40

Springer Series in Chemical Physics
Edited by Fritz Peter Schäfer

Springer Series in Chemical Physics
Editors: V. I. Goldanskii R. Gomer F. P. Schäfer J. P. Toennies

Volume 40 **High Resolution Spectroscopy of Transient Molecules**
Editor: E. Hirota

Volume 41 **High Resolution Spectral Atlas of Nitrogen Dioxide 559–597 nm**
By K. Uehara and H. Sasada

Volumes 1–39 are listed on the back inside cover

Eizi Hirota

High-Resolution Spectroscopy of Transient Molecules

With Contributions by
Y. Endo K. Kawaguchi S. Saito
T. Suzuki C. Yamada

With 80 Figures

Springer-Verlag
Berlin Heidelberg New York Tokyo

Professor Eizi Hirota, Ph.D.
The Institute for Molecular Science,
Okazaki 444, Japan

Series Editors

Professor Vitalii I. Goldanskii
Institute of Chemical Physics
Academy of Sciences
Kosygin Street 4
SU 117334 Moscow, USSR

Professor Robert Gomer
The James Franck Institute
The University of Chicago
5640 Ellis Avenue
Chicago, IL 60637, USA

Professor Dr. Fritz Peter Schäfer
Max-Planck-Institut für
Biophysikalische Chemie
D-3400 Göttingen-Nikolausberg
Fed. Rep. of Germany

Professor Dr. J. Peter Toennies
Max-Planck-Institut für Strömungsforschung
Böttingerstraße 6–8
D-3400 Göttingen
Fed. Rep. of Germany

ISBN 3-540-15302-0 Springer-Verlag Berlin Heidelberg New York Tokyo
ISBN 0-387-15302-0 Springer-Verlag New York Heidelberg Berlin Tokyo

Typesetting: Daten- und Lichtsatz-Service, Würzburg
Offsetprinting and bookbinding: Graphischer Betrieb Konrad Triltsch, Würzburg

2153/3020-543210

Preface

It is a great challenge in chemistry to clarify every detail of reaction processes. In older days chemists mixed starting materials in a flask and took the resultants out of it after a while, leaving all the intermediate steps uncleared as a sort of black box. One had to be content with only changing temperature and pressure to accelerate or decelerate chemical reactions, and there was almost no hope of initiating new reactions. However, a number of new techniques and new methods have been introduced and have provided us with a clue to the examination of the black box of chemical reaction. Flash photolysis, which was invented in the 1950s, is such an example; this method has been combined with high-resolution electronic spectroscopy with photographic recording of the spectra to provide a large amount of precise and detailed data on transient molecules which occur as intermediates during the course of chemical reactions.

In 1960 a fundamentally new light source was devised, i.e., the laser. When the present author and coworkers started high-resolution spectroscopic studies of transient molecules at a new research institute, the Institute for Molecular Science in Okazaki in 1975, the time was right to exploit this new light source and its microwave precursor in order to shed light on the black box. These light sources have allowed them to develop new spectroscopic means of high resolution and high sensitivity, which have proved extremely useful in unraveling molecular properties of many transients which play unique roles in the black box. The knowledge obtained from these studies is expected to lead to a new type of chemistry and to open new areas in molecular science and related fields; it will be of considerable significance in a wide range of fields from molecular structure even to semiconductor technology. This book describes primarily the achievements obtained during the past ten years by the high-resolution molecular spectroscopy group at the Institute for Molecular Science.

The greatest part of the work presented in this book was contributed by my coworkers, but any errors in it must be entirely ascribed to me. I should like explicitly to mention the names of the other members of the group: Shuji Saito, Chikashi Yamada, Yasuki Endo, Kentarou Kawaguchi, and Tetsuo Suzuki. We have benefitted much from collaborations and discussions with many visiting scientists, in particular, Drs. T. Amano, R. J. Butcher, J. E. Butler, R. F. Curl, J. T. Hougen, M. Kakimoto, K. Matsumura, A. R. W. McKellar, T. A. Miller, K. Nagai, N. Ohashi, K. Takagi, and M. Tanimoto. I should also like to mention that T. Amano and I started our studies of free radicals in Professor Yonezo Morino's laboratory more than twenty years ago. I owe much to Professor Morino for his continuous encouragement since this earlier stage of the study. I also wish to express my deep gratitude to the former Director General, Dr. Hideo Akamatu, Professor Emeritus of the IMS, and to the present Director General, Professor Saburo Nagakura; both of them have helped make it possible to undertake the research described in this book. I am also grateful to Dr. Helmut Lotsch of Springer-Verlag for reading and commenting on the manuscript. Finally most of the manuscript was typed by Miss Masako Teshima; without her assistance this book would not have been completed so quickly.

Okazaki, Japan, January, 1985 *Eizi Hirota*

Contents

1. Introduction 1

1.1 Historical Background of Free Radical Studies by High-Resolution Spectroscopy 1

1.2 Significance of High-Resolution Spectra of Transient Molecules 3

2. Theoretical Aspects of High-Resolution Molecular Spectra 5

2.1 Molecular Rotation 5

2.1.1 Angular Momentum 5

2.1.2 Rotational Energy of a Rigid Body 8

2.2 Vibration-Rotation Interaction 11

2.3 Fine and Hyperfine Structures 17

2.3.1 Electron Spin and Electron Orbital Angular Momenta 17

2.3.2 Fine-Structure Hamiltonian 19

2.3.3 Hamiltonian Matrix Elements Appropriate for Hund's Case (a) 20

2.3.4 Λ-Type Doubling in $^2\Pi$ States 22

2.3.5 Hamiltonian Matrix Elements Appropriate for Hund's Case (b) 24

2.3.6 Rotational Structure of a ^2E Molecule 28

2.3.7 Hyperfine Interaction 33

2.3.8 Matrix Elements of the Hyperfine Interaction Appropriate for Case $(a)_\beta$ 34

2.3.9 Matrix Elements of the Hyperfine Interaction Appropriate for Case $(b)_{\beta J}$ 39

2.3.10 Hyperfine Interactions Caused by More than One Nucleus 41

2.3.11 Hyperfine Interactions in a C_{3v} Symmetric Top Molecule 42

2.4 Vibronic Interaction Including the Renner-Teller Effect 47

2.4.1 Born-Oppenheimer Approximation and Vibronic Interaction 47

2.4.2 Renner Effect on a Linear Polyatomic Molecule in a Π Electronic State 48

2.4.3 Rotational Energy Levels of a Linear Triatomic Molecule Affected by Vibronic Interactions . . . 57
2.4.4 Hyperfine Structure in a $^2\Pi$ Polyatomic Molecule Affected by Rovibronic Interactions . . 61
2.5 Zeeman and Stark Effects 62
2.5.1 Zeeman Effect of a Molecule Without Orbital Angular Momentum 64
2.5.2 Zeeman Effect of a Molecule with Electron Orbital Angular Momentum 70
2.5.3 Stark Effect 71

3. Experimental Details 74
3.1 Microwave Spectrometer 75
3.1.1 Requirements to be Met by a Microwave Spectrometer for the Study of Transient Molecules 75
3.1.2 Historical Survey of Microwave Studies of Transient Molecules 76
3.1.3 High Sensitivity Millimeter-Wave Spectrometer to Study Transient Molecules 77
3.2 Infrared Laser Spectrometers 81
3.2.1 Diode Laser Spectrometer 82
3.2.2 Infrared Laser Magnetic Resonance Spectrometer . 90
3.2.3 Far-Infrared Laser Magnetic Resonance Spectrometer 93
3.2.4 Difference Frequency Laser Spectrometer 96
3.3 Dye Laser Spectroscopic System 98
3.3.1 CW Dye Laser 98
3.3.2 Doppler-Limited Excitation Spectroscopy Using a CW Dye Laser as a Source 99
3.3.3 Intermodulated Fluorescence Spectroscopy . . . 102
3.4 Double Resonance Spectroscopy 103
3.4.1 Microwave or Radio Frequency Optical Double Resonance 104
3.4.2 An MODR Spectrometer to Study Transient Molecules 105
3.4.3 Infrared-Optical Double Resonance 108
3.5 Generation of Transient Molecules 112

4. Individual Molecules 119
4.1 Diatomic Free Radicals 119
4.1.1 $^2\Sigma$ and $^3\Sigma$ Diatomic Molecules 119
4.1.2 $^2\Pi$ Diatomic Molecules 129
4.1.3 Molecules in Δ States 137
4.2 Linear Polyatomic Molecules 139
4.2.1 Molecules in Σ Electronic States 139

4.2.2 Molecules in Π Electronic States 141
4.3 Nonlinear XY_2- and XYZ-Type Triatomic Free Radicals 150
4.3.1 Nonlinear XYZ-Type Free Radicals of C_s Symmetry 150
4.3.2 Nonlinear XY_2-Type Free Radicals of C_{2v} Symmetry 156
4.3.3 Molecular Structure and Anharmonic Potential Constants of Nonlinear XYZ-Type Molecules of C_s Symmetry 164
4.4 Symmetric Top and Other Polyatomic Free Radicals . . 169
4.4.1 The Methyl Radical CH_3 169
4.4.2 The Trifluoromethyl Radical CF_3 172
4.4.3 Structures and Internal Motions of the Methyl Radical and Its Derivatives 175
4.4.4 The NO_3 Radical 179
4.4.5 The Methoxy Radical CH_3O 180
4.5 Fine and Hyperfine Interactions in Free Radicals 184
4.5.1 Hyperfine Interaction Constants of Diatomic Molecules 184
4.5.2 Spin-Rotation Interaction in Nonlinear Polyatomic Molecules 189
4.5.3 Hyperfine Interaction in Nonlinear Polyatomic Molecules 192
4.6 Molecules in Metastable Electronic States 194
4.6.1 LEF and MODR Study of H_2CS in the $\tilde{a}\,^3A_2$ State 194
4.6.2 Perturbations in the Singlet-Singlet Transition of a Few Simple Carbenes 198

5. Applications and Future Prospects 201
5.1 Applications to Chemical Reactions 201
5.2 Applications to Atmospheric Chemistry 206
5.3 Applications to Astronomy 207
5.3.1 HCCN 209
5.3.2 Phosphorus-Containing Compounds 209
5.4 Future Developments 210
5.4.1 Possible Improvements of Spectroscopic Techniques 210
5.4.2 Future Trends in High-Resolution Spectroscopic Studies of Transient Molecules 211

References 213

Subject Index 229

1. Introduction

1.1 Historical Background of Free Radical Studies by High-Resolution Spectroscopy

The concept of free radicals has long roots in chemistry; according to *Herzberg* [1.1] it already existed in the 19th century. An identification of a chemically stable free radical triphenylmethyl in the early 20th century gave impetus to the study of free radicals. However, it remained a quite difficult task for chemists to characterize free radicals, because most of them appeared only for a short period of time. A number of simple free radicals, mostly diatomic, were later identified in flames and electrical discharges, by recording and analyzing their emission spectra, e.g., CH, OH, CN, and C_2. The advent of new quantum mechanics provided spectroscopy with a sound basis, and brought about remarkable progress in spectroscopic studies of diatomic free radicals in the late 1920s and early 1930s, as compiled by *Herzberg* in [1.2]. After World War II, *Herzberg* and his students started high-resolution studies of transient molecules; a new technique of flash photolysis was introduced to generate such species, and emission and absorption spectra were recorded by large grating spectrometers [1.3].

Development of short-wavelength techniques such as radar during the war brought about new spectroscopic methods, i.e., electron spin resonance (ESR), or more properly called electron paramagnetic resonance (EPR) and zero-field microwave spectroscopy. The ESR has been applied primarily to free radicals trapped in low-temperature inert matrices, whereas *Carrington* and his associates have successfully applied EPR to a number of diatomic and triatomic free radicals in the gas phase [1.4]. The hyperfine coupling constants and g factors obtained from these investigations constitute sources of information on transient molecules that is not obtainable from traditional optical spectroscopy. Microwave spectroscopy has remarkably improved the precision with which we can determine molecular parameters, and has been applied to quite a large number of molecules, most of which were, however, chemically stable [1.5, 6]. In 1955 *Townes* and co-workers [1.7] succeeded in detecting transitions between Λ doublets of the OH radical. Nine years later, *Powell* and *Lide* [1.8] and *Winnewisser* et al. [1.9] reported the detection of a second example, SO, by this technique. It must, however, be admitted that the microwave investigation of free radicals did not appear to be very promising in the 1960s, because of the difficulty in detecting the low-energy photon of microwaves.

In the infrared (IR) region an important contribution was made in the 1950s by *Pimentel* and *Herr* [1.10], who set up rapid scan spectrometers and observed the vibrational spectra of a few free radicals. However, the resolution was quite low (a few cm^{-1}), and furthermore, the spectrometers did not seem to be sensitive enough to detect many short-lived transients. One should not overlook other important contributions of IR spectroscopy to free radical studies, namely a number of matrix isolation studies by *Milligan* and *Jacox* [1.11] and *Raymond* and *Andrews* [1.12] among others. Absence of rotational structures in the matrix spectra precludes the possibility of determining molecular structures from the observed spectra. Nevertheless, some qualitative information has been obtained on structures from the isotopic shifts of the vibrational frequencies. The matrix spectral data as recently compiled by *Jacox* [1.13] have been of great use in observing gas phase IR spectra.

The invention of the laser has brought about a drastic change in the spectroscopy of free radicals, although progress was rather slow in the beginning. This is perhaps due to the fact that free radical studies need versatile light sources, but most lasers developed earlier were rather limited in many respects, including wavelength coverage. An epoch-making step was taken in 1968 by *Evenson* and co-workers [1.14], who developed a spectroscopic method for investigating paramagnetic molecules; it was called laser magnetic resonance or LMR, apparently influenced by NMR, but here the light source was a far-infrared (FIR) laser and the spectra observed corresponded to electric-dipole transitions between rotational levels of paramagnetic molecules. The LMR bears similarity to NMR only in how it scans the spectrum, i.e., magnetic field scanning. It is rather closer, or essentially identical to EPR [1.15]. Far-infrared (FIR) LMR is known as one of the most sensitive spectroscopic methods of high resolution. Later, LMR was extended to the IR region, where CO_2, N_2O, and CO lasers are used as sources. Infrared LMR seems, however, to be gradually being replaced by tunable laser spectroscopy, such as diode laser spectroscopy, mainly because of the limitation in wavelength coverage. Two kinds of tunable laser sources have also been developed in 2–4 μm, namely the color-center laser and the difference frequency laser (Sect. 3.2). The former is also being extended to about 1 μm.

In the visible region the laser spectroscopic study of unstable molecules was initiated with the observation of fluorescence induced by an Ar^+ laser; e.g., BaO $A^1\Sigma \rightarrow X^1\Sigma$ by *Sakurai* et al. [1.16] and LiNa $^1\Pi \rightarrow {}^1\Sigma$ by *Hessel* [1.17]. The development of the dye laser has drastically changed spectroscopy in the visible and uv regions; quite a large number of reports have already appeared on applications of this light source to molecular spectroscopy. Special reference should be made to microwave optical double resonance (MODR) experiments carried out by *Curl* and co-workers on NH_2 [1.18] and also by *Harris* and collaborators on BaO [1.19] (Sect. 3.4). The intermodulated fluorescence

(IMF) technique developed by *Sorem* and *Schawlow* [1.20] was also applied to a few free radicals such as BO_2 [1.21] and NH_2 [1.22].

1.2 Significance of High-Resolution Spectra of Transient Molecules

Chemistry is essentially a science of material transformation, based largely upon empirical approaches, by which many new compounds have been synthesized. It is questionable, however, whether a chemical reaction is really so simple that it can adequately be expressed by $A + B \rightarrow C + D$, for example. During the course of a chemical reaction, at least one chemical bond is surely broken and new bonds may be formed in place of the old one. Thus, we may suppose that intermediate species with "incomplete" chemical bonds exist. This is in fact the case; quite a large number of such intermediate species, referred to as free radicals, have now been detected. Although their existence was presumed on the basis of various, often indirect, pieces of evidence, the details of their structures and properties have long remained unclear, because of the inherent high reactivities and short lifetimes caused by their "incomplete" chemical bonds. This book describes some recent responses to the challenge of unraveling the nature of simple transient molecules using spectroscopic methods of high resolution and high sensitivity, such as microwave and laser spectroscopy.

The book starts by describing the fundamental principles upon which molecular spectra are based, paying special attention to complications caused by unpaired electrons. Experimental techniques then follow; the spectroscopic systems and the methods of generating transient species which have been employed are discussed in some detail, because the success or failure of free radical studies depends critically on how judiciously the experimental methods are selected and/or combined when acting on particularly difficult molecules. All reaction intermediates must be unambiguously identified in order to understand fully the details of the reaction mechanism. High-resolution spectra of these species allow us not only to confirm the presence of these species in the reaction system, but also to monitor their distributions over quantum states; chemical reactions may often result in selective population of molecules in certain levels.

Many transient molecules generated by chemical reactions and some other means are free radicals. One may be struck by how dramatically the types of spectroscopic information on free radicals differ from that on ordinary molecules. This information is examined in the next chapters. The "incomplete" chemical bonds let the free radicals play unique roles in the theory of molecular structure, by providing clear examples of various types of fine and hyperfine interactions between unpaired electrons and nuclei with finite spins present in the molecule. The unpaired electrons endow the molecules with paramagnetism generated by their unquenched orbital and/or spin angular momenta; the magnetic moments are thus of the order of a Bohr magneton. The magnetic moments interact with those, if any, of nuclei in the molecule, splitting its spectral lines into multiplets referred to as the magnetic hyperfine structure. This is completely absent in a singlet "sigma" molecule without any electron spin and/or orbital angular momenta. The hyperfine structure often gives unique information on the electronic structure of the molecule. The free radicals thus play a specially important role in the study of molecular structure.

As one might expect, high-resolution spectra of free radicals provide extremely reliable means of monitoring these species in various environments. For example, in the study of chemical reactions, the observed spectrum makes it possible to view selectively a particular species in the presence of many others, and one can thus get very specific information on chemical systems by means of high-resolution spectra. This sort of application forms the subject of the last chapter. Another example is radioastronomy, where laboratory microwave data have been employed to identify chemical species, to evaluate Doppler shifts, and to calculate column densities of interstellar molecules. It is interesting to note that about 40% of the identified species are free radicals which exist in space as "stable" species, presumably because of low density there. In a few cases radioastronomical observations even preceded laboratory detection, but the significance of interstellar spectra observed by radiotelescopes is fully appreciated only when they are compared with laboratory data. Radioastronomical observations are now being extended to the IR region.

Chemical processes in the Earth's atmosphere have attracted much attention because of air pollution caused by exhausts from automobiles, aeroplanes, and factories. Molecular spectroscopy has already contributed much to monitoring chemically stable species in the atmosphere, and short-lived species such as OH, HO_2, ClO, SH, and NO_3 will be observable in the near future, based on laboratory data.

Plasmas are one of the most efficient sources of transient molecules (Sect. 3.5). It is also indispensable in semiconductor fabrication; amorphous silicon is produced by electrical discharge in silane or disilane, and high-resolution spectra of intermediate species, if known, will be of potential use in diagnosing chemical conditions for making electronic devices.

In the future, free radicals will attract much more attention from many more fields than before, because they always occur in chemical processes. The demand for high-resolution spectra will increase accordingly; they are an indispensable means in unambiguously identifying and monitoring free radicals in various systems and environments.

2. Theoretical Aspects of High-Resolution Molecular Spectra

A high-resolution molecular spectrum denotes here a spectrum which exhibits well-resolved rotational structures or it corresponds to the rotational transition of a molecule (i.e., the pure rotational spectrum). Although direct transitions among fine and/or hyperfine structure component levels are rarely discussed, they are certainly classified as high-resolution spectra.

Molecule rotation has been thoroughly discussed by many authors and a large number of excellent textbooks have already been published [2.1–3]. However, in view of a great diversity of notation and phase convention employed for physical quantities, a brief description is given below on a rigid rotation of the molecule (Sect. 2.1), which is then followed by a few complicated aspects such as the vibration-rotation interaction (Sect. 2.2), fine and hyperfine structure interaction (Sect. 2.3), and other vibronic interactions (Sect. 2.4). The effect of the external field, either electric or magnetic, is also discussed in Sect. 2.5.

2.1 Molecular Rotation

2.1.1 Angular Momentum

In discussing the rotation of a molecule, it is appropriate to introduce two Cartesian coordinate systems, namely, the space-fixed coordinate system and the molecule-fixed coordinate system, designated as

$F = X, Y, Z$ for the space-fixed coordinates and

$g = x, y, z$ for the molecule-fixed coordinates.

The two systems are taken to share their origin in common which coincides with the center of gravity of the molecule. They are transformed to each other by a unitary matrix whose elements are the direction cosines Φ_{Fg}:

$$g = \sum_F \Phi_{Fg} F, \tag{2.1.1}$$

$$F = \sum_F \Phi_{Fg} g. \tag{2.1.2}$$

The angular momentum of the molecular rotation is an axial vector, and its components are tranformed as (2.1.1, 2)

$$J_g = \sum_F \Phi_{Fg} J_F, \tag{2.1.3}$$

$$J_F = \sum_g \Phi_{Fg} J_g. \tag{2.1.4}$$

The angular momentum is given in units of $\hbar$ throughout the present book. Then the commutation relations between the space-fexed components of $\boldsymbol{J}$ are given by

$$J_X J_Y - J_Y J_X = \mathrm{i} J_Z, \tag{2.1.5}$$

$$J_Y J_Z - J_Z J_Y = \mathrm{i} J_X, \tag{2.1.6}$$

$$J_Z J_X - J_X J_Z = \mathrm{i} J_Y, \tag{2.1.7}$$

and the square of the total angular momentum

$$\boldsymbol{J}^2 = J_X^2 + J_Y^2 + J_Z^2 \tag{2.1.8}$$

commutes with any of its components:

$$\boldsymbol{J}^2 J_F - J_F \boldsymbol{J}^2 = 0. \tag{2.1.9}$$

Because Φ_{Fg} ($F = X$, Y, and Z) is a component of a unit vector, the commutation relations of J_F with $\Phi_{F'g}$ are

$$J_X \Phi_{Yg} - \Phi_{Yg} J_X = \mathrm{i} \Phi_{Zg}, \tag{2.1.10}$$

$$\Phi_{Xg} J_Y - J_Y \Phi_{Xg} = \mathrm{i} \Phi_{Zg}, \tag{2.1.11}$$

$$J_Y \Phi_{Zg} - \Phi_{Zg} J_Y = \mathrm{i} \Phi_{Xg}, \tag{2.1.12}$$

$$\Phi_{Yg} J_Z - J_Z \Phi_{Yg} = \mathrm{i} \Phi_{Xg}, \tag{2.1.13}$$

$$J_Z \Phi_{Xg} - \Phi_{Xg} J_Z = \mathrm{i} \Phi_{Yg}, \tag{2.1.14}$$

$$\Phi_{Zg} J_X - J_X \Phi_{Zg} = \mathrm{i} \Phi_{Yg}, \tag{2.1.15}$$

$$J_F \Phi_{Fg} - \Phi_{Fg} J_F = 0. \tag{2.1.16}$$

A straightforward calculation using (2.1.3, 5–7, 10–16) leads to the commutation relations between the molecule-fixed components of the angular momentum:

$$J_x J_y - J_y J_x = -\mathrm{i} J_z, \tag{2.1.17}$$

$$J_y J_z - J_z J_y = -\mathrm{i} J_x, \tag{2.1.18}$$

$$J_z J_x - J_x J_z = -\mathrm{i} J_y. \tag{2.1.19}$$

Note that the right-hand sides of (2.1.17–19) have the factor $-$ i, instead of + i as in (2.1.5–7). This reversed sign is often referred to as the anomalous sign [2.4]. When the components of the angular momentum along the rotating axes are considered, the commutation relations must have the anomalous sign. Equations (2.1.8, 9) hold also for J_g:

$$\boldsymbol{J}^2 = J_x^2 + J_y^2 + J_z^2, \tag{2.1.20}$$

$$\boldsymbol{J}^2 J_g - J_g \boldsymbol{J}^2 = 0, \tag{2.1.21}$$

and the commutation relations of J_g with $\Phi_{Fg'}$ are similar to (2.1.10–16) except for replacement of i by $-$ i:

$$J_x \Phi_{Fy} - \Phi_{Fy} J_x = -\mathrm{i}\Phi_{Fz}, \tag{2.1.22}$$

$$\Phi_{Fx} J_y - J_y \Phi_{Fx} = -\mathrm{i}\Phi_{Fz}, \tag{2.1.23}$$

$$J_y \Phi_{Fz} - \Phi_{Fz} J_y = -\mathrm{i}\Phi_{Fx}, \tag{2.1.24}$$

$$\Phi_{Fy} J_z - J_z \Phi_{Fy} = -\mathrm{i}\Phi_{Fx}, \tag{2.1.25}$$

$$J_z \Phi_{Fx} - \Phi_{Fz} J_z = -\mathrm{i}\Phi_{Fy}, \tag{2.1.26}$$

$$\Phi_{Fz} J_x - J_x \Phi_{Fz} = -\mathrm{i}\Phi_{Fy}, \tag{2.1.27}$$

$$J_g \Phi_{Fg} - \Phi_{Fg} J_g = 0. \tag{2.1.28}$$

The commutation relations, between the components of the angular momentum, (2.1.5–7, 9) and (2.1.17–19, 21), allow us to choose a set of eigenfunctions which simultaneously diagonalizes $\boldsymbol{J}^2$ and J_Z or $\boldsymbol{J}^2$ and J_z such that

$$\langle J, k, M \,|J^2|\, J, k, M\rangle = J(J+1), \tag{2.1.29}$$

$$\langle J, k, M \,|J_Z|\, J, k, M\rangle = M, \tag{2.1.30}$$

$$\langle J, k, M \,|J_z|\, J, k, M\rangle = k, \tag{2.1.31}$$

where $|M| \leqq J$ and $|k| \leqq J$. The nonvanishing matrix elements of J_X and J_Y or J_x and J_y are given by

$$\langle J, k, M \pm 1 \,|J_X|\, J, k, M\rangle = [J(J+1) - M(M \pm 1)]^{1/2}/2, \tag{2.1.32}$$

$$\langle J, K, M \pm 1 \,|J_Y|\, J, k, M\rangle = \mp\, \mathrm{i}[J(J+1) - M(M \pm 1)]^{1/2}/2, \tag{2.1.33}$$

$$\langle J, k \pm 1, M \,|J_x|\, J, k, M\rangle = [J(J+1) - k(k \pm 1)]^{1/2}/2, \tag{2.1.34}$$

$$\langle J, k \pm 1, M \,|J_y|\, J, k, M\rangle = \pm\, \mathrm{i}[J(J+1) - k(k \pm 1)]^{1/2}/2, \tag{2.1.35}$$

where the phase is chosen following *Condon* and *Shortley* [2.1.5–8]. The ladder operators defined by

$$J_{F\pm} = J_X \pm \mathrm{i} J_Y, \tag{2.1.36}$$

$$J_{g\pm} = J_x \pm \mathrm{i} J_y \tag{2.1.37}$$

are often more convenient than J_X and J_Y or J_x and J_y and have nonvanishing matrix elements

$$\langle J, k, M \pm 1 \,|J_{F\pm}|\, J, k, M\rangle = [J(J+1) - M(M \pm 1)]^{1/2}, \tag{2.1.38}$$

$$\langle J, k \mp 1, M \,|J_{g\pm}|\, J, k, M\rangle = [J(J+1) - k(k \mp 1)]^{1/2}. \tag{2.1.39}$$

2.1.2 Rotational Energy of a Rigid Body

The rotation of a molecule is well approximated in most cases by a model which consists of mass points (i.e., atoms) rigidly connected with each other by sticks without any masses. The ith mass point with mass m_i is located at $\boldsymbol{R}_i = (X_i, Y_i, Z_i)$ and the center of mass of the whole molecule at $\boldsymbol{R} = \sum m_i \boldsymbol{R}_i / M$, where M denotes the total mass. The position vector of the ith mass point relative to the center of mass is designated by $\boldsymbol{r}_i$, i.e., $\boldsymbol{R}_i = \boldsymbol{R} + \boldsymbol{r}_i$ and $\sum m_i \boldsymbol{r}_i = 0$. The velocity of the ith mass point relative to the space-fixed coordinate is given by

$$V_i = \dot{\boldsymbol{R}} + \dot{\boldsymbol{r}}_i = \dot{\boldsymbol{R}} + \boldsymbol{\omega} \times \boldsymbol{r}_i + \boldsymbol{v}_i, \tag{2.1.40}$$

where $\boldsymbol{\omega}$ denotes the angular velocity of the molecule-fixed axis system relative to the space-fixed coordinate and $\boldsymbol{v}_i$ the velocity of the ith mass point relative to the molecule-fixed coordinate, which vanishes when the molecule is rigid. The kinetic energy of a rigid molecule is then given by

$$T = (1/2) \sum_i m_i V_i^2 = (1/2)\, M \dot{\boldsymbol{R}}^2 + (1/2) \sum_i m_i (\boldsymbol{\omega} \times \boldsymbol{r}_i)^2. \tag{2.1.41}$$

The first term represents the translation energy of the entire molecule and will not be considered further. The second term, the rotational energy T_{R}, is rewritten as

$$T_{\mathrm{R}} = I_{xx}\omega_x^2 + I_{yy}\omega_y^2 + I_{zz}\omega_z^2, \tag{2.1.42}$$

where the x, y, and z axes are taken such that $\sum m_i x_i y_i = \sum m_i y_i z_i = \sum m_i z_i x_i = 0$. The x, y, and z axes are then referred to as the principal axes of inertia and

$$I_{xx} = \sum m_i (y_i^2 + z_i^2),$$

$$I_{yy} = \sum m_i (z_i^2 + x_i^2), \quad \text{and}$$

$$I_{zz} = \sum m_i (x_i^2 + y_i^2)$$

as the principal moments of inertia.

The symmetric top is defined to be a molecule for which two of the three principal moments of inertia are equal: $I_{xx} = I_{yy} + I_{zz}$. The z axis is then referred to as the symmetry axis. The rotational energy (2.1.42) may be transformed to an expression in which the angular velocity is replaced by the angular momentum and the moment of inertia by the rotational constant:

$$\hbar J_g = T_R/\partial \omega_g, \tag{2.1.43}$$

$$B = h/(8\pi^2 I_{xx}), \tag{2.1.44}$$

$$A \text{ or } C = h/(8\pi^2 I_{zz}). \tag{2.1.45}$$

It is customary to use the notations A, B, and C such that $A \geqq B \geqq C$. Therefore, A should be used in (2.1.45) when $I_{zz} < I_{xx}$ and the molecule is classified as a prolate symmetric top, otherwise C must be used, namely, for an oblate symmetric top. When the angular momentum (2.1.43) is taken to be a quantum-mechanical operator, the rotational Hamiltonian of the symmetric top is obtained as

$$\boldsymbol{H}_R/h = B\boldsymbol{J}^2 + (A - B)\, J_z^2 \quad \text{(prolate)}, \tag{2.1.46}$$

$$= B\boldsymbol{J}^2 + (C - B)\, J_z^2 \quad \text{(oblate)}. \tag{2.1.47}$$

It is easy to confirm that $\boldsymbol{H}_R$ commutes with $\boldsymbol{J}^2$, J_z, and J_Z, and thus the basis set $|J, k, M\rangle$ mentioned above and defined by (2.1.29–31) is also the eigenfunction of $\boldsymbol{H}_R$. In fact,

$$\langle J, k, M\, |H_R/h|\, J', k', M'\rangle = (E_R/h)\, \delta_{JJ'}\, \delta_{kk'}\, \delta_{MM'}$$

$$= BJ(J+1) + (A - B)\, k^2 \quad \text{(prolate)}, \tag{2.1.48}$$

$$= BJ(J+1) + (C - B)\, k^2 \quad \text{(oblate)}. \tag{2.1.49}$$

The quantum number k represents the magnitude of the rotation about the symmetry axis, and, since the eigenvalue E_R involves k only in the form of k^2, the right-handed and left-handed rotations of the same magnitude correspond to the same eigenvalue. In other words, each eigenstate with a finite k is doubly degenerate, and this degeneracy is referred to as K degeneracy. For the prolate top the levels with larger $|k|$ values are higher, whereas the reverse is true for the oblate top.

Because the linear molecule has no rotational freedom about the molecular axis it may be regarded as a symmetric top with $k = 0$. Its rotational wave function is known to be the spherical harmonics $Y_{JM}(\theta, \phi) \equiv |J, 0, M\rangle$. Using (2.1.38, 39) the symmetric-top rotational wave function may be constructed from that of a linear molecule:

$$|J, \pm |k|, \pm |M|\rangle = N\, J_{g\mp}^{|k|}\, J_{F\pm}^{|M|}\, |J, 0, 0\rangle, \tag{2.1.50}$$

where N denotes the normalization factor. Equation (2.1.50) is useful in discussing the symmetry properties of the symmetric-top rotational wave function, as explained in Table 2.1.

Tabelle 2.1. Symmetry properties of the symmetric-top rotational wave function

	C_2^x	C_2^y	C_2^z
J_x	J_x	$-J_x$	$-J_x$
J_y	$-J_y$	J_y	$-J_y$
J_z	$-J_z$	$-J_z$	J_z
$J_{g\pm}$	$J_{g\mp}$	$-J_{g\mp}$	$-J_{g\pm}$
$\|J, 0, M\rangle$	$(-1)^J \|J, 0, M\rangle$	$(-1)^J \|J, 0, M\rangle$	$\|J, 0, M\rangle$
$\|J, k, M\rangle$	$(-1)^J \|J, -k, M\rangle$	$(-1)^{J+k} \|J, -k, M\rangle$	$(-1)^k \|J, k, M\rangle$
$S(J, k, M, \gamma)$	$(-1)^{J+\gamma} S(J, k, M, \gamma)$	$(-1)^{J+k+\gamma} S(J, k, M, \gamma)$	$(-1)^k S(J, k, M, \gamma)$

It is often convenient to take a linear combination

$$S(J, k, M, \gamma) = (1/\sqrt{2})\,[|J, k, M\rangle + (-1)^\gamma |J, -k, M\rangle], \tag{2.1.51}$$

which is referred to as the Wang wave function, where $\gamma = 0$ or 1. The symmetry properties of $S(J, k, M, \gamma)$ are also included in Table 2.1.

The rotational Hamiltonian for an asymmetric rotor is given by

$$\boldsymbol{H}_\mathrm{R}/h = AJ_a^2 + BJ_b^2 + CJ_c^2, \tag{2.1.52}$$

where the molecule-fixed coordinate axes are designated by a, b, and c rather than x, y, and z, following the convention $A > B > C$. As (2.1.52) shows, $\boldsymbol{H}_\mathrm{R}$ is invariant under the four symmetry operations E, C_2^a, C_2^b, and C_2^c. These operations form a group referred to as V (called the four group), and its character table is reproduced in Table 2.2. There are four species A, B_a, B_b, and B_c, and each of the eigenstates of (2.1.52) must belong to one of these species. Although the eigenfunction and eigenvalue of (2.1.52) cannot be given in simple analytical forms, the eigenfunction may be expanded in terms of the Wang symmetric-top wave functions. When the molecule under consideration is close to a prolate symmetric top, (2.1.52) may be written as

$$\boldsymbol{H}_\mathrm{R}/h = [(B + C)/2]\,\boldsymbol{J}^2 + [A - (B + C)/2]\,J_a^2 + [(B - C)/2]\,(J_b^2 - J_c^2), \tag{2.1.53}$$

Table 2.2. Character table of the four group and the symmetry of the asymmetric rotor rotational energy level

V	E	C_2^a	C_2^b	C_2^c	K_a	K_c
A	1	1	1	1	e	e
B_a	1	1	-1	-1	e	o
B_b	1	-1	1	-1	o	o
B_c	1	-1	-1	1	o	e

whereas, when the oblate-top limit is more appropriate, $\boldsymbol{H}_\mathrm{R}$ is given by

$$\begin{aligned}\boldsymbol{H}_\mathrm{R}/h = {} & [(A+B)/2]\boldsymbol{J}^2 + [C-(A+B)/2]\,J_c^2 \\ & + [(A-B)/2]\,(J_a^2 - J_b^2). \end{aligned} \tag{2.1.54}$$

In both cases, the last term gives rise to matrix elements of $\Delta k = \pm 2$. Furthermore, the $\langle k\,|\,k'\rangle$ element is equal to the $\langle -k\,|\,-k'\rangle$ element. It is thus easily proved that the eigenstate is classified according to whether k is even or odd and also to the sum or difference of $|k\rangle$ and $|-k\rangle$, namely to the parity γ. Referring to Tables 2.1, 2, it is easily verified that each species of the four group is represented by a linear combination of $S(J, k, M, \gamma)$ and the four species correspond to the parities of k and γ. Table 2.2 further shows that it is sufficient to examine the parity of each eigenstate when it is subjected to the two operations C_2^a and C_2^c. Because the parity is reserved irrespective of the degree of asymmetry of the molecule, we may use the k quantum numbers in the symmetric-top limits to label each rotational eigenstate of an asymmetric top; K_a and K_c are such numbers in the prolate and oblate limits, respectively, as the suffixes a and c indicate. Each rotational level of an asymmetric rotor is thus expressed as $J_{K_a K_c}$. [The asymmetry of the molecule is often measured by Ray's parameter κ defined by $(2B-A-C)/(A-C)$, which takes the values of -1 and $+1$ in the two symmetric-top limits. Therefore, K_a and K_c are often written as K_{-1} and K_{+1}.]

2.2 Vibration-Rotation Interaction

Although a rigid body closely approximates a molecule in respect to its overall rotation, a real molecule vibrates intramolecularly, namely the chemical bond and the valence angle holding atoms in the molecule are by no means rigid, resulting in the interaction between the vibrational and rotational motions. The position vector $\boldsymbol{r}_i$ of the ith mass points thus no longer has constant magnitude, but is given by

$$\boldsymbol{r}_i = \boldsymbol{r}_i^e + \delta\boldsymbol{r}_i. \tag{2.2.1}$$

The present discussion shall be restricted to the intramolecular vibration of small amplitude well approximated by the harmonic oscillator, leaving the large-amplitude motions like internal rotation and inversion as special cases to be discussed individually. The center-of-mass condition $\sum_i m_i \boldsymbol{r}_i = 0$ requires $\sum_i m_i \delta\boldsymbol{r}_i = 0$, because $\sum m_i \boldsymbol{r}_i^e = 0$. Differentiation of this condition with respect to time leads to $\sum_i m_i \boldsymbol{v}_i = 0$. The "molecule-fixed" axis system is now defined by

$$\sum_i m_i \boldsymbol{r}_i^e \times \boldsymbol{r}_i = 0 \quad \text{or} \quad \sum_i m_i \boldsymbol{r}_i^e \times \delta\boldsymbol{r}_i = 0, \tag{2.2.2}$$

which, when differentiated with time, gives

$$\sum_i m_i \boldsymbol{r}_i^e \times \boldsymbol{v}_i = 0, \tag{2.2.3}$$

referred to as the Eckart condition. The $3N-6$ ($3N-5$ for a linear molecule) internal degrees of freedom are best represented by the $3N-6$ (or $3N-5$) normal coordinates Q_r, and $\delta \boldsymbol{r}_i$ may be expanded in Q_r:

$$\sqrt{m_i}\, \delta \boldsymbol{r}_i = \sum_i \boldsymbol{l}_{ir} Q_r. \tag{2.2.4}$$

The time derivative of (2.2.4) is $\sqrt{m_i}\, \boldsymbol{v}_i = \sum \boldsymbol{l}_{ir} \dot{Q}_r$, and the vibrational kinetic energy is given by

$$T_{\mathrm{V}} = (1/2) \sum_i m_i v_i^2 = (1/2) \sum_i \dot{Q}_r^2, \tag{2.2.5}$$

where the orthonormal condition for Q_i

$$\sum_i \boldsymbol{l}_{ir} \cdot \boldsymbol{l}_{is} \delta_{rs} \tag{2.2.6}$$

is employed. The center-of-mass condition and (2.2.2) may also be written as

$$\sum_i \sqrt{m_i}\, \boldsymbol{l}_{ir} = 0, \tag{2.2.7}$$

$$\sum_i \sqrt{m_i} \boldsymbol{r}_i^e \times \boldsymbol{l}_{ir} = 0. \tag{2.2.8}$$

The $\boldsymbol{l}$ matrix defined by (2.2.4) is of $3N \times (3N-6)$ or $3N \times (3N-5)$ dimension, but may be made a $3N \times 3N$ squared unitary matrix by adding the following submatrix of $3N \times 6$ dimension formed by referring to (2.2.7, 8):

$$\begin{bmatrix}
(m_1/M)^{1/2} & 0 & 0 & 0 & -z_1^e\sqrt{m_1/I_{yy}^e} & y_1^e\sqrt{m_1/I_{zz}^e} \\
(m_2/M)^{1/2} & 0 & 0 & 0 & -z_2^e\sqrt{m_2/I_{yy}^e} & y_2^e\sqrt{m_2/I_{zz}^e} \\
\vdots & \vdots & \vdots & \vdots & \vdots & \vdots \\
(m_N/M)^{1/2} & 0 & 0 & 0 & -z_N^e\sqrt{m_N/I_{yy}^e} & y_N^e\sqrt{m_N/I_{zz}^e} \\
0 & (m_1/M)^{1/2} & 0 & z_1^e\sqrt{m_1/I_{xx}^e} & 0 & -x_1^e\sqrt{m_1/I_{zz}^e} \\
0 & (m_2/M)^{1/2} & 0 & z_2^e\sqrt{m_2/I_{xx}^e} & 0 & -x_2^e\sqrt{m_2/I_{zz}^e} \\
\vdots & \vdots & \vdots & \vdots & \vdots & \vdots \\
0 & (m_N/M)^{1/2} & 0 & z_N^e\sqrt{m_N/I_{xx}^e} & 0 & -x_N^e\sqrt{m_N/I_{zz}^e} \\
0 & 0 & (m_1/M)^{1/2} & -y_1^e\sqrt{m_1/I_{xx}^e} & x_1^e\sqrt{m_1/I_{yy}^e} & 0 \\
0 & 0 & (m_2/M)^{1/2} & -y_2^e\sqrt{m_2/I_{xx}^e} & x_2^e\sqrt{m_2/I_{yy}^e} & 0 \\
\vdots & \vdots & \vdots & \vdots & \vdots & \vdots \\
0 & 0 & (m_N/M)^{1/2} & -y_N^e\sqrt{m_N/I_{xx}^e} & x_N^e\sqrt{m_N/I_{yy}^e} & 0
\end{bmatrix}. \tag{2.2.9}$$

It then follows that

$$\sum_r l_{ir}^{(x)} l_{jr}^{(x)} = \delta_{ij} - \sqrt{m_i m_j}\,(1/M + z_i^e z_j^e/I_{yy}^e + y_i^e y_j^e/I_{zz}^e), \tag{2.2.10}$$

$$\sum_r l_{ir}^{(x)} l_{jr}^{(y)} = (\sqrt{m_i m_j}/I_{zz}^e)\, x_j^e y_i^e, \tag{2.2.11}$$

and similar equations obtained from (2.2.10, 11) through cyclic permutation of x, y, and z also hold [2.9]. The moments of inertia are now functions of the normal coordinates as follows:

$$\boldsymbol{I} = \boldsymbol{I}^e + \sum_r \boldsymbol{a}_r Q_r + \sum_r \sum_s \boldsymbol{A}_{rs} Q_r Q_s, \tag{2.2.12}$$

where the coefficients $\boldsymbol{a}_r$ and $\boldsymbol{A}_{rs}$ may be expressed in terms of $\boldsymbol{r}_i^e$ and $\boldsymbol{l}_{ir}$.

It should be pointed out that the Eckart condition is not a unique way of defining the molecule-fixed coordinate system; it is a matter of convenience. It largely eliminates the coupling between the vibration and rotation, but not completely. A term

$$2\boldsymbol{\omega} \cdot \sum_i m_i (\delta \boldsymbol{r}_i \times \boldsymbol{v}_i) = 2\boldsymbol{\omega} \cdot \sum_r \sum_s \zeta_{rs} Q_r \dot{Q}_s \tag{2.2.13}$$

remains, where

$$\zeta_{rs} = \sum_i \boldsymbol{l}_{ir} \times \boldsymbol{l}_{is} = -\zeta_{sr}. \tag{2.2.14}$$

This term is referred to as the Coriolis term, and ζ_{rs} is called the Coriolis coupling constant.

When (2.1.40) with $\boldsymbol{v}_i$ retained is substituted into (2.1.41) and the translation energy is dropped, the kinetic energy for the vibration and rotation then are

$$T_V + T_R = (\hbar^2/2) \sum_s p_s^2 + (\hbar^2/2)\,(\boldsymbol{J} - \boldsymbol{p})^\dagger\, \mu (\boldsymbol{J} - \boldsymbol{p}), \tag{2.2.15}$$

where p_s denotes the momentum conjugate to Q_s, μ the inverse of the moment of inertia tensor (2.2.12) with the correction term arising from the Coriolis interaction $-\sum_r \sum_s \sum_t \zeta_{rt} \zeta_{st} Q_r Q_s$, and $\boldsymbol{p}$ is given by

$$\boldsymbol{p} = \sum_r \sum_s \zeta_{sr} Q_s p_r. \tag{2.2.16}$$

Equation (2.2.15) is added to the vibrational potential function given below to obtain the vibrational and rotational Hamiltonian:

$$V/hc = (1/2) \sum_r \omega_r q_r^2 + (1/3!) \sum_r \sum_s \sum_t \phi_{rst} q_r q_s q_t$$
$$+ (1/4!) \sum_r \sum_s \sum_t \sum_u \phi_{rstu} q_r q_s q_t q_u + \ldots, \tag{2.2.17}$$

where q_r denotes the dimensionless normal coordinate defined by $q_r = (2\pi c \omega_r/\hbar)^{1/2} Q_r$. The eigenvalue problem is solved by using the perturbation method, where the zeroth-order basis function is a product of the harmonic oscillators for the normal modes and the rigid body rotational wave function [2.10–12].

The first-order correction is nonvanishing only for the symmetric-top and spherical-top molecules in excited states of degenerate modes and is referred to as the first-order Coriolis term. For the (prolate) symmetric-top it is given by

$$-2(A\zeta)_t k l_t + \eta_{tJ} J(J+1) k l_t + \eta_{tK} k^3 l_t, \tag{2.2.18}$$

where l_t denotes the vibrational angular momentum quantum number taking $\pm v_t, \pm(v_t - 2), \ldots, \pm 1$ or 0 and the second and third terms represent higher-order (third-order) corrections.

Among the second-order correction terms the most important one is the change $\boldsymbol{\alpha}_r$ in the rotational constant upon the excitation of the rth vibrational mode, defined by

$$\boldsymbol{B}_{\mathrm{v}} = \boldsymbol{B}_{\mathrm{e}} - \sum_r \boldsymbol{\alpha}_r (v_r + d_r/2), \tag{2.2.19}$$

where v_r and d_r denote the vibrational quantum number and the degree of degeneracy of the rth normal mode, respectively. The detailed expressions for $\boldsymbol{\alpha}_r$ are given below. For an asymmetric top

$$\alpha_r^B = -(2B^2/\omega_r)\Big[\sum_{\xi} 3(a_r^{(b\xi)})^2/(4I_{\xi\xi}^e) + \sum_s (\zeta_{r,s}^{(b)})^2 (3\omega_r^2 + \omega_s^2)/(\omega_r^2 - \omega_s^2)$$
$$+ \pi(c/h)^{1/2} \sum_s \phi_{rrs} a_s^{(bb)} (\omega_r/\omega_s^{3/2})\Big], \tag{2.2.20}$$

and similar expressions apply to α_r^A and α_r^C after replacing b with a and c, respectively. For a symmetric top the formulas are different for α^A (or α^C) and α^B and also for α_r and α_t, where r and t denote the nondegenerate and degenerate vibrations, respectively; namely

$$\alpha_r^A = -(2A^2/\omega_r)\,[3(a_r^{(zz)})^2/(4I_{aa}^e) + \sum_s (\zeta_{r,s}^{(z)})^2 (3\omega_r^2 + \omega_s^2)/(\omega_r^2 - \omega_s^2)$$
$$+ \pi(c/h)^{1/2} \sum_s \phi_{rrs} a_s^{(zz)} (\omega_r/\omega_s^{3/2})], \tag{2.2.21}$$

$$\alpha_t^A = -(2A^2/\omega_t)\,[3(a_{t1}^{(xz)})^2/(4I_{bb}^e) + \sum_{t'} (\zeta_{t1,t'2}^{(z)})^2 (3\omega_t^2 + \omega_{t'}^2)/(\omega_t^2 - \omega_{t'}^2)$$
$$+ \pi(c/h)^{1/2} \sum_s \phi_{st1t1} a_s^{(zz)} (\omega_t/\omega_s^{3/2})] \tag{2.2.22}$$

$$\alpha_r^B = -(2B^2/\omega_r)\,[3[(a_r^{(xx)})^2 + (a_r^{(xy)})^2]/(4I_{bb}^e)$$
$$+ \sum_t [(\zeta_{r,t1}^{(y)})^2 + (\zeta_{r,t1}^{(x)})^2](3\omega_r^2 + \omega_t^2)/(\omega_r^2 - \omega_t^2)$$
$$+ \pi(c/h)^{1/2} \sum_s \phi_{rrs} a_s^{(xx)} (\omega_r/\omega_s^{3/2})], \tag{2.2.23}$$

$$\alpha_t^B = -(2B^2/\omega_t)\,[3(a_{t1}^{(xz)})^2/(8I_{aa}^e) + 3(a_{t1}^{(xx)})^2/(4I_{bb}^e)$$
$$+ (1/2)\sum_r [(\zeta_{r,t1}^{(y)})^2 + (\zeta_{r,t1}^{(x)})^2](3\omega_t^2 + \omega_r^2)/(\omega_t^2 - \omega_r^2)$$
$$+ \sum_{t'} [(\zeta_{t1,t'1}^{(y)})^2 + (\zeta_{t1,t'1}^{(x)})^2](3\omega_t^2 + \omega_{t'}^2)/(\omega_t^2 - \omega_{t'}^2)$$
$$+ \pi(c/h)^{1/2} \sum_r \phi_{rt1t1} a_r^{(xx)} (\omega_t/\omega_r^{3/2})]. \tag{2.2.24}$$

Here s denotes the nondegenerate mode and t' the doubly degenerate mode, and the symmetry axis is taken to coincide with the z axis and the x and y axes are perpendicular to it. Note that α_r contains the cubic potential constants.

Another important second-order correction is the centrifugal distortion effect on the rotational energy [2.13]. The lowest-order expression of it is proportional to J^4 and is given by

$$\boldsymbol{H}_{\mathrm{cent}}/h = (1/4) \sum_\alpha \sum_\beta \sum_\gamma \sum_\delta \tau_{\alpha\beta\gamma\delta} J_\alpha J_\beta J_\gamma J_\delta, \tag{2.2.25}$$

where the centrifugal distortion constant $\tau_{\alpha\beta\gamma\delta}$ is expressed as

$$\tau_{\alpha\beta\gamma\delta} = -(\hbar^4/h)/(I_{\alpha\alpha}^e I_{\beta\beta}^e I_{\gamma\gamma}^e I_{\delta\delta}^e) \sum_r a_r^{(\alpha\beta)} a_r^{(\gamma\delta)}/(8\pi^2 c^2 \omega_r^2) \tag{2.2.26}$$

and α, β, γ, and δ are the principal axes. In view of its definition (2.2.26), only 21 $\tau_{\alpha\beta\gamma\delta}$ are independent; there are 3 of the $\tau_{\alpha\alpha\alpha\alpha}$ type, 3 of the $\tau_{\alpha\alpha\beta\beta}$ type, 3 of the $\tau_{\alpha\beta\alpha\beta}$ type, 3 of the $\tau_{\alpha\alpha\beta\gamma}$ type, 3 of the $\tau_{\alpha\beta\alpha\gamma}$ type, and 6 of the $\tau_{\alpha\alpha\alpha\beta}$ type. For a molecule of orthorhombic symmetry (C_{2v}, D_2, D_{2h}, and higher symmetry) only the former three types are nonvanishing. For a lower-symmetry molecule other types of $\tau_{\alpha\beta\gamma\delta}$ exist, but their contributions are second order when the rigid-body rotational wave functions are used as bases and may be neglected in most cases. Since the $\tau_{\alpha\beta\alpha\beta}$ terms can be absorbed in the $\tau_{\alpha\alpha\beta\beta}$ terms because of the commutation relations (2.1.17–19), (2.2.25) is simplified to

$$\boldsymbol{H}_{\mathrm{cent}}/h = (1/4) \sum_\alpha \sum_\beta \tau'_{\alpha\alpha\beta\beta} J_\alpha^2 J_\beta^2, \quad \text{where} \tag{2.2.27}$$

$$\tau'_{\alpha\alpha\beta\beta} = \tau_{\alpha\alpha\beta\beta} + 2\tau_{\alpha\beta\alpha\beta}(\alpha \neq \beta), \tag{2.2.28}$$
$$\tau'_{\alpha\alpha\alpha\alpha} = \tau_{\alpha\alpha\alpha\alpha}.$$

This simplification also modifies the rotational constants slightly:

$$A \to A' = A + (1/4)\,(3\,\tau_{bcbc} - 2\,\tau_{caca} - 2\,\tau_{abab}), \tag{2.2.29}$$

$$B \to B' = B + (1/4)\,(3\,\tau_{caca} - 2\,\tau_{abab} - 2\,\tau_{bcbc}), \tag{2.2.30}$$

$$C \to C' = C + (1/4)\,(3\,\tau_{abab} - 2\,\tau_{bcbc} - 2\,\tau_{caca}). \tag{2.2.31}$$

Equation (2.2.27) contains six τ contains, but *Watson* [2.13] has shown that only five of the six constants can be determined from the observed spectrum. An expression appropriate for a near prolate symmetric rotor is reproduced here:

$$\begin{aligned} \boldsymbol{H}_{\mathrm{cent}}/h = &- \Delta_J \boldsymbol{J}^4 - \Delta_{JK} \boldsymbol{J}^2 J_a^2 - \Delta_K J_a^4 \\ &- 2\,\delta_J \boldsymbol{J}^2 (J_b^2 - J_c^2) - \delta_K [J_a^2 (J_b^2 - J_c^2) + (J_b^2 - J_c^2)\, J_a^2], \end{aligned} \tag{2.2.32}$$

and the tranformation from (2.2.27) to (2.2.32) again introduces additional corrections to A', B', and C' as follows:

$$A' \to A'' = A' + 16\,R_6 \tag{2.2.33}$$

$$B' \to B'' = B' - 16\,R_6 (A - C)/(B - C) \tag{2.2.34}$$

$$C' \to C'' = C' + 16\,R_6 (A - B)/(B - C), \tag{2.2.35}$$

where $R_6 = (\tau_{bbbb} + \tau_{cccc} - 2\,\tau'_{bbcc})/64$.

Other effects of the vibration-rotation interaction are much less important and contribute little to the molecular spectra. There are a few exceptions, namely the cases where two rotational levels are in near degeneracy and are connected by a vibration-rotation matrix element.

A good example is a C_{3v} symmetric top, in which rotational levels with $k - l_t = 0$ or $3n$ (n denoting an integer) consist of pairs of levels, one belonging to A_1 and the other to A_2 symmetry, and may thus be split, whereas other levels with $k - l_t = 3n \pm 1$ belong to E symmetry and are not split. In a $v_t = 1$ state, where $l_t = \pm 1$, the $k = \pm 1$, $l_t = \pm 1$ levels are degenerate to the first order, as (2.2.18) indicates, but are split in the second order by the $\Delta k = \pm 2$, $\Delta l_t = \pm 2$ matrix elements (the so-called 2,2 interaction), resulting in a doublet referred to as the l-type doublet. The splitting is given by

$$\Delta E_l/h = q_t J(J + 1), \tag{2.2.36}$$

where the l-type doubling constant q_t is

$$\begin{aligned} q_t = (2\,B^2/\omega_t)\,[&- 3\,(a_{t1}^{(yz)})^2/(4\,I_{aa}^e) \\ &+ \sum_r [(\zeta_{r,t1}^{(y)})^2 - (\zeta_{r,t1}^{(x)})^2]\,(3\,\omega_t^2 + \omega_r^2)/(\omega_t^2 - \omega_r^2) \\ &+ 2\,\pi (c/h)^{1/2} \sum_{t'} \phi_{t1t1t'1}\, a_{t'1}^{(yy)} (\omega_t/\omega_{t'}^{3/2})]. \end{aligned} \tag{2.2.37}$$

As expected from molecular symmetry, the interaction may be classified according to the value of $\Delta(k - l_t) = 0, \pm 3, \pm 6, \ldots$. One of the ± 3 cases, i.e., $\Delta k = \pm 1$, $\Delta l_t = \mp 2$, is often referred to as the 2, -1 interaction [2.14].

When two vibrational states come close, the rotational energy level structure of each state may show large deviations from that discussed above. There are two main interactions; one is a purely vibrational effect referred to as the Fermi interaction and the other is due to the Coriolis interaction. The Fermi interaction is essentially a mixing of the two vibrational states caused by anharmonic terms of the vibrational potential function. What frequently happens is that a stretching (s) fundamental state ($v_s = 1$) mixes with the first overtone state ($v_b = 2$) of a bending (b) mode. These two states are connected by the cubic anharmonic potential term $(1/2)\, \phi_{sbb} q_s q_b^2$. Consequent to mixing, one state will acquire the character of the other state; the rotational constants are then weighted averages of the original ones.

The most important Coriolis interaction is that between two fundamental vibrational states, although higher-order interactions have also been reported in quite a large number of cases. The interaction perturbs the rotational levels, as indicated by (2.2.20–24); when the resonance is exact, these expressions for α_r cease to be valid.

2.3 Fine and Hyperfine Structures

When a molecule involves one or more unpaired electrons, as free radicals do, the rotational structure of its spectrum shows splittings referred to as the fine and hyperfine structures. The resultant total electron spin angular momentum $\boldsymbol{S}$ and electron orbital angular momentum $\boldsymbol{L}$ couple with each other and with the molecular rotation, resulting in fine structures of the spectrum. If the molecule further contains at least one nucleus with nonzero nuclear spin, the nuclear spin angular momentum $\boldsymbol{I}$ couples with $\boldsymbol{S}$ and/or $\boldsymbol{L}$ to produce additional structures in the spectrum, called the magnetic hyperfine structures. The nuclear spin angular momentum generates hyperfine structures also in the spectrum of molecules without unpaired electrons, through either the nuclear electric quadrupole interaction (provided that $I \geq 1$) or the nuclear spin-overall rotation interaction. This section deals with these interactions and presents matrix elements needed in analyzing the observed spectra.

2.3.1 Electron Spin and Electron Orbital Angular Momenta

In a well-isolated electronic state, the total electron spin angular momentum $\boldsymbol{S}$ is defined with the eigenvalue $S = 0, 1/2, 1, \ldots$, and, according to this value, the electronic state is classified as singlet, doublet, triplet, and so on. The electron orbital angular momentum $\boldsymbol{L}$ is, on the other hand, usually not well specified in molecules, in sharp contrast with the case of atoms, in which the electrostatic field acting on electrons is approximately central so that $\boldsymbol{L}^2$ is nearly conserved. For diatomic and linear polyatomic molecules, however, the projection of $\boldsymbol{L}$ along the molecular axis is often well defined with the eigenvalue Λ. The electronic state of a linear molecule is thus classified by the value of Λ; the states with $\Lambda = 0, \pm 1, \pm 2, \pm 3, \ldots$ are called $\Sigma, \Pi, \Delta, \Phi, \ldots$. The orbital angular momentum is nearly or completely quenched in most of the symmetric and asymmetric top molecules, so that these molecules normally behave like linear molecules in Σ states.

Hund proposed five extreme cases of angular momentum coupling, of which two are by far the most important. One of them is called Hund's case (a), in which both the orbital and spin angular momenta are strongly coupled with the molecular axis, so their components Λ and Σ along the axis and the sum $\Omega = \Lambda + \Sigma$ are well defined. The end-over-end rotation angular momentum $\boldsymbol{R}$ couples with Ω to form the total angular momentum $\boldsymbol{J}$. The other important scheme is Hund's case (b) where the orbital angular momentum is still strongly coupled to the molecular axis, but the electron spin is decoupled so that Σ is no longer defined. The rotational angular momentum $\boldsymbol{R}$ is added to Λ to form $\boldsymbol{N}$, which is then combined with $\boldsymbol{S}$ to give the total angular momentum $\boldsymbol{J}$:

$$\boldsymbol{J} = \boldsymbol{N} + \boldsymbol{S}. \tag{2.3.1}$$

Most molecules with finite Λ values, like many diatomic and linear polyatomic molecules, are best treated by Hund's case (a), while molecules in Σ states and also most nonlinear molecules closely approximate Hund's case (b). An exception is the CH_3O radical in the $\tilde{X}\,^2E$ state, where the orbital angular momentum is not completely quenched [2.15]. The bending vibration in a linear polyatomic molecule, when excited, generates the vibrational angular momentum $\boldsymbol{l}$, which, coupled with the electron orbital angular momentum, splits excited vibrational states, a phonemenon known as the Renner-Teller effect, Sect. 2.4.

An appropriate base function for Hund's case (a) is a decoupled representation given by $|J \Omega M_J\rangle\, |S \Sigma\rangle\, |n, \Lambda\rangle$. The first wave function is the symmetric top rotational wave function which diagonalizes $\boldsymbol{J}^2$, J_z, and J_Z with the eigenvalues $J(J+1)$, Ω, and M_J. The second term represents the spin part, which is an eigenfunction of $\boldsymbol{S}^2$ and S_z with the eigenvalues $S(S+1)$ and Σ. The last part denotes the electronic wave function, which is an eigenfunction of L_z with the eigenvalue Λ; n stands for other parameters specifying the electronic state. The nonvanishing off-diagonal matrix elements of $\boldsymbol{S}$ are

$$\langle S \Sigma \pm 1 |\, S_\pm\, | S \Sigma \rangle = [S(S+1) - \Sigma(\Sigma \pm 1)]^{1/2}, \tag{2.3.2}$$

where

$$S_\pm = S_x \pm \mathrm{i} S_y .$$

It is a little more complicated to evaluate the matrix elements in Hund's case (b). An appropriate rotational wave function is again that for a symmetric top, but J and Ω are replaced by N and K, respectively, i.e. $|NKM\rangle$. It should be noted that the molecule-fixed components of $\boldsymbol{S}$ do not commute with $\boldsymbol{N}$. The overall base function may be written in a coupled representation as $|NKSJM_J\rangle$. The case (b) function is related to the case (a) basis set by a transformation:

$$|NKSJM_J\rangle = \sum_{\Sigma,P} (-1)^{-J+P+2S}(2N+1)^{1/2}\begin{pmatrix} J & N & S \\ P & -K & -\Sigma \end{pmatrix} \times |JPM_J\rangle\,|S\Sigma\rangle, \qquad (2.3.3)$$

where the axial component of J is designated by P, rather than Ω, to encompass cases of nonlinear molecules.

2.3.2 Fine-Structure Hamiltonian

The unpaired electrons introduce three kinds of fine-structure interactions, spin-orbit, spin-rotation, and spin-spin interactions. When the electronic orbital angular momentum is not quenched, the spin-orbit interaction is by far the most important among the three. The simplest expression for this interaction is given by

$$\boldsymbol{H}_{\mathrm{SO}} = A_{\mathrm{SO}}\boldsymbol{L}\cdot\boldsymbol{S}, \qquad (2.3.4)$$

as in the case of atoms [2.5]. (All Hamiltonians in this chapter are expressed in units of either MHz or cm^{-1}.) For diatomic and linear polyatomic molecules for which L_z is well defined, i.e., molecules which closely approximate Hund's case (a), (2.3.4) may be divided into two parts:

$$\boldsymbol{H}_{\mathrm{SO}} = A_{\mathrm{SO}}[L_z S_z + (L_+ S_- + L_- S_+)/2], \qquad (2.3.5)$$

of which the first gives the first-order term, while the second part connects electronic states which differ in Λ by ± 1. The latter is combined with other Hamiltonian operators including itself which are linear in $L_\pm$ to generate a large number of second-order terms; the Λ-type doubling is an important example of such terms, Sect. 2.3.4. For linear molecules in $^2\Sigma$ states and nonlinear molecules, the first-order term is missing, leaving the second-order contributions as the most important.

The second interaction is called the spin-rotation interaction, which is expressed as [2.4, 16]

$$\boldsymbol{H}_{\mathrm{SR}} = (1/2)\sum_{\alpha,\beta} \varepsilon_{\alpha\beta}(N_\alpha S_\beta + S_\beta N_\alpha). \qquad (2.3.6)$$

Since the molecular rotation generates a weak magnetic field, the unpaired electron spin magnetic moment will interact with it, resulting in the spin-rotation interaction. However, this classical mechanism has been known to be relatively unimportant and outweighed by the second-order term generated by a combination of the spin-orbit interaction and the electronic Coriolis interaction [the last term of (2.3.11), or more appropriately for Hund's case (b), the term in $N_+ L_- + N_- L_+$] [2.17]. For linear molecules $\boldsymbol{N}$ is always perpendicular to the molecular axis, and (2.3.6) simplifies to

$$\boldsymbol{H}_{\mathrm{SR}} = \gamma \boldsymbol{N} \cdot \boldsymbol{S} . \tag{2.3.7}$$

The third interaction is referred to as the spin-spin interaction, which occurs only for cases with $S \geqq 1$, namely for molecules involving more than one unpaired electron. The Hamiltonian for this interaction is given by [2.18]

$$\boldsymbol{H}_{\mathrm{SS}} = \alpha_0 (3 S_z^2 - \boldsymbol{S}^2) + \beta_0 (S_x^2 - S_y^2) . \tag{2.3.8}$$

Only the first term exists in linear molecules except for those in $^3\Pi$ states [2.19]. In terms of classical mechanics, this interaction may be regarded as the dipole-dipole interaction between two unpaired electron spin magnetic moments. It has, however, been known that the second-order contribution of the spin-orbit interaction is again substantial, especially when the molecule involves heavy elements [2.17].

In analyzing high-resolution molecular spectra, the centrifugal corrections to these interactions must be taken into account.

2.3.3 Hamiltonian Matrix Elements Appropriate for Hund's Case (a)

The discussion is limited to diatomic and linear polyatomic molecules. The rotational Hamiltonian is given by

$$\boldsymbol{H}_{\mathrm{ROT}} = B_{\mathrm{v}} \boldsymbol{R}^2 - D_{\mathrm{v}} \boldsymbol{R}^4 + \dots . \tag{2.3.9}$$

Upon replacing $\boldsymbol{R}$ with $\boldsymbol{J} - \boldsymbol{L} - \boldsymbol{S}$, the first term reduces to

$$\boldsymbol{H}_{\mathrm{ROT}} = B_{\mathrm{v}} [(J_x - L_x - S_x)^2 + (J_y - L_y - S_y)^2] + \dots . \tag{2.3.10}$$

The terms $2 B_{\mathrm{v}} (L_x S_x + L_y S_y)$ may be combined with the spin-orbit interaction Hamiltonian (2.3.4). When $B(L_x^2 + L_y^2)$ is omitted, (2.3.10) reduces to

$$\begin{aligned} \boldsymbol{H}_{\mathrm{ROT}} = {} & B_{\mathrm{v}} (\boldsymbol{J}^2 - J_z^2 + \boldsymbol{S}^2 - S_z^2) \\ & - B_{\mathrm{v}} (J_+ S_- + J_- S_+) - B_{\mathrm{v}} (J_+ L_- + J_- L_+) + \dots . \end{aligned} \tag{2.3.11}$$

The second term connects two states which differ in Σ by ± 1 and thus causes spin uncoupling when the rotation is excited; Hund's case (a) then moves to Hund's case (b) [2.20]. The last term is referred to as the electronic Coriolis term; it has matrix elements between two electronic states which differ in Λ by ± 1. As mentioned above, it is combined with the spin-orbit interaction term

$(A_{SO} + 2B_v)(L_+S_- + L_-S_+)/2$ to generate second-order terms. The degeneracy of pair levels in a $^2\Pi$ molecule, $\Lambda = 2\Sigma = \pm 1$ or $\Lambda = -2\Sigma = \pm 1$, is lifted by these second-order terms, a phenomenon referred to as Λ-type doubling [2.21]. The same mechanism also contributes to the spin-rotation interaction in $^2\Sigma$ molecules.

The effective Hamiltonian for diatomic and linear polyatomic molecules is thus given by

$$\boldsymbol{H}_{\text{eff}} = \boldsymbol{H}_{\text{ROT}} + \boldsymbol{H}_{\text{SO}} + \boldsymbol{H}_{\text{SS}} + \boldsymbol{H}_{\text{SR}} + \boldsymbol{H}_{\text{SOCD}}, \tag{2.3.12}$$

where

$$\boldsymbol{H}_{\text{ROT}} = B_v(\boldsymbol{J}^2 - J_z^2 + \boldsymbol{S}^2 - S_z^2) - B_v(J_+S_- + J_-S_+) - D_v(\boldsymbol{J} - \boldsymbol{L} - \boldsymbol{S})^4, \tag{2.3.13}$$

$$\boldsymbol{H}_{\text{SO}} = A_{\text{SO}} L_z S_z, \tag{2.3.14}$$

$$\boldsymbol{H}_{\text{SS}} = (2/3)\,\lambda_v(3S_z^2 - \boldsymbol{S}^2) + \eta\, L_z S_z[S_z^2 - (3\boldsymbol{S}^2 - 1)/5], \tag{2.3.15}$$

$$\boldsymbol{H}_{\text{SR}} = \gamma_v(\boldsymbol{J} - \boldsymbol{S}) \cdot \boldsymbol{S}, \quad \text{and} \tag{2.3.16}$$

$$\boldsymbol{H}_{\text{SOCD}} = (1/2)\,A_{Dv}[(\boldsymbol{J} - \boldsymbol{L} - \boldsymbol{S})^2 L_z S_z + L_z S_z(\boldsymbol{J} - \boldsymbol{L} - \boldsymbol{S})^2] + (1/3)\,\lambda_D[(\boldsymbol{J} - \boldsymbol{L} - \boldsymbol{S})^2(3S_z^2 - \boldsymbol{S}^2) + (3S_z^2 - \boldsymbol{S}^2)(\boldsymbol{J} - \boldsymbol{L} - \boldsymbol{S})^2]. \tag{2.3.17}$$

As noted above, $\boldsymbol{H}_{\text{SS}}$ is absent in doublet states. The α_0 constant of (2.3.8) has been replaced by $(2/3)\,\lambda_v$, because λ_v is more common in the discussion of the spectra of diatomic molecules [2.22]. *Brown* et al. [2.23] have introduced the η term in (2.3.15) in fitting the spectra in quartet and higher multiplet spin states. The operators $(\boldsymbol{J} - \boldsymbol{L} - \boldsymbol{S})^4$ in (2.3.13) and $(\boldsymbol{J} - \boldsymbol{L} - \boldsymbol{S})^2$ in (2.3.17) may be reduced as for $\boldsymbol{H}_{\text{ROT}}$, (2.3.11); since these terms represent small corrections, their matrix elements off-diagonal in Λ need not be retained in most cases.

Table 2.3. Matrix elements of $\boldsymbol{H}_{\text{eff}}$ for a $^2\Pi$ electronic state [a]

$$\langle ^2\Pi_{1/2}, J, \pm |\, H_{\text{eff}}\, | ^2\Pi_{1/2}, J, \pm \rangle = -A_{\text{SO}}/2 + o_v^* + (p_v^* + q_v^*)/2 + D_v + (B_v - A_{Dv}/2 - D_v + q_v^*/2)\,X^2 - D_v X^4 \mp (-1)^{J-S}(q_v + p_v/2)\,X$$

$$\langle ^2\Pi_{3/2}, J, \pm |\, H_{\text{eff}}\, | ^2\Pi_{3/2}, J, \pm \rangle = A_{\text{SO}} + q_v^*/2 - 3D_v + (B_v + A_{Dv}/2 + 3D_v + q_v^*/2)(X^2 - 2) - D_v X^4$$

$$\langle ^2\Pi_{3/2}, J, \pm |\, H_{\text{eff}}\, | ^2\Pi_{1/2}, J, \pm \rangle = -[B_v - \gamma_v/2 + p_v^*/4 + q_v^*/2 - 2D_v(X^2 - 1) \mp (-1)^{J-S}(q_v/2)\,X](X^2 - 1)^{1/2}$$

[a] $X = J + 1/2$

The matrix elements of $\boldsymbol{H}_{\mathrm{eff}}$ are evaluated using the wave function $|J\Omega M_J\rangle\,|S\Sigma\rangle\,|n,\Lambda\rangle$ as a basis, and are summarized in Table 2.3 for the $^2\Pi$ electronic state.

2.3.4 Λ-Type Doubling in $^2\Pi$ States

In Sect. 2.3.2 we ignored terms involving the operators $L_\pm$ which connect electronic states different in Λ by ± 1, as already mentioned. When treated by second-order perturbation theory, these terms contribute to the matrix elements diagonal in Λ and modify the meanings of molecular parameters present in these matrix elements. A more conspicuous effect of the off-diagonal terms is to lift the degeneracy between the $+\Lambda$ and $-\Lambda$ states; the pair levels thus split are referred to as Λ-type doublets. All levels with finite Λ and $\Omega = \Lambda + \Sigma$ are degenerate and may be split by Λ-type doubling, but the effect is most marked in Π molecules, so we shall restrict our discussion to $^2\Pi$ molecules only.

The Λ-type doubling may be regarded as a consequence of breaking the axial symmetry of the Π molecule; the doubly degenerate levels are split into two, one symmetric and the other antisymmetric with respect to reflection on a plane including the molecular axis z. The operation is chosen to be σ_{xz}. Here the phase of *Lepard* [2.24] is adopted:

$$\sigma_{xz}\,|J\Omega\rangle = (-1)^{J-\Omega}\,|J, -\Omega\rangle, \tag{2.3.18}$$

$$\sigma_{xz}\,|S\Sigma\rangle = (-1)^{S-\Sigma}\,|S, -\Sigma\rangle, \quad \text{and} \tag{2.3.19}$$

$$\sigma_{xz}\,|n,\Lambda\rangle = (-1)^{\Lambda+s}\,|n, -\Lambda\rangle, \tag{2.3.20}$$

where $s = 0$ for all electronic states except Σ^- for which $s = 1$. The total base wave function for a doublet state is thus transformed by σ_{xz} as

$$\begin{aligned}\sigma_{xz}\,|J\Omega S\Sigma n\Lambda\rangle &= (-1)^{J-\Omega+S-\Sigma+\Lambda+s}\,|J, -\Omega, S, -\Sigma, n, -\Lambda\rangle \\ &= (-1)^{J-S+s}\,|J, -\Omega, S, -\Sigma, n, -\Lambda\rangle.\end{aligned} \tag{2.3.21}$$

Therefore, the following linear combination of the original basis functions

$$|^2\,|\Lambda|^s, J, \pm\rangle = (1/\sqrt{2})\,[|J\Omega S\Sigma n\Lambda\rangle \pm (-1)^{J-S+s}\,|J-\Omega S-\Sigma n-\Lambda\rangle] \tag{2.3.22}$$

will have $\pm$ parity with respect to the σ_{xz} operation.

The terms in the Hamiltonian which cause Λ doubling are yielded as follows:

$$\boldsymbol{H}_\Lambda = (1/2)\,(A_{\mathrm{SO}} + 2B_{\mathrm{v}})\,(L_+S_- + L_-S_+) - B_{\mathrm{v}}(J_+L_- + J_-L_+), \tag{2.3.23}$$

the nonvanishing matrix elements of which are given, in terms of the original base functions, by

$$\langle J\Omega S\Sigma \mp 1\, n'\Lambda \pm 1 |\, H_\Lambda | J\Omega S\Sigma n\Lambda\rangle$$
$$= [S(S+1) - \Sigma(\Sigma \mp 1)]^{1/2} \langle n'\Lambda \pm 1 |\, L_\pm (B_v + A_{SO}/2)\, | n\Lambda\rangle, \quad (2.3.24\,a)$$

and

$$\langle J\Omega \pm 1\, S\Sigma n'\Lambda \pm 1 |\, H_\Lambda\, | J\Omega S\Sigma n\Lambda\rangle$$
$$= -\,[J(J+1) - \Omega(\Omega \pm 1)]^{1/2} \langle n'\Lambda \pm 1 |\, L_\pm B_v\, | n\Lambda\rangle. \quad (2.3.24\,b)$$

When only a $^2\Pi$ and a $^2\Sigma$ state are taken into account, the matrix elements (2.3.24) may be expressed in terms of three parameters $\mathcal{O}$, ζ, and η, defined by [2.25]

$$\langle \Omega = 1/2, \Lambda = 1 |\, H_\Lambda\, | \Omega = 1/2, \Lambda = 0\rangle$$
$$= \langle \Lambda = 1 |\, L_+ (B_v + A_{SO}/2)\, | \Lambda = 0\rangle \equiv \mathcal{O}, \quad (2.3.25\,a)$$

$$\langle \Omega = -1/2, \Lambda = -1 |\, H_\Lambda\, | \Omega = -1/2, \Lambda = 0\rangle \equiv (-1)^s \mathcal{O}, \quad (2.3.25\,b)$$

$$\langle \Omega = 1/2, \Lambda = 1 |\, H_\Lambda\, | \Omega = -1/2, \Lambda = 0\rangle$$
$$= -(J + 1/2) \langle \Lambda = +1 |\, L_+ B_v\, | \Lambda = 0\rangle \equiv -\zeta, \quad (2.3.25\,c)$$

$$\langle \Omega = -1/2, \Lambda = -1 |\, H_\Lambda\, | \Omega = 1/2, \Lambda = 0\rangle \equiv -(-1)^s \zeta, \quad (2.3.25\,d)$$

$$\langle \Omega = 3/2, \Lambda = 1 |\, H_\Lambda\, | \Omega = 1/2, \Lambda = 0\rangle$$
$$= -[(J + 1/2)^2 - 1]^{1/2} \langle \Lambda = 1 |\, L_+ B_v\, | \Lambda = 0\rangle \equiv -\eta, \quad (2.3.25\,e)$$

and

$$\langle \Omega = -3/2, \lambda = -1 |\, H_\Lambda\, | \Omega = -1/2, \Lambda = 0\rangle \equiv -(-1)^s \eta. \quad (2.3.25\,f)$$

When the symmetrized basis set (2.3.22) is used, these matrix elements are summarized as

$$\langle ^2\Pi_{1/2}, J, \pm |\, H_\Lambda\, | ^2\Sigma^s_{1/2}, J, \pm\rangle = \mathcal{O} \mp (-1)^{J-S+s} \zeta, \quad (2.3.26\,a)$$

and

$$\langle ^2\Pi_{3/2}, J, \pm |\, H_\Lambda\, | ^2\Sigma^s_{1/2}, J, \pm\rangle = -\eta. \quad (2.3.26\,b)$$

It should be noted that a great simplification takes place in going from (2.3.25) to (2.3.26), thanks to the fact that states of different parity are not mixed by $\boldsymbol{H}_\Lambda$.

The matrix elements effective for a $^2\Pi$ state may then be calculated by applying a Van Vleck transformation to (2.3.26):

$$\langle ^2\Pi_{1/2}, J, \pm |\, H_\Lambda\, | ^2\Pi_{1/2}, J, \pm\rangle$$
$$= -\sum_i [\mathcal{O}_i \bar{\mathcal{O}}_i + \zeta_i \bar{\zeta}_i \mp (-1)^{J-S+s} (\mathcal{O}_i \bar{\zeta}_i + \zeta_i \bar{\mathcal{O}}_i)]/\Delta E_i, \quad (2.3.27\,a)$$

$$\langle ^2\Pi_{3/2}, J, \pm |\, H_\Lambda\, | ^2\Pi_{3/2}, J, \pm\rangle = -\sum_i \eta_i \bar{\eta}_i / \Delta E_i, \quad (2.3.27\,b)$$

and

$$\langle {}^2\Pi_{3/2}, J, \pm | H_\Lambda | {}^2\Pi_{1/2}, J, \pm \rangle$$
$$= -\sum_i [-\eta_i \bar{\mathcal{O}}_i \pm (-1)^{J-S+s} \eta_i \bar{\zeta}_i]/\Delta E_i, \qquad (2.3.27\,\mathrm{c})$$

where $\Delta E_i = E_{\Sigma i} - E_\Pi$ and the bar above the constant denotes the conjugate complex. After removing the J-dependent factors from ζ and η, the three parameters are replaced by the following constants [2.21]:

$$o_v^* = -\sum_i |\langle \Lambda = 1 | L_+ A_{SO}/2 | \Lambda = 0 \rangle|^2/\Delta E_i, \qquad (2.3.28\,\mathrm{a})$$

$$p_v^* = -4\sum_i \langle \Lambda = 1 | L_+ A_{SO}/2 | \Lambda = 0 \rangle \langle \Lambda = 0 | L_- B_v | \Lambda = 1 \rangle/\Delta E_i, \qquad (2.3.28\,\mathrm{b})$$

$$q_v^* = -2\sum_i |\langle \Lambda = 1 | L_+ B_v | \Lambda = 0 \rangle|^2/\Delta E_i, \qquad (2.3.28\,\mathrm{c})$$

$$p_v = -4\sum_i (-1)^s \langle \Lambda = 1 | L_+ A_{SO}/2 | \Lambda = 0 \rangle \langle \Lambda = 0 | L_- B_v | \Lambda = 1 \rangle/\Delta E_i, \qquad (2.3.28\,\mathrm{d})$$

and

$$q_v = -2\sum_i (-1)^s |\langle \Lambda = 1 | L_+ B_v | \Lambda = 0 \rangle|^2/\Delta E_i. \qquad (2.3.28\,\mathrm{e})$$

As shown in Table 2.3, p_v and q_v cause Λ doubling, whereas the remaining three only modify other parameters: o_v^* contributes to the separation between ${}^2\Pi_{1/2}$ and ${}^2\Pi_{3/2}$, which is essentially determined by A_{SO}, q_v^* enters in the matrix primarily in the form $B_v - q_v^*/2$, and p_v^* replaces the spin-rotation interaction constant γ_v with $\gamma_v - p_v^*/2$. Therefore, o_v^*, p_v^*, and q_v^* are not determinable constants. It is interesting to note that both p_v and q_v contain a sign factor $(-1)^s$, which means that contributions of Σ^+ and Σ^- states to p_v and q_v are cancelled. This fact is especially important in interpreting the observed p_v and q_v constants using their second-order expressions (2.3.28 d, e).

2.3.5 Hamiltonian Matrix Elements Appropriate for Hund's Case (b)

Because most of the symmetric and asymmetric top molecules belong to Hund's case (b), this section deals primarily with such nonlinear molecules. Hund's case (b) diatomic and linear polyatomic molecules are readily treated by simply ignoring the component of the rotational angular momentum $\boldsymbol{R}$ along the molecular axis.

In Hund's case (b) the spin-rotation interaction is by far the most important of the fine-structure interactions, and the matrix elements necessary for this interaction are readily obtained using the spherical irreducible tensor method [2.6]. Some appropriate tensor relations are given in Table 2.4.

Table 2.4. Useful relations for spherical tensors

1) *The spherical tensor notation of a vector* $\boldsymbol{r}$

$$r_0 = z\,,$$
$$r_\pm = \mp 2^{-1/2}(x \pm \mathrm{i}\,y).$$

2) *Coupling of two tensor operators*

$$X^l_m = (-1)^{l_1 - l_2 - m}(2l+1)^{1/2} \sum_{m_1,\, m_2} \begin{pmatrix} l_1 & l_2 & l \\ m_1 & m_2 & m \end{pmatrix} T^{l_1}_{m_1} U^{l_2}_{m_2}\,.$$

3) *Scalar product of two tensors*

$$\boldsymbol{T} \cdot \boldsymbol{U} = \sum_q (-1)^q\, T^k_q\, U^k_{-q}\,.$$

4) *The Wigner-Eckart theorem*

$$\langle r', j', m'|\, T^k_q \,|r, j, m\rangle = (-1)^{j'-m'} \begin{pmatrix} j' & k & j \\ -m' & q & m \end{pmatrix} \langle r', j' \,||\, T^k \,||\, r, j\rangle,$$

where $\langle r', j' \,||\, T^k \,||\, r, j\rangle$ is called the reduced matrix element.

5) *Computation of the reduced matrix elements.*
Choose m, m', and q so as to give a trivial value for $\langle r', j, m'|\, T^k_q \,|r, j, m\rangle$ and divide it by

$$(-1)^{j-m'} \begin{pmatrix} j & k & j \\ -m' & q & m \end{pmatrix}.$$

For example,

$$\langle S \,||\, T^1(\boldsymbol{S}) \,||\, S\rangle = [S(S+1)(2S+1)]^{1/2}, \quad \text{or}$$

$$\langle J'K' \,||\, \mathscr{D}^k_{\cdot q}(\theta, \phi) \,||\, JK\rangle = [(2J'+1)(2J+1)]^{1/2}(-1)^{J'-K'} \begin{pmatrix} J' & k & J \\ -K' & q & K \end{pmatrix}.$$

6) *Matrix elements of coupled tensor operators.*
i) Two noncommuting tensor operators operating on the same system:

$$\langle r', j' \,||\, \boldsymbol{X}^K \,||\, r, j\rangle = (2K+1)^{1/2}(-1)^{K+j+j'} \sum_{r'',\, j''} \begin{Bmatrix} k_1 & k_2 & K \\ j & j' & j'' \end{Bmatrix}$$
$$\times \langle r', j' \,||\, T^{k_1} \,||\, r'', j''\rangle \langle r'', j'' \,||\, T^{k_2} \,||\, r, j\rangle.$$

ii) Tensor operators operating on different systems:

$$\langle r', j'_1, j'_2, J' \,||\, \boldsymbol{X}^K \,||\, r, j_1, j_2, J\rangle = \sum_{r''} \langle r', j'_1 \,||\, T^{k_1} \,||\, r'', j_1\rangle \langle r'', j'_2 \,||\, U^{k_2} \,||\, r, j_2\rangle$$
$$\times [(2J'+1)(2J+1)(2K+1)]^{1/2} \begin{Bmatrix} j'_1 & j_1 & k_1 \\ j'_2 & j_2 & k_2 \\ J' & J & K \end{Bmatrix}.$$

Table 2.4. (continued)

iii) Scalar product of two commuting tensor operators:

$$\langle r', j_1', j_2', J', M' | \boldsymbol{T}^k \cdot \boldsymbol{U}^k | r, j_1, j_2, J, M \rangle = \delta_{J'J} \delta_{M'M} (-1)^{j_1 + j_2' + J} \begin{Bmatrix} J & j_2' & j_1' \\ k & j_1 & j_2 \end{Bmatrix}$$

$$\times \sum_{r''} \langle r', j_1' || T^k || r'', j_1 \rangle \langle r'', j_2' || U^k || r, j_2 \rangle.$$

iv) Single operator in coupled scheme.

If a tensor operator $\boldsymbol{T}^k$ is operating only on part 1,

$$\langle r', j_1', j_2', J' || T^k || r, j_1, j_2, J \rangle = \delta_{j_2' j_2} (-1)^{j_1' + j_2 + J + k} [(2J' + 1)\ (2J + 1)]^{1/2} \begin{Bmatrix} j_1' & J' & j_2 \\ J & j_1 & k \end{Bmatrix}$$

$$\times \langle r', j_1' || T^k || r, j_1 \rangle.$$

Similarly, for U^k operating only on part 2,

$$\langle r', j_1', j_2', J' || U^k || r, j_1, j_2, J \rangle = \delta_{j_1' j_1} (-1)^{j_1 + j_2 + J' + k} [(2J' + 1)\ (2J + 1)]^{1/2} \begin{Bmatrix} j_2' & J' & j_1 \\ J & j_2 & k \end{Bmatrix}$$

$$\times \langle r', j_2' || U^k || r, j_2 \rangle.$$

The spin-rotation Hamiltonian $\boldsymbol{H}_{\mathrm{SR}}$ (2.3.6) is rewritten in terms of spherical tensors as

$$\boldsymbol{H}_{\mathrm{SR}} = (1/2) \sum_{k=0}^{2} [T^k(\boldsymbol{\varepsilon}) \cdot T^k(\boldsymbol{N}, \boldsymbol{S}) + T^k(\boldsymbol{N}, \boldsymbol{S}) \cdot T^k(\boldsymbol{\varepsilon})], \tag{2.3.29}$$

where

$$T_p^k(\boldsymbol{N}, \boldsymbol{S}) = (-1)^p (2k + 1)^{1/2} \sum_{p_1, p_2} T_{p_1}^1(\boldsymbol{N})\, T_{p_2}^1(\boldsymbol{S}) \begin{pmatrix} 1 & 1 & k \\ p_1 & p_2 & -p \end{pmatrix} \tag{2.3.30}$$

and $T^k(\boldsymbol{\varepsilon})$ represents an irreducible tensor of rank $k = 0, 1$, and 2 of the spin-rotation coupling constant. The components of the tensors are related to the Cartesian components as follows:

$$T_{+1}^1(\boldsymbol{X}) = -(1/\sqrt{2})\,(X_x + \mathrm{i} X_y), \tag{2.3.31 a}$$

$$T_0^1(\boldsymbol{X}) = X_z, \quad \text{and} \tag{2.3.31 b}$$

$$T_{-1}^1(\boldsymbol{X}) = (1/\sqrt{2})\,(X_x - \mathrm{i} X_y), \tag{2.3.31 c}$$

where $\boldsymbol{X} = \boldsymbol{N}$ or $\boldsymbol{S}$, and

$$T_0^0(\boldsymbol{\varepsilon}) = -(1/\sqrt{3})\,(\varepsilon_{xx} + \varepsilon_{yy} + \varepsilon_{zz}), \tag{2.3.32 a}$$

$$T_0^1(\boldsymbol{\varepsilon}) = -(1/\sqrt{2})\,\mathrm{i}(\varepsilon_{xy} - \varepsilon_{yx}), \tag{2.3.32 b}$$

$$T_{\pm 1}^1(\boldsymbol{\varepsilon}) = (1/2)\,[(\varepsilon_{xz} - \varepsilon_{zx}) \pm \mathrm{i}(\varepsilon_{yz} - \varepsilon_{zy})], \tag{2.3.32 c}$$

$$T_0^2(\varepsilon) = (1/\sqrt{6})\,(2\varepsilon_{zz} - \varepsilon_{xx} - \varepsilon_{yy}), \tag{2.3.32 d}$$

$$T_{\pm 1}^2(\varepsilon) = \mp (1/2)\,[(\varepsilon_{xz} + \varepsilon_{zx}) \pm \mathrm{i}(\varepsilon_{yz} + \varepsilon_{zy})], \tag{2.3.32 e}$$

and

$$T_{\pm 2}^2(\varepsilon) = (1/2)\,[(\varepsilon_{xx} - \varepsilon_{yy}) \pm \mathrm{i}(\varepsilon_{xy} + \varepsilon_{yx})]. \tag{2.3.32 f}$$

See [2.26]. The spin-rotation interaction tensor consists of nine components, as shown by (2.3.32), but, as pointed out by *Brown* and *Sears* [2.27], only six of them are independent even in a molecule without any symmetry (i.e., C_1). The number of independent components is smaller for higher symmetry of the molecule; four are nonvanishing for a C_s molecule, while only the diagonal elements exist for molecules of orthorhombic symmetry.

The matrix elements of the Hamiltonian (2.3.29) are evaluated using a Hund's case (b) function $|N K S J M_J\rangle$ as a basis set. The rotational part of this function is the symmetric top wave function discussed in Sect. 2.1.2 with $\boldsymbol{N}$ replacing $\boldsymbol{J}$, i.e., $|N K M\rangle$, which simultaneously diagonalizes $\boldsymbol{N}^2$ and N_z and may be expressed by the rotation matrix of rank N:

$$|N K M\rangle = [(2N + 1)/8\pi]^{1/2}\,\mathscr{D}_{MK}^{(N)*}(\omega). \tag{2.3.33}$$

The anomalous sign mentioned in Sect. 2.1.1 poses a problem in dealing with operators expressed in a molecule-fixed coordinate, as in the Hamiltonian (2.3.29). The difficulty may be avoided by transforming all the molecule-fixed components (q) of the angular momenta involved to the space-fixed components (p); the transformation is easily written down using the rotation matrix [2.26]:

$$T_p^k(\varepsilon) = \sum_q \mathscr{D}_{pq}^{(k)*}(\omega)\, T_q^k(\varepsilon). \tag{2.3.34}$$

Making use of the tensor relations summarized in Table 2.4, the matrix elements of $\boldsymbol{H}_{\mathrm{SR}}$ (2.3.29) are given by

$$\begin{aligned}
&\langle N'K'SJ'M_J'|\, H_{\mathrm{SR}}\, |N K S J M_J\rangle \\
&= \delta_{J'J}\,\delta_{M_F' M_F} \sum_{k=0}^{2} (2k+1)^{1/2}\,[S(S+1)(2S+1)(2N+1)(2N'+1)]^{1/2} \\
&\quad \times (-1)^{J+S+N'} \begin{Bmatrix} N & S & J \\ S & N' & 1 \end{Bmatrix} \\
&\quad \times (1/2)\Bigg\{(-1)^k [N(N+1)(2N+1)]^{1/2} \begin{Bmatrix} 1 & 1 & k \\ N' & N & N \end{Bmatrix} \\
&\qquad + [N'(N'+1)(2N'+1)]^{1/2} \begin{Bmatrix} 1 & 1 & k \\ N & N' & N' \end{Bmatrix}\Bigg\} \\
&\quad \times \sum_q (-1)^{N'-K'} \begin{pmatrix} N' & k & N \\ -K' & q & K \end{pmatrix} T_q^k(\varepsilon).
\end{aligned} \tag{2.3.35}$$

Equation (2.3.35) shows that the spin-rotation has matrix elements of $\Delta N = 0, \pm 1$ and $\Delta K = 0, \pm 1$, and ± 2 in the most general cases. For a C_s symmetry molecule, the spin-rotation interaction connects levels of the same parity, so that the energy matrix will be factorized into two parts, one with even parity and the other with odd parity. However, the presence of $\Delta N = \pm 1$ matrix elements makes the structure of the energy matrix considerably different from that for an ordinary molecule without unpaired electrons [2.18]. For C_s symmetry, $\Delta K = \pm 1$ elements do not allow any further factorization of the matrix, whereas they are missing for orthorhombic molecules so that the matrix is divided into even K and odd K parts.

Brown and *Sears* [2.27] have also examined the centrifugal distortion terms for the spin-rotation interaction. Following *Watson* [2.13], they derived two types of expressions, one called the asymmetric or A reduction, which is more appropriate for asymmetric rotors, and the other referred to as the symmetric or S reduction, more suitable for symmetric or nearly symmetric tops. All the matrix elements for a C_s molecule are given in Table 2.5 separately for two types of reductions.

A similar treatment may be applied to the spin-spin interaction in a triplet state. Here we present only the result [2.28]:

$$\begin{aligned}\langle N'K'SJ'M_J'|\, H_{SS}\, |NKSJM_J\rangle \\ = -(1/2)\sqrt{30}\,\delta_{J'J}\delta_{M_F'M_F}\, g_S^2\beta^2(-1)^{N+J+1}\begin{Bmatrix} J & 1 & N' \\ 2 & N & 1 \end{Bmatrix} \\ \times \sum_q (-1)^{N'-K'}[(2N+1)(2N'+1)]^{1/2}\begin{pmatrix} N' & 2 & N \\ -K' & q & K \end{pmatrix} T_q^2(C),\end{aligned} \tag{2.3.36}$$

where $T_q^2(C)$ represents the averages of spherical harmonics over the vibronic coordinates [2.26], namely

$$T_q^2(C) = \sum_{i>j} \langle (4\pi/5)^{1/2}\, Y_{2q}(\theta,\phi)/r_{ij}^3\rangle, \tag{2.3.37}$$

with i and j denoting the unpaired electrons. The two coupling constants appearing in (2.3.8) are related to $T_q^2(C)$ as

$$\alpha_0 = -(1/2)\, g_S^2\beta^2\, T_0^2(C) \quad \text{and} \tag{2.3.38 a}$$

$$\beta_0 = -(1/2)\sqrt{6}\, g_S^2\beta^2\, T_{\pm 2}^2(C). \tag{2.3.38 b}$$

2.3.6 Rotational Structure of a ^2E Molecule

This section is devoted to a special case of the symmetric top molecule in a degenerate electronic state. An example so far investigated in detail is the methoxy radical CH_3O in the ^2E ground electronic state. Because of the electronic degeneracy, the molecule is expected to be Jahn-Teller distorted

Table 2.5. Matrix elements of the spin-rotation interaction and its centrifugal distortion for an asymmetric top molecule in a doublet electronic state[a, b]

1) *Asymmetric reduction*

$$\langle NKSJ|H_{SR}^{asym}|NKSJ\rangle = -[\Gamma(NSJ)/2N(N+1)]\{\varepsilon_{aa}K^2 + (\varepsilon_{bb}+\varepsilon_{cc}) \times [N(N+1)-K^2]/2 + \Delta_K^s K^4 + (\Delta_{NK}^s + \Delta_{KN}^s)K^2N(N+1) + \Delta_N^s N^2(N+1)^2\}$$

$$\langle NK\pm 1, SJ|H_{SR}^{asym}|NKSJ\rangle = -[\Gamma(NSJ)(2K\pm 1)/4N(N+1)] \times [N(N+1)-K(K\pm 1)]^{1/2}(\varepsilon_{ab}+\varepsilon_{ba})/2$$

$$\langle NK\pm 2, SJ|H_{SR}^{asym}|NKSJ\rangle = -[\Gamma(NSJ)/4N(N+1)]\{[N(N+1)-K(K\pm 1)] \times [N(N+1)-(K\pm 1)(K\pm 2)]\}^{1/2}\{(\varepsilon_{bb}-\varepsilon_{cc})/2 + 2\delta_N^s N(N+1) + \delta_K^s[K^2+(K\pm 2)^2]\}$$

$$\langle N-1, KSJ|H_{SR}^{asym}|NKSJ\rangle = -\phi(NSJ)(K/2N)(N^2-K^2)^{1/2} \times \{\varepsilon_{aa}-(\varepsilon_{bb}+\varepsilon_{cc})/2 + \Delta_K^s K^2 + \Delta_{NK}^s N^2\}$$

$$\langle N-1, K\pm 1, SJ|H_{SR}^{asym}|NKSJ\rangle = -\phi(NSJ)[(N\pm 2K+1)/4N] \times [(N\mp K)(N\mp K-1)]^{1/2}(\varepsilon_{ab}+\varepsilon_{ba})/2$$

$$\langle N-1, K\pm 2, SJ|H_{SR}^{asym}|NKSJ\rangle = -[\phi(NSJ)/4N][(N\mp K)(N\mp K-1)(N\mp K-2) \times (N\pm K+1)]^{1/2}\{\pm(\varepsilon_{bb}-\varepsilon_{cc})/2 + \delta_K^s[K(N\pm K)+(K\pm 2)(N\pm K+2)]\}.$$

2) *Symmetric reduction*

$$\langle NKSJ|H_{SR}^{sym}|NKSJ\rangle = -[\Gamma(NSJ)/2N(N+1)]\{\varepsilon_{aa}K^2 + (\varepsilon_{bb}+\varepsilon_{cc}) \times [N(N+1)-K^2]/2 + D_K^s K^4 + (D_{NK}^s + D_{KN}^s)K^2N(N+1) + D_N^s N^2(N+1)^2\}$$

$$\langle NK\pm 1, SJ|H_{SR}^{sym}|NKSJ\rangle = -[\Gamma(NSJ)(2K\pm 1)/4N(N+1)] \times [N(N+1)-K(K\pm 1)]^{1/2}(\varepsilon_{ab}+\varepsilon_{ba})/2$$

$$\langle NK\pm 2, SJ|H_{SR}^{sym}|NKSJ\rangle = -[\Gamma(NSJ)/4N(N+1)]\{[N(N+1)-K(K\pm 1)] \times [N(N+1)-(K\pm 1)(K\pm 2)]\}^{1/2}\{(\varepsilon_{bb}-\varepsilon_{cc})/2 + 2d_1^s N(N+1)\}$$

$$\langle NK\pm 4, SJ|H_{SR}^{sym}|NKSJ\rangle = -[\Gamma(NSJ)/2N(N+1)]\,d_2^s\{[N(N+1)-K(K\pm 1)] \times [N(N+1)-(K\pm 1)(K\pm 2)][N(N+1)-(K\pm 2)(K\pm 3)] \times [N(N+1)-(K\pm 3)(K\pm 4)]\}^{1/2}$$

$$\langle N-1, KSJ|H_{SR}^{sym}|NKSJ\rangle = -\phi(NSJ)(K/2N)(N^2-K^2)^{1/2} \times \{\varepsilon_{aa}-(\varepsilon_{bb}+\varepsilon_{cc})/2 + D_K^s K^2 + D_{NK}^s N^2\}$$

$$\langle N-1, K\pm 1, SJ|H_{SR}^{sym}|NKSJ\rangle = -\phi(NSJ)[(N\pm 2K+1)/4N] \times [(N\mp K)(N\mp K-1)]^{1/2}(\varepsilon_{ab}+\varepsilon_{ba})/2$$

$$\langle N-1, K\pm 2, SJ|H_{SR}^{sym}|NKSJ\rangle = \mp[\phi(NSJ)/4N][(N\mp K-1)(N\pm K+1) \times (N\mp K-2)(N\mp K)]^{1/2}(\varepsilon_{bb}-\varepsilon_{cc})/2$$

$$\langle N-1, K\pm 4, SJ|H_{SR}^{sym}|NKSJ\rangle = \mp[\phi(NSJ)/2N]\,d_2^s\{(N\mp K)[N^2-(K\pm 1)^2] \times [N^2-(K\pm 2)^2][N^2-(K\pm 3)^2](N\mp K-4)\}^{1/2}$$

[a] Applicable to molecules with at least C_s symmetry. For orthorhombic molecules, matrix elements with $\Delta K = \pm 1$ are missing.

[b] $\Gamma(NSJ) = N(N+1)+S(S+1)-J(J+1)$ and $\phi(NSJ) = [(N-J+S)(N+J+S+1) \times (S+J-N+1)(N+J-S)/(2N+1)(2N-1)]^{1/2}$. Note that $\phi(NSJ)=1$ for $S=1/2$

from C_{3v} symmetry. As briefly mentioned in Sect. 2.4, the Jahn-Teller effect will primarily lower the molecular symmetry and also split vibronically excited levels which are otherwise degenerate [2.29]. Although to a much lesser extent, the rotational spectrum in the ground vibronic state exhibits the effect of Jahn-Teller interaction when it is observed by high-resolution spectroscopic means such as microwave spectroscopy. The details are discussed below.

The angular momenta which we take into account include the rotational angular momentum $\boldsymbol{R}$, the electronic orbital angular momentum $\boldsymbol{L}$, and the electron spin angular momentum $\boldsymbol{S}$, just as in the case of a $^2\Pi$ diatomic molecule. In fact, the rotational Hamiltonian of an ^2E molecule is an extension of that of a $^2\Pi$ molecule. One important difference is that the axial component of $\boldsymbol{R}$ is present in the symmetric top molecule and makes the energy level structure much more complicated than in $^2\Pi$ diatomic molecules. The orbital angular momentum, if not quenched, will be strongly coupled to the molecular symmetry axis as in $^2\Pi$ diatomic cases, making a Hund's case (a) type function $|JPM_J\rangle|S\Sigma\rangle|\Lambda\rangle$ a natural choice of the base function, where P, Σ, and Λ stand for the projections of $\boldsymbol{J}$, $\boldsymbol{S}$, and $\boldsymbol{L}$ on the molecular symmetry axis. We employ the following coupling scheme

$$\boldsymbol{N} = \boldsymbol{R} + \boldsymbol{L} \quad \text{and} \quad \boldsymbol{J} = \boldsymbol{N} + \boldsymbol{S}. \tag{2.3.39}$$

Brown [2.30] has discussed the rotational Hamiltonian for a symmetric top in a ^2E electronic state, assuming the spin-orbit Hamiltonian of the form $a\boldsymbol{L}\cdot\boldsymbol{S}$, which may be divided into two parts $a_{\parallel}L_zS_z$ and $a_{\perp}(L_xS_x + L_yS_y)$ by introducing two different coupling constants. (The spin-orbit coupling constant in this section shall be designated by a, retaining A for the rotational constant associated with the symmetry axis.) As in Sect. 2.3.3, the symmetric top rotational Hamiltonian is given by

$$\begin{aligned} \boldsymbol{H}_{\mathrm{SYM}} &= (A - B)(N_z - L_z - G_z)^2 + B(\boldsymbol{N} - \boldsymbol{L} - \boldsymbol{G})^2 \\ &= AN_z^2 + B(N_x^2 + N_y^2) - 2AN_z(L_z + G_z) - 2B(N_xL_x + N_yL_y) \\ &\quad + (\text{terms independent of } \boldsymbol{N}), \end{aligned} \tag{2.3.40}$$

where $\boldsymbol{R}$ has been replaced by $\boldsymbol{N} - \boldsymbol{L} - \boldsymbol{G}$, and $\boldsymbol{G}$ stands for the vibrational angular momentum which is absent in diatomic molecules. The third and fourth terms of (2.3.40) represent the vibronic Coriolis interaction and connect different electronic states, causing higher-order interactions [2.31]. Replacing $\boldsymbol{N}$ with $\boldsymbol{J} - \boldsymbol{S}$ gives the $-B(J_+S_- + J_-S_+)$ term which leads to spin uncoupling as in the $^2\Pi$ molecule. There are additional terms proportional to the squares of $\boldsymbol{N}$, but they are not included in (2.3.40); they are either of the $(\mu_{xx} - \mu_{yy} \pm 2\mathrm{i}\mu_{xy})N_{\mp}^2$ type or of $(\mu_{zx} \pm \mathrm{i}\mu_{zy})(N_zN_{\mp} + N_{\mp}N_z)$, where $\mu_{gg'}$ denotes the component of the μ tensor in (2.2.15) [2.31, 32]. *Hougen* [2.31] has given the following expression for these terms:

$$\begin{aligned} \boldsymbol{H}_{\mathrm{PERT}} &= h_1(\mathscr{L}_-^2N_+^2 + \mathscr{L}_+^2N_-^2) \\ &\quad + h_2[\mathscr{L}_-^2(N_zN_- + N_-N_z) + \mathscr{L}_+^2(N_zN_+ + N_+N_z)], \end{aligned} \tag{2.3.41}$$

where $\mathscr{L}_{\pm}^2$ represent artificial ladder operators which convert one component of the vibronic E state into the other:

$$\mathscr{L}_{\pm}^2 |\Lambda = \mp 1\rangle = |\Lambda = \pm 1\rangle. \tag{2.3.42}$$

The h_1 and h_2 terms are analogs of the (2, 2) and (2, $-$ 1) interaction terms in the theory of vibration-rotation coupling briefly mentioned in Sect. 2.2. The h_1 term causes Λ-type doubling similar to that of $^2\Pi$ molecules. As indicated by their origins, these terms distort the molecule from C_{3v} symmetry.

As in other cases, the spin-rotation interaction is derived for most parts through a second-order perturbation treatment of the spin-orbit interaction and the electronic Coriolis interaction and again consists of terms both diagonal and off-diagonal in Λ as follows:

$$\begin{aligned} \boldsymbol{H}_{\mathrm{SR}} = {} & \varepsilon_{zz} N_z S_z + (\varepsilon_{xx} + \varepsilon_{yy})(N_+ S_- + N_- S_+)/4 \\ & + \varepsilon_1 (\mathscr{L}_-^2 N_+ S_+ + \mathscr{L}_+^2 N_- S_-) \\ & + \varepsilon_{2a}[\mathscr{L}_-^2 (N_z S_- + S_- N_z) + \mathscr{L}_+^2 (N_z S_+ + S_+ N_z)] \\ & + \varepsilon_{2b}[\mathscr{L}_-^2 (N_- S_z + S_z N_-) + \mathscr{L}_+^2 (N_+ S_z + S_z N_+)]. \end{aligned} \tag{2.3.43}$$

The first two terms are identical to (2.3.29, 30) specified for an orthorhombic molecule with the ε components given by (2.3.32 a, d) only, while the other three terms in ε_1, ε_{2a}, and ε_{2b} are typical of the ^2E molecules and give matrix elements of $\Delta\Lambda = \pm 2$.

It is convenient to take a linear combination of the base function, as for $^2\Pi$ molecules discussed in Sect. 2.3.4, in order to distinguish levels according to their parity:

$$\begin{aligned} |JPS\Sigma; \pm\rangle = {} & (1/\sqrt{2})\,[|\Lambda = 1\rangle\, |S\Sigma\rangle\, |JPM_J\rangle \\ & \pm (-1)^{J-P+S-\Sigma}\, |\Lambda = -1\rangle\, |S -\Sigma\rangle\, |J -PM_J\rangle]. \end{aligned} \tag{2.3.44}$$

Care should be taken of the signs of the quantum numbers specifying each state. The P and Σ in the symmetrized wave function are identical to those for the $\Lambda = +1$ function on the right-hand side of (2.3.44). Therefore, the symmetrized wave function with $\Sigma = 1/2$ corresponds to the $^2E_{3/2}$ spin state for which $\Lambda = \pm 1$ and $\Sigma = \pm 1/2$, whereas that with $\Sigma = -1/2$ corresponds to the $^2E_{1/2}$ state for which $\Lambda = \pm 1$ and $\Sigma = \mp 1/2$. When $P - \Sigma = 3n + 1$, the symmetrized wave functions with + and $-$ signs belong to A_1 and A_2 symmetry, respectively, and thus the pair states may be split; the splitting is largest for $K = P - \Sigma = 1$. On the other hand, the pair levels with $P - \Sigma + 3n + 1$ belong to E symmetry and will degenerate unless the hyperfine interactions are considered.

The matrix elements of $\boldsymbol{H}_{\mathrm{SO}}$, $\boldsymbol{H}_{\mathrm{SYM}}$, $\boldsymbol{H}_{\mathrm{PERT}}$, and $\boldsymbol{H}_{\mathrm{SR}}$ are evaluated by substituting $\boldsymbol{J} - \boldsymbol{S}$ for $\boldsymbol{N}$. The Jahn-Teller mixing among the vibrational states with the same $j = l - \Lambda/2$ value makes the expectation values of L_z and $(L_z + G_z)$ different from those for a vibronic state which is free from the Jahn-Teller effect.

According to *Child* and *Longuet-Higgins* [2.33], the matrix elements of L_z and $(L_z + G_z)$ are

$$L_z \,|\Lambda\rangle = \zeta_e d \,|\Lambda\rangle \quad \text{and} \tag{2.3.45 a}$$

$$(L_z + G_z)\,|\Lambda\rangle = [(\zeta_e - \zeta_s/2)\, d + j\zeta_s]\,|\Lambda\rangle, \tag{2.3.45 b}$$

where ζ_s denotes the first-order Coriolis coupling constant of a Jahn-Teller active doubly degenerate mode s, and d introduced by *Child* and *Longuet-Higgins* [2.33] represents the degree of Jahn-Teller coupling, $d = 1$ and 0 for the weak and strong coupling limits, respectively. Strictly speaking, (2.3.45 b) is applicable only to X_3-type molecules, so that it will be replaced here by [2.30]

$$(L_z + G_z)\,|\Lambda\rangle = \Lambda\zeta_t\,|\Lambda\rangle. \tag{2.3.45 c}$$

All the matrix elements are listed in Table 2.6, where the centrifugal distortion terms not discussed here are also included. For details, see [2.34, 35].

Table 2.6. Matrix elements of the Hamiltonian for a ^2E symmetric top molecule

1) *Diagonal elements*

$$\begin{aligned}
\langle J P S \Sigma; \pm | H | J P S \Sigma; \pm \rangle &= a\zeta_e d\Sigma + a_D \zeta_e d (P - \Sigma)^2 \Sigma \\
&- 2A\zeta_t (P - \Sigma) + \eta_e \zeta_t (P - \Sigma)[J(J+1) - 2P\Sigma + 3/4] + \eta_K \zeta_t (P - \Sigma)^3 \\
&+ A(P - \Sigma)^2 + B[J(J+1) - P^2 + 1/2] \\
&- D_N (J + 1/2)^2 [(J + 1/2)^2 - 4P\Sigma + 2] \\
&- D_{NK} (P - \Sigma)^2 [J(J+1) - 2P\Sigma + 3/4] \\
&- D_K (P - \Sigma)^4 \\
&+ \varepsilon_{aa} (P - \Sigma)\Sigma - (\varepsilon_{bb} + \varepsilon_{cc})/4.
\end{aligned}$$

2) *The spin uncoupling term*

$$\begin{aligned}
&\langle J P + 1, S\Sigma' = 1/2; \pm | H | J P S \Sigma = -1/2; \pm \rangle \\
&= -\{\eta_e \zeta_t (P + 1/2) + B - (\varepsilon_{bb} + \varepsilon_{cc})/4 \\
&\quad - 2D_N (J + 1/2)^2 - D_{NK} (P + 1/2)^2\}[J(J+1) - P(P+1)]^{1/2}.
\end{aligned}$$

3) (2, 2) *interaction terms*

$$\begin{aligned}
&\langle J, -P + 3, \Sigma' = 1/2; \pm | H | J P \Sigma = 1/2; \pm \rangle \\
&= \mp (-1)^{J-P+S-\Sigma} h_{1N} [J(J+1) - P(P-1)]^{1/2} \\
&\quad \times [J(J+1) - (P-1)(P-2)]^{1/2} [J(J+1) - (P-2)(P-3)]^{1/2}
\end{aligned}$$

$$\begin{aligned}
&\langle J, -P + 2, \Sigma' = 1/2; \pm | H | J P \Sigma = -1/2; \pm \rangle \\
&= \pm (-1)^{J-P+S-\Sigma} \{h_1 + h_{1N} [J(J+1) + P + 3/4] \\
&\quad + (h_{1K}/2)[(P + 1/2)^2 + (3/2 - P)^2]\} \\
&\quad \times [J(J+1) - P(P-1)]^{1/2} [J(J+1) - (P-1)(P-2)]^{1/2}
\end{aligned}$$

Table 2.6. (continued)

$$\langle J, -P+1, \Sigma' = -1/2; \pm | H | J P \Sigma = -1/2; \pm \rangle$$
$$= \mp (-1)^{J-P+S-\Sigma} \{2h_1 + h_{1N}[3J(J+1) - P(P-1) + 3/2]$$
$$+ h_{1K}[(P+1/2)^2 + (3/2-P)^2] - \varepsilon_1\}[J(J+1) - P(P-1)]^{1/2}.$$

4) (2, − 1) *interaction terms*

$$\langle J, -P-2, \Sigma' = -1/2; \pm | H | J P \Sigma = -1/2; \pm \rangle$$
$$= \pm (-1)^{J-P+S-\Sigma} 2h_{2N}(P+1)[J(J+1) - P(P+1)]^{1/2}$$
$$\times [J(J+1) - (P+1)(P+2)]^{1/2}$$

$$\langle J, -P-1, \Sigma' = 1/2; \pm | H | J P \Sigma = -1/2; \pm \rangle$$
$$= + (-1)^{J-P+S-\Sigma} \{2(P+1)\{h_2 + h_{2N}[J(J+1) + P + 7/4]$$
$$+ (h_{2K}/2)[(P+1/2)^2 + (P+3/2)^2]\} - \varepsilon_{2b}\}$$
$$\times [J(J+1) - P(P+1)]^{1/2}$$

$$\langle J, -P, \Sigma' = 1/2; \pm | H | J P \Sigma = 1/2; \pm \rangle$$
$$= \pm (-1)^{J-P+S-\Sigma} 2P\{h_2 + h_{2N}[2J(J+1) - P^2 + 3/4]$$
$$+ (h_{2K}/2)[(P-1/2)^2 + (P+1/2)^2] - \varepsilon_{2a}\}.$$

5) (2, − 4) *interaction terms*

$$\langle J, -P-4, \Sigma' = 1/2; \pm | H | J P \Sigma = -1/2; \pm \rangle$$
$$= \pm (-1)^{J-P+S-\Sigma} h_4 [J(J+1) - P(P+1)]^{1/2}$$
$$\times [J(J+1) - (P+1)(P+2)]^{1/2} [J(J+1) - (P+2)(P+3)]^{1/2}$$
$$\times [J(J+1) - (P+3)(P+4)]^{1/2}$$

$$\langle J, -P-3, \Sigma' = 1/2; \pm | H | J P \Sigma = 1/2; \pm \rangle$$
$$= \mp (-1)^{J-P+S-\Sigma} 4h_4 [J(J+1) - P(P+1)]^{1/2}$$
$$\times [J(J+1) - (P+1)(P+2)]^{1/2} [J(J+1) - (P+2)(P+3)]^{1/2}$$

2.3.7 Hyperfine Interaction

All interactions involving the nuclear spin angular momentum $\boldsymbol{I}$ are called the hyperfine interaction. The splittings caused by this interaction are usually small and are resolved only with spectroscopic methods of resolution better than a few MHz. The most important hyperfine interaction for open-shell molecules is magnetic, caused by the coupling between unpaired electrons and nuclear spins in the molecule. The nuclear electric quadrupole hyperfine interaction, i.e., the electrostatic interaction between a nucleus with I larger than 1/2 and surrounding electrons, makes equally substantial contributions to the spectra of both free radicals and a large number of closed-shell molecules. It should be noted that, in contrast with the quadrupole interaction, all nuclei with finite spin quantum numbers are involved in the magnetic hyperfine interaction. The nuclear spin-rotation and nuclear spin-spin interactions are orders of magnitude smaller, and may be neglected in most microwave investigations, except for a few cases where nuclei with large magnetic moments such as P and F are involved.

There are two different types of magnetic hyperfine interaction; one is between the electron orbital angular momentum $\boldsymbol{L}$ and the nuclear spin angular momentum $\boldsymbol{I}$ and the other between the electron spin $\boldsymbol{S}$ and the nuclear spin $\boldsymbol{I}$. The first type may be expressed by the following Hamiltonian:

$$\boldsymbol{H}_{IL} = a\boldsymbol{I} \cdot \boldsymbol{L}. \tag{2.3.46}$$

On the other hand, the second type is further divided into two parts, the Fermi contact interaction $\boldsymbol{H}_{\mathrm{F}}$ and the dipole-dipole interaction $\boldsymbol{H}_{\mathrm{DD}}$, each Hamiltonian expressed as

$$\boldsymbol{H}_{\mathrm{F}} = b_{\eta}\boldsymbol{I} \cdot \boldsymbol{S}, \quad \text{and} \tag{2.3.47}$$

$$\boldsymbol{H}_{\mathrm{DD}} = \boldsymbol{S} \cdot \boldsymbol{T} \cdot \boldsymbol{I}, \tag{2.3.48}$$

where $\boldsymbol{T}$ denotes a second-rank traceless tensor with five independent components. The Fermi contact interaction constant b_{η} is sometimes designated as a_{F}. The expression for the nuclear quadrupole interaction is identical to that for a closed-shell molecule, but the matrix elements are much more complicated for free radicals than for molecules without unpaired electrons. The nuclear spin-rotation interaction Hamiltonian is identical to that for the electron spin-rotation interaction (2.3.6), except that $\boldsymbol{S}$ is replaced by $\boldsymbol{I}$:

$$\boldsymbol{H}_{\mathrm{NSR}} = (1/2) \sum_{\alpha,\beta} C_{\alpha\beta}(N_{\alpha}I_{\beta} + I_{\beta}N_{\alpha}). \tag{2.3.49}$$

For most molecules which do not involve heavy atoms, the hyperfine energy is much smaller than others, making it appropriate to couple the nuclear spin angular momentum with the total angular momentum $\boldsymbol{J}$ excluding the nuclear spin to form the real total angular momentum $\boldsymbol{F}$:

$$\boldsymbol{F} = \boldsymbol{J} + \boldsymbol{I}, \tag{2.3.50}$$

as long as there is no external field. This coupling scheme is referred to as case $(\mathrm{a})_{\beta}$ or case $(\mathrm{b})_{\beta J}$ [2.1]. The basis function is accordingly of the form

$$|\ldots JIFM_F\rangle. \tag{2.3.51}$$

2.3.8 Matrix Elements of the Hyperfine Interaction Appropriate for Case $(\mathrm{a})_{\beta}$

The hyperfine Hamiltonians are rewritten using spherical tensor notations as [2.36]

$$\boldsymbol{H}_{\mathrm{HF}} = \boldsymbol{H}_{IL} + \boldsymbol{H}_{\mathrm{F}} + \boldsymbol{H}_{\mathrm{DD}} + \boldsymbol{H}_{Q}, \tag{2.3.52}$$

where

$$H_{IL} = 2 g_N \beta \beta_N r^{-3} T^1(\boldsymbol{I}) \cdot T^1(\boldsymbol{L}), \tag{2.3.53}$$

$$H_{\mathrm{F}} = (8\pi/3)\, g_S g_N \beta \beta_N |\psi(0)|^2\, T^1(\boldsymbol{I}) \cdot T^1(\boldsymbol{S}), \tag{2.3.54}$$

$$H_{\mathrm{DD}} = -\sqrt{10}\, g_S g_N \beta \beta_N T^1(\boldsymbol{I}) \cdot T^1(\boldsymbol{S}, C^2), \quad \text{and} \tag{2.3.55}$$

$$H_Q = T^2(\boldsymbol{Q}) \cdot T^2(\nabla E), \tag{2.3.56}$$

where

$$T^1(\boldsymbol{I}) \cdot T^1(\boldsymbol{S}, C^2) = -\sum_{p,q} \sqrt{3} \begin{pmatrix} 1 & 1 & 2 \\ p & q-p & -q \end{pmatrix} \times T^1_p(\boldsymbol{I})\, T^1_{q-p}(\boldsymbol{S})\, C^2_{-q}(\theta, \phi)\, r^{-3}, \tag{2.3.57}$$

$$T^2(\boldsymbol{Q}) = \sum_n e_n r_n^2\, C^2(\theta_n, \phi_n), \quad \text{and} \tag{2.3.58}$$

$$T^2(\nabla E) = -\sum_i e_i\, C^2(\theta_i, \phi_i)\, r_i^{-3}. \tag{2.3.59}$$

In (2.3.58, 59), the suffixes n and i denote the charged particles within and outside the nucleus, respectively, and $C^2_q(\theta, \phi)$ is related to the second-rank spherical harmonics as [2.26]

$$C^2_q(\theta, \phi) = (4\pi/5)^{1/2}\, Y_{2q}(\theta, \phi). \tag{2.3.60}$$

As for the fine-structure interactions, it is convenient to transform the components of the angular momenta in the molecule-fixed coordinate to those in the space-fixed coordinate, see (2.3.34). The transformation may be accomplished as follows:

$$T^k_p(\boldsymbol{A}) = \sum_q \mathscr{D}^{k\,*}_{p-q}(\theta, \phi)\, T^k_{-q}(\boldsymbol{A}), \tag{2.3.61}$$

where p and q stand for the space-fixed and molecule-fixed components, respectively.

Using the case $(a)_\beta$ $|J\Omega S\Sigma\Lambda; IFM_F\rangle$ function as the basis set, the matrix elements of the four Hamiltonians in (2.3.52) are evaluated as

$$\begin{aligned} &\langle J'\Omega' S\Sigma'\Lambda'; IF'M_F'|\, H_{IL}\, |J\Omega S\Sigma\Lambda; IFM_F\rangle \\ &= 2 g_N \beta \beta_N \delta_{FF'} \delta_{M_F M_F'} (-1)^{I+J'+F} \begin{Bmatrix} I & J' & F \\ J & I & 1 \end{Bmatrix} \langle I\| T^1(\boldsymbol{I}) \| I\rangle \\ &\quad \times \sum_q \langle J'\Omega'|\, \mathscr{D}^1_{\cdot q}(\theta, \phi)\, |J\Omega\rangle \langle S\Sigma'|S\Sigma\rangle \langle \Lambda'|\, T^1_q(\boldsymbol{L})\, r^{-3}\, |\Lambda\rangle, \end{aligned} \tag{2.3.62}$$

where

$$\langle J'\Omega'|\,\mathscr{D}^{1}_{\cdot q}(\theta,\phi)\,|J\Omega\rangle = (-1)^{J'-\Omega'}[(2J+1)(2J'+1)]^{1/2}\begin{pmatrix} J' & 1 & J \\ -\Omega' & q & \Omega \end{pmatrix},$$

$$\langle S\Sigma'|S\Sigma\rangle = \delta_{\Sigma'\Sigma'} \quad \text{and} \tag{2.3.63}$$

$$\langle I|\,|T^1(\mathbf{I})|\,|I\rangle = [I(I+1)(2I+1)]^{1/2}. \tag{2.3.64}$$

If we are interested only in terms diagonal in Λ, only the $q=0$ term needs to be retained in (2.3.62):

$$\langle\Lambda|\,T^1_0(\mathbf{L})\,r^{-3}\,|\Lambda\rangle = \langle\Lambda|\,r^{-3}\,|\Lambda\rangle\,\Lambda. \tag{2.3.65}$$

Therefore, (2.3.62) reduces to

$$\begin{aligned}
&\langle J'\Omega'S\Sigma'\Lambda';IF'M_F'|\,H_{IL}\,|J\Omega S\Sigma\Lambda;IFM_F\rangle \\
&\quad = 2g_N\beta\beta_N\delta_{FF'}\delta_{M_FM_F'}\delta_{\Sigma\Sigma'}\delta_{\Omega\Omega'} \\
&\quad\quad \times(-1)^{I+J'+F}\begin{Bmatrix} I & J' & F \\ J & I & 1 \end{Bmatrix}[I(I+1)(2I+1)]^{1/2} \\
&\quad\quad \times(-1)^{J'-\Omega'}[(2J+1)(2J'+1)]^{1/2}\begin{pmatrix} J' & 1 & J \\ -\Omega' & 0 & \Omega \end{pmatrix}\Lambda\,\langle\Lambda|\,r^{-3}\,|\Lambda\rangle.
\end{aligned} \tag{2.3.66}$$

The matrix elements of the Fermi and dipole-dipole interactions are calculated similarly, giving

$$\begin{aligned}
&\langle J'\Omega'S\Sigma'\Lambda';IF'M_F'|\,H_F\,|J\Omega S\Sigma\Lambda;IFM_F\rangle \\
&\quad = (8\pi/3)\,g_Sg_N\beta\beta_N\,|\psi(0)|^2\,\delta_{FF'}\delta_{M_FM_F'}(-1)^{I+J'+F}\begin{Bmatrix} I & J' & F \\ J & I & 1 \end{Bmatrix} \\
&\quad\quad \times[I(I+1)(2I+1)]^{1/2}\sum_q(-1)^{J'-\Omega'} \\
&\quad\quad \times[(2J'+1)(2J+1)]^{1/2}\begin{pmatrix} J' & 1 & J \\ -\Omega' & q & \Omega \end{pmatrix} \\
&\quad\quad \times(-1)^{S-\Sigma'}[S(S+1)(2S+1)]^{1/2}\begin{pmatrix} S & 1 & S \\ -\Sigma' & q & \Sigma \end{pmatrix},
\end{aligned} \tag{2.3.67}$$

and

$$\begin{aligned}
&\langle J'\Omega'S\Sigma'\Lambda';IF'M_F'|\,H_{DD}\,|J\Omega S\Sigma\Lambda;IFM_F\rangle \\
&\quad = \sqrt{30}\,g_Sg_N\beta\beta_N\delta_{FF'}\delta_{M_FM_F'}(-1)^{I+J'+F}\begin{Bmatrix} I & J' & F \\ J & I & 1 \end{Bmatrix} \\
&\quad\quad \times[I(I+1)(2I+1)]^{1/2}\sum_{q,\mu}(-1)^{J'-\Omega'} \\
&\quad\quad \times[(2J'+1)(2J+1)]^{1/2}\begin{pmatrix} J' & 1 & J \\ -\Omega' & q & \Omega \end{pmatrix}
\end{aligned}$$

$$\times (-1)^q \begin{pmatrix} 1 & 2 & 1 \\ \mu & q-\mu & -q \end{pmatrix} (-1)^{S-\Sigma'} \begin{pmatrix} S & 1 & S \\ -\Sigma' & \mu & \Sigma \end{pmatrix}$$
$$\times \langle \Lambda' || C^2_{q-\mu}(\theta, \phi) || \Lambda \rangle. \quad (2.3.68)$$

Using the Wigner-Eckart theorem, equation 5 in Table 2.4, the $T^2(\boldsymbol{Q})$ tensor (2.3.58) may be expressed in terms of the $T^2(\boldsymbol{I})$ tensor as

$$\langle I M_I | \; T^2(\boldsymbol{Q}) \; | I M_I \rangle = f \langle I M_I | \; T^2(\boldsymbol{I}) \; | I M_I \rangle, \quad (2.3.69)$$

where f denotes a proportionality factor and all matrix elements off-diagonal in the nuclear states are neglected. It is customary to define the nuclear quadrupole moment by $eQ/2 = \langle II | \; T^2_0(\boldsymbol{Q}) \; | II \rangle$, which replaces f in (2.369), and the reduced matrix element of $T^2(\boldsymbol{Q})$ is given in terms of eQ as

$$\langle I || T^2(\boldsymbol{Q}) || I \rangle = \{(I+1)(2I+1)(2I+3)/[I(2I-1)]\}^{1/2} eQ/2. \quad (2.3.70)$$

The matrix elements of the nuclear quadrupole interaction are then given by

$$\langle J' \Omega' S \Sigma' \Lambda'; I F' M'_F | \; H_Q \; | J \Omega S \Sigma \Lambda; I F M_F \rangle$$
$$= \delta_{FF'} \delta_{M_F M'_F} \delta_{\Sigma\Sigma'} (-1)^{I+J'+F} \begin{Bmatrix} I & J' & F \\ J & I & 2 \end{Bmatrix}$$
$$\times \{(I+1)(2I+1)(2I+3)/[I(2I-1)]\}^{1/2} eQ/2$$
$$\times \sum_q (-1)^{J'-\Omega'} [(2J'+1)(2J+1)]^{1/2} \begin{pmatrix} J' & 2 & J \\ -\Omega' & q & \Omega \end{pmatrix}$$
$$\times \langle \Lambda' | \; T^2_q(\nabla E) \; | \Lambda \rangle. \quad (2.3.71)$$

These results may be specialized for a $^2\Pi$ molecule, a representative example of Hund's case (a) molecules:

$$\langle ^2\Pi_\Omega; I J' F; \pm | \; H_{\mathrm{HF}} \; | ^2\Pi_\Omega; I J F; \pm \rangle$$
$$= G(J'JIF)(-1)^{J'-\Omega} \begin{pmatrix} J' & 1 & J \\ -\Omega & 0 & \Omega \end{pmatrix} [a\Lambda + (b+c)\Sigma]$$
$$+ Q(J'JIF)(-1)^{J'-\Omega} \begin{pmatrix} J' & 2 & J \\ -\Omega & 0 & \Omega \end{pmatrix} eQq_1/4, \quad (2.3.72\,\mathrm{a})$$

and

$$\langle ^2\Pi_{3/2}; I J' F; \pm | \; H_{\mathrm{HF}} \; | ^2\Pi_{1/2}; I J F; \pm \rangle$$
$$= - G(J'JIF)(-1)^{J'-3/2} \begin{pmatrix} J' & 1 & J \\ -3/2 & 1 & 1/2 \end{pmatrix} b/\sqrt{2}, \quad (2.3.72\,\mathrm{b})$$

when only terms diagonal in Λ are retained. Among the off-diagonal terms, those with $\Delta\Lambda = \pm 2$ are important, since they contribute to Λ doublets to first order:

$$\langle {}^2\Pi_{1/2}; IJ'F; \pm | H_{\mathrm{HF}} | {}^2\Pi_{1/2}; IJF; \pm \rangle$$
$$= \pm G(J'JIF) \begin{pmatrix} J' & 1 & J \\ 1/2 & -1 & 1/2 \end{pmatrix} d/\sqrt{2}, \tag{2.3.73 a}$$

and

$$\langle {}^2\Pi_{3/2}; IJ'F; \pm | H_Q | {}^2\Pi_{1/2}; IJF; \pm \rangle$$
$$= \mp Q(J'JIF) \begin{pmatrix} J' & 2 & J \\ 3/2 & -2 & 1/2 \end{pmatrix} eQq_2/(4\sqrt{6}). \tag{2.3.73 b}$$

In (2.3.72, 73) $G(J'JIF)$ and $Q(J'JIF)$ are defined as

$$G(J'JIF) = [I(I+1)(2I+1)(2J'+1)(2J+1)]^{1/2}$$
$$\times (-1)^{J'+I+F} \begin{Bmatrix} I & J' & F \\ J & I & 1 \end{Bmatrix}, \tag{2.3.74 a}$$

and

$$Q(J'JIF) = \{(I+1)(2I+1)(2I+3)(2J'+1)(2J+1)/[(2I-1)]\}^{1/2}$$
$$\times (-1)^{J'+I+F} \begin{Bmatrix} I & J' & F \\ J & I & 2 \end{Bmatrix}. \tag{2.3.74 b}$$

The parameters a, b, c, and d which appear in (2.3.72, 73) stand for the hyperfine coupling constants first introduced by *Frosch* and *Foley* [2.37], defined as

$$a = 2 g_N \beta \beta_N \langle \Lambda = 1 | r^{-3} | \Lambda = 1 \rangle, \tag{2.3.75 a}$$

$$b = b_\eta - c/3, \tag{2.3.75 b}$$

$$b_\eta = (8\pi/3) g_S g_N \beta \beta_N |\psi(0)|^2, \tag{2.3.75 c}$$

$$c = (3/2) g_S g_N \beta \beta_N \langle (3\cos^2\theta - 1)/r^3 \rangle$$
$$= 3 g_S g_N \beta \beta_N \langle \Lambda = 1 | C_0^2(\theta, \phi) r^{-3} | \Lambda = 1 \rangle, \quad \text{and} \tag{2.3.75 d}$$

$$d = (3/2) g_S g_N \beta \beta_N \langle \sin^2\theta / r^3 \rangle$$
$$= -\sqrt{6} g_S g_N \beta \beta_N \langle \Lambda = 1 | C_2^2(\theta, \phi) r^{-3} | \Lambda = -1 \rangle. \tag{2.3.75 e}$$

It should be noted that the phase of the electronic wave function $|\Lambda\rangle$ is chosen such that the integral $\langle \Lambda = \pm 1 | \exp(\pm 2i\phi) | \Lambda = \mp 1 \rangle = -1$ [2.38]. This choice requires the introduction of the minus sign in the definition of the d constant, to make it consistent with Frosch and Foley's definition. The two field gradient constants, in (2.3.72 a, 73 b) are defined as

$$q_1 = -e \langle (3\cos^2\theta - 1)/r^3 \rangle = 2 \langle \Lambda = 1 | T_0^2(\nabla E) | \Lambda = 1 \rangle, \tag{2.3.76 a}$$

and

$$q_2 = -3e \langle \sin^2\theta / r^3 \rangle = -\sqrt{24} \langle \Lambda = 1 | T_2^2(\nabla E) | \Lambda = -1 \rangle. \tag{2.3.76 b}$$

2.3.9 Matrix Elements of the Hyperfine Interaction Appropriate for Case $(b)_{\beta J}$

The matrix elements of the hyperfine interaction in case $(b)_{\beta J}$ are derived similarly. Symmetric and asymmetric top molecules, however, have no nuclear spin-electron orbital interaction. There are, in general, five independent parameters for each of the dipole-dipole and the nuclear electric quadrupole hyperfine interactions. The physical meaning of such parameters would be clearer when they are given in a Cartesian coordinate system instead of in the spherical coordinate system. If we define the coupling constant by

$$T_q^2(C) = \langle n|\, C_q^2(\theta, \phi)\, r^{-3}\, |n\rangle, \tag{2.3.77}$$

its five components are related to those in the Cartesian coordinate system as

$$g_S g_N \beta \beta_N T_0^2(C) = T_{zz}/2 = -(T_{xx} + T_{yy})/2, \tag{2.3.78 a}$$

$$g_S g_N \beta \beta_N T_{\pm 1}^2(C) = \mp (1/\sqrt{6})\,(T_{xz} \pm \mathrm{i}\, T_{yz}), \tag{2.3.78 b}$$

$$g_S g_N \beta \beta_N T_{\pm 2}^2(C) = (1/\sqrt{24})\,(T_{xx} - T_{yy} \pm 2\mathrm{i}\, T_{xy}). \tag{2.3.78 c}$$

Similar relations hold for the quadrupole coupling tensor:

$$eQ\, T_0^2(\nabla E) = \chi_{zz}/2, \tag{2.3.79 a}$$

$$eQ\, T_{\pm 1}^2(\nabla E) = \mp (1/\sqrt{6})\,(\chi_{xz} \pm \mathrm{i}\chi_{yz}), \tag{2.3.79 b}$$

$$eQ\, T_{\pm 2}^2(\nabla E) = (1/\sqrt{24})\,(\chi_{xx} - \chi_{yy} \pm 2\mathrm{i}\chi_{xy}). \tag{2.3.79 c}$$

The matrix elements of the hyperfine Hamiltonian are obtained as follows:

$$\begin{aligned}
&\langle N'K'SJ'IF'M_F'|\, H_{\mathrm{HF}}\, |NKSJIFM_F\rangle \\
&\quad = \delta_{M_F M_F'} \delta_{FF'} \delta_{NN'} \delta_{KK'} (-1)^{N+S+J'} (-1)^{J+I+F+1} \\
&\qquad \times [(2J'+1)(2J+1)]^{1/2} [S(S+1)(2S+1)I(I+1)(2I+1)]^{1/2} \\
&\qquad \times \begin{Bmatrix} I & J' & F \\ J & I & 1 \end{Bmatrix} \begin{Bmatrix} S & J' & N \\ J & S & 1 \end{Bmatrix} b_\eta \\
&\qquad - \delta_{M_F M_F'} \delta_{FF'} \sqrt{30}\, g_S g_N \beta \beta_N (-1)^{J+I+F} \\
&\qquad \times [(2J'+1)(2J+1)(2N'+1)(2N+1)]^{1/2} \\
&\qquad \times \begin{Bmatrix} I & J' & F \\ J & I & 1 \end{Bmatrix} \begin{Bmatrix} N' & N & 2 \\ S & S & 1 \\ J' & J & 1 \end{Bmatrix} \sum_q (-1)^{N'-K'} \begin{pmatrix} N' & 2 & N \\ -K' & q & K \end{pmatrix} T_q^2(C) \\
&\qquad + \delta_{M_F M_F'} \delta_{FF'} (eQ/2) (-1)^{J+I+F} (-1)^{N'+S+J} \\
&\qquad \times \{(I+1)(2I+1)(2I+3)/[I(I+1)]\}^{1/2}
\end{aligned}$$

$$\times [(2J'+1)(2J+1)(2N'+1)(2N+1)]^{1/2}$$
$$\times \begin{Bmatrix} I & J' & F \\ J & I & 2 \end{Bmatrix} \begin{Bmatrix} N' & J' & S \\ J & N & 2 \end{Bmatrix}$$
$$\times \sum_q (-1)^{N'-K'} \begin{pmatrix} N' & 2 & N \\ -K' & q & K \end{pmatrix} T_q^2(\nabla E). \tag{2.3.80}$$

For diatomic and linear polyatomic molecules in $^2\Sigma$ states, only $K = 0$ elements need to be retained in (2.3.80). The hyperfine coupling constants in (2.3.80) are related to the Frosch and Foley b and c constants as follows:

$$b_\eta = b + c/3, \tag{2.3.81 a}$$

$$g_S g_N \beta \beta_N T_0^2(C) = c/3. \tag{2.3.81 b}$$

The matrix elements of the nuclear spin-rotation interaction Hamiltonian may be derived similarly to the electron spin-rotation interaction. The result is

$$\langle N'K'SJ'IF'M_F'|\, H_{\mathrm{NSR}}\, |NKSJIFM_F\rangle$$
$$= (-1)^{J+I+F}[I(I+1)(2I+1)]^{1/2} \begin{Bmatrix} I & J' & F \\ J & I & 1 \end{Bmatrix}$$
$$\times (-1)^{N+S+J+1}[(2J'+1)(2J+1)(2N'+1)(2N+1)]^{1/2}$$
$$\times \begin{Bmatrix} N & J & S \\ J' & N' & 1 \end{Bmatrix} (1/2) \sum_{k=0,2} (2k+1)^{1/2}$$
$$\times \left\{ (-1)^k [N(N+1)(2N+1)]^{1/2} \begin{Bmatrix} k & 1 & 1 \\ N & N' & N \end{Bmatrix} \right.$$
$$\left. + [N'(N'+1)(2N'+1)]^{1/2} \begin{Bmatrix} k & 1 & 1 \\ N & N' & N' \end{Bmatrix} \right\}$$
$$\times \sum_q (-1)^{N'-K'} \begin{pmatrix} N' & k & N \\ -K' & q & K \end{pmatrix} T_q^k(C), \tag{2.3.82}$$

where

$$T_0^0(C) = -(1/\sqrt{3})(C_{aa} + C_{bb} + C_{cc}), \tag{2.3.83 a}$$

$$T_0^2(C) = \sqrt{6}(2C_{aa} - C_{bb} - C_{cc}), \quad \text{and} \tag{2.3.83 b}$$

$$T_{\pm 2}^2(C) = (C_{bb} - C_{cc})/2, \tag{2.3.83 c}$$

with $C_{\alpha\alpha}$ denoting the $\alpha\alpha$ component of the nuclear spin-rotation coupling constant. The off-diagonal elements like C_{ab} do not vanish when the molecular symmetry is lower than orthorhombic, but their contributions are usually very small.

2.3.10 Hyperfine Interactions Caused by More than One Nucleus

The discussions has so far been limited to molecules which involve only one nucleus with spin. This section deals with cases where more than one nucleus is involved in hyperfine interactions. When two such nuclei are equivalent, we must take into account the symmetry of the molecule in deriving the hyperfine matrix elements. These cases will be treated later. For the rest each nucleus is considered individually.

Suppose there are two nonequivalent nuclei with nuclear spins I_1 and I_2. They may be coupled to the rotational angular momentum $\boldsymbol{J}$ successively to form the total angular momentum $\boldsymbol{F}_1 = \boldsymbol{J} + \boldsymbol{I}_1$ and $\boldsymbol{F} = \boldsymbol{F}_1 + \boldsymbol{I}_2$, where the nucleus with larger hyperfine coupling constants is called 1. The hyperfine matrix elements for nucleus 1 are identical to those given in Sects. 2.3.7–9, provided that F is replaced by F_1; they are thus diagonal in F_1 and F. The matrix elements for the second nucleus are obtained by multiplying the relevant matrix elements by the factor

$$(-1)^{J'+F_1+I_1+k}[(2F_1'+1)(2F_1+1)]^{1/2}\begin{Bmatrix} I_1 & J' & F_1' \\ k & F_1 & J \end{Bmatrix}, \tag{2.3.84}$$

where $k = 1$ and 2 for the magnetic and the nuclear electric quadrupole hyperfine interactions, respectively. As an example, the matrix elements of the Fermi interaction for the second nucleus are derived as [2.39]

$$\begin{aligned} &\langle N'K'SJ'I_1F_1'I_2F'M_F'|H_F(2)|NKSJI_1F_1I_2FM_F\rangle \\ &= \delta_{M_FM_F'}\delta_{FF'}\delta_{NN'}\delta_{KK'}(-1)^{2F_1+F+2J'+S+N+I_1+I_2+2} \\ &\quad\times[(2F_1'+1)(2F_1+1)(2J'+1)(2J+1)S(S+1)(2S+1) \\ &\quad\times I_2(I_2+1)(2I_2+1)]^{1/2} \\ &\quad\times\begin{Bmatrix} F & I_2 & F_1' \\ 1 & F_1 & I_2 \end{Bmatrix}\begin{Bmatrix} I_1 & J' & F_1' \\ 1 & F_1 & J \end{Bmatrix}\begin{Bmatrix} N & J' & S \\ 1 & S & J \end{Bmatrix} b_\eta\,. \end{aligned} \tag{2.3.85}$$

This procedure may be easily extended to multinuclear systems; the matrix elements are obtained by succesive application of (2.3.84) with appropriate replacement of the quantum numbers involved.

When the hyperfine interaction is much larger for nucleus 1 than for nucleus 2, F_1 becomes a good quantum number; the hyperfine structure will have a form as if there are $2I_1 + 1$ levels widely split by the first nucleus each of which consists of $2I_2 + 1$ sublevels with narrower splittings caused by the second nucleus. The matrix elements diagonal in F_1 give the essential parts of the hyperfine structure. On the other hand, when the hyperfine interactions of two nuclei are not much different in magnitude, F_1 ceases to be a good quantum number; a large matrix including matrix elements off-diagonal in F_1 needs to be diagonalized to reproduce the hyperfine levels precisely. Care must be taken in corresponding the calculated with the observed spectra [2.40].

When a molecule contains two equivalent nuclei like XH_2- and XYH_2-type molecules which involve two protons, the hyperfine interaction is described in terms of the total nuclear spin operator $\boldsymbol{I}_0 = \boldsymbol{I}_1 + \boldsymbol{I}_2$. The two nuclei have the identical spin quantum number $I = I_1 = I_2$, then the total spin quantum number may take the values $2I, 2I-1, \ldots, 1, 0$. When a C_2 operation exchanges two nuclei, $(2I+1)(I+1)$ of the $(2I+1)^2$ total spin functions are symmetric and the rest $(2I+1)I$ are antisymmetric. The spin state with $I_0 = 2I$ is obviously symmetric, the $I_0 = 2I - 1$ state is antisymmetric, and so on. (Note that each state specified by I_0 is $2I_0 + 1$-fold degenerate). For example, for $I = 1/2$, the $I_0 = 1$ state is symmetric and the $I_0 = 0$ state antisymmetric. The symmetric and antisymmetric spin functions must be combined with rovibronic wave functions to make the overall wave functions either symmetric or antisymmetric depending on whether the equivalent nuclei in the molecule obey Bose or Fermi statistics.

The PH_2 and NH_2 radicals are examples containing two equivalent protons, which are exchanged by the C_2 rotation about the b axis. Since the proton is a Fermi particle with $I = 1/2$, the $I_0 = 1$ and 0 spin states must be combined with K_a, K_c = even, even or odd, odd rotational levels and with K_a, K_c = even, odd or odd, even rotational levels, respectively, in the ground vibronic state which belongs to B_2 symmetry. Therefore, the proton hyperfine structure is absent in the latter group of rotational levels, leaving only the phosphorus and nitrogen hyperfine structures, respectively, in PH_2 and NH_2. On the other hand, in the former group of rotational levels, both the proton and phosphorus/nitrogen hyperfine interactions are nonvanishing, making the overall hyperfine structures very complicated [2.41].

The matrix elements for the two equivalent nuclei are identical to those for a single nucleus provided that the averages of the coupling constants such as $b_{\eta 0} = (b_{\eta 1} + b_{\eta 2})/2$ and $\boldsymbol{T}_0 = (\boldsymbol{T}_1 + \boldsymbol{T}_2)/2$ are used in place of those for a single nucleus.

The "difference" spin angular momentum operator $\boldsymbol{I}_- = \boldsymbol{I}_1 - \boldsymbol{I}_2$ remains to be discussed, but, since it gives matrix elements only between rotational levels of different K, its contributions may usually be ignored.

2.3.11 Hyperfine Interactions in a C_{3v} Symmetric Top Molecule

The present discussion is restricted to C_{3v} molecules with three equivalent nuclei which are not subjected to the nuclear quadrupole interaction. Two typical examples will be presented: CF_3 in $^2A_1$ and CH_3O in ^2E vibronic states [2.42, 34]. The magnetic hyperfine Hamiltonian may written as

$$\boldsymbol{H}_{\mathrm{HF}} = \sum_{i=1}^{3} (a_{\mathrm{F}i} \boldsymbol{S} \cdot \boldsymbol{I}_i + \boldsymbol{S} \cdot \boldsymbol{T}_i \cdot \boldsymbol{I}_i), \tag{2.3.86}$$

where $a_{\mathrm{F}i}$ stands for the Fermi contact parameter in this section and $\boldsymbol{T}_i$ for the dipole-dipole interaction tensor of the ith nucleus. The C_{3v} molecular symmetry makes it appropriate to take linear combinations of the $\boldsymbol{I}_i$ operators and accordingly of the coupling constants as follows:

$$\boldsymbol{I}_0 = \boldsymbol{I}_1 + \boldsymbol{I}_2 + \boldsymbol{I}_3, \tag{2.3.87 a}$$

$$\boldsymbol{I}_\pm = \boldsymbol{I}_1 + \mathrm{e}^{\pm 2\pi i/3}\,\boldsymbol{I}_2 + \mathrm{e}^{\pm 4\pi i/3}\,\boldsymbol{I}_3, \tag{2.3.87 b}$$

$$a_{\mathrm{F0}} = (a_{\mathrm{F1}} + a_{\mathrm{F2}} + a_{\mathrm{F3}})/3, \tag{2.3.88 a}$$

$$a_{\mathrm{F\pm}} = (a_{\mathrm{F1}} + \mathrm{e}^{\pm 2\pi i/3}\,a_{\mathrm{F2}} + \mathrm{e}^{\pm 4\pi i/3}\,a_{\mathrm{F3}})/3, \tag{2.3.88 b}$$

$$\boldsymbol{T}_0 = (\boldsymbol{T}_1 + \boldsymbol{T}_2 + \boldsymbol{T}_3)/3, \quad \text{and} \tag{2.3.89 a}$$

$$\boldsymbol{T}_\pm = (\boldsymbol{T}_1 + \mathrm{e}^{\pm 2\pi i/3}\,\boldsymbol{T}_2 + \mathrm{e}^{\pm 4\pi i/3}\,\boldsymbol{T}_3)/3. \tag{2.3.89 b}$$

The hyperfine Hamiltonian (2.3.86) is then rewritten as

$$\boldsymbol{H}_{\mathrm{HF}} = \sum_{\alpha=0,\pm} (a_{\mathrm{F}\alpha}\,\boldsymbol{S}\cdot\boldsymbol{I}_{-\alpha} + \boldsymbol{S}\cdot\boldsymbol{T}_\alpha\cdot\boldsymbol{I}_{-\alpha}), \tag{2.3.90}$$

and its matrix elements may be evaluated using the spherical tensor method. When the case (b)$_\beta$ function $|n\,N\,K\,S\,J\,\Gamma\,I_0\,F\,M_F\rangle$ is used as the basis, the matrix elements are given by

$$\begin{aligned}
&\langle n\,N'\,K'\,S\,J'\,\Gamma'\,I_0'\,F'\,M_F'|H_{\mathrm{HF}}|n\,N\,K\,S\,J\,\Gamma\,I_0\,F\,M_F\rangle \\
&\quad = \delta_{FF'}\,\delta_{M_F M_F'}\,\delta_{NN'}\,\delta_{KK'}(-1)^{N+S+J'}(-1)^{J+I_0'+F+1} \\
&\qquad \times [(2J'+1)(2J+1)S(S+1)(2S+1)]^{1/2}\begin{Bmatrix} I_0' & J' & F \\ J & I_0 & 1 \end{Bmatrix}\begin{Bmatrix} S & J' & N \\ J & S & 1 \end{Bmatrix} \\
&\qquad \times \langle \Gamma'; I_0' \| T^1(\boldsymbol{I}_{-\alpha}) \| \Gamma; I_0\rangle \langle n'|a_{F\alpha}|n\rangle \\
&\qquad\quad - g_S\,g_N\,\beta\,\beta_N\,\delta_{FF'}\,\delta_{M_F M_F'}\sqrt{30}(-1)^{J+I_0'+F} \\
&\qquad \times [S(S+1)(2S+1)(2J'+1)(2J+1)(2N'+1)(2N+1)]^{1/2} \\
&\qquad \times \begin{Bmatrix} I_0' & J' & F \\ J & I_0 & 1 \end{Bmatrix}\begin{Bmatrix} N' & N & 2 \\ S & S & 1 \\ J' & J & 1 \end{Bmatrix}\langle \Gamma'; I_0' \| T^1(\boldsymbol{I}_{-\alpha}) \| \Gamma; I_0\rangle \\
&\qquad \times \sum_q (-1)^{N'-K'}\begin{pmatrix} N' & 2 & N \\ -K' & q & K \end{pmatrix}\langle n'|T_q^2(C_\alpha)|n\rangle,
\end{aligned} \tag{2.3.91}$$

where n denotes the vibronic part of the wave function and Γ the symmetry (A_1, A_2, or E) of the nuclear spin function. The dipole-dipole interaction tensors listed in (2.3.89) are expressed in the spherical coordinate system in (2.3.91); the transformation from this system to the Cartesian coordinate system is discussed below.

The nuclear spin function $|\Gamma; I_0, M_{I0}\rangle$ may be expressed in terms of $|m_1, m_2, m_3\rangle$, which is a product of the spin functions of the three nuclei specified by the magnetic quantum numbers, as done by *Hougen* [2.31]. There is one-to-one correspondence between the spin functions of largest I_0 and M_{I0} (i.e., $I_0 = M_{I0} = 3I$) and of largest m_i (i.e., I), and they are obviously totally symmetric, A_1. Spin functions which have the same I_0 but different M_{I0} may be obtained by successively applying $I_x - \mathrm{i}I_y$ to $|n; I_0, M_{I0} = I_0\rangle$. The spin functions with smaller I_0 values are obtained consecutively by imposing orthogonality conditions on linear combinations of functions which have the identical M_{I0} value. For three nuclei of $I = I/2$, the largest I_0 value is 3/2; there are four states of $I_0 = 3/2$ which belong to A_1, as noted above. Four other states all have $I_0 = I/2$, and form two pairs of $E_\pm$ symmetry. The $M_{I0} = I_0$ functions of A_1 and $E_\pm$ symmetry are given by

$$|A_1; 3/2, 3/2\rangle = |1/2, 1/2, 1/2\rangle \quad \text{and} \tag{2.3.92 a}$$

$$|E_\pm; 1/2, 1/2\rangle = (1/\sqrt{3})\,[|-1/2, 1/2, 1/2\rangle + \mathrm{e}^{\pm 2\pi\mathrm{i}/3}|1/2, -1/2, 1/2\rangle + \mathrm{e}^{\pm 4\pi\mathrm{i}/3}|1/2, 1/2, -1/2\rangle], \tag{2.3.92 b}$$

respectively. The spin states of three nuclei with $I = 1$ are classified into five groups: $|A_1; 3, M_{I0}\rangle$, $|A_1; 1, M_{I0}\rangle$, $|E_\pm; 2, M_{I0}\rangle$, $|E_\pm; 1, M_{I0}\rangle$, and $|A_2; 0, 0\rangle$.

The reduced matrix element $\langle\Gamma'; I_0' \| T^1(\boldsymbol{I}_\alpha) \| \Gamma; I_0\rangle$ may be evaluated by using the Wigner-Eckart theorem. For three nuclei with $I = 1/2$, the nonvanishing matrix elements are [2.42, 31]

$$\langle\Gamma; I_0 \| T^1(\boldsymbol{I}_0) \| \Gamma; I_0\rangle = [I_0(I_0+1)(2I_0+1)]^{1/2}, \tag{2.3.93 a}$$

$$\langle E_\pm; I_0 \| T^1(\boldsymbol{I}_\mp) \| E_\mp; I_0\rangle = -2[I_0(I_0+1)(2I_0+1)]^{1/2}, \tag{2.3.93 b}$$

and

$$\langle E_\pm; 1/2 \| T^1(\boldsymbol{I}_\pm) \| A_1; 3/2\rangle = \sqrt{6}\,. \tag{2.3.93 c}$$

Symmetry consideration of the vibronic wave function $|n\rangle$ is useful in determining what integrals are nonvanishing among $\langle n'|a_{\mathrm{F}\alpha}|n\rangle$ and $\langle n'|T_q^2(C_\alpha)|n\rangle$. In a nondegenerate vibronic state, only terms with $\alpha = 0$ are present among the Fermi term integrals, making the matrix elements identical to those for a single nucleus. For the dipole-dipole interaction, the symmetry argument requires $q - \alpha = 0, \pm 3n$, with n denoting an integer for the matrix elements to be nonvanishing; the operators $\boldsymbol{I}_\pm$ give matrix elements between levels with $\Delta K = \pm 1$ or ± 2. Selection rules for degenerate vibronic states are discussed later.

Because the total wave function must belong to either A_1 or A_2, the following combinations of the rovibronic wave functions and nuclear spin functions are permissible for three $I = 1/2$ nuclei [2.31];

$$|^{evsr}A_1\rangle|^{n}A_1\rangle, \quad |^{evsr}A_2\rangle|^{n}A_1\rangle \quad \text{or}$$

$$(1/\sqrt{2})\,[|^{evsr}E_+\rangle|^{n}E_-\rangle \pm |^{evsr}E_-\rangle|^{n}E_+\rangle],$$

where $|^{evsr}\Gamma\rangle$ and $|^{n}\Gamma\rangle$ denote the rovibronic and nuclear spin functions, respectively. In a nondegenerate vibronic state, $|^{evsr}A_1\rangle$ and $|^{evsr}A_2\rangle$ correspond to levels with $K = 3n$, whereas $|^{evsr}E_\pm\rangle$ to those with $K \neq 3n$. Because the matrix elements with $\Delta K = \pm 2$ connect $|^{n}E_+\rangle$ and $|^{n}E_-\rangle$, the $K = 1$ levels may be split into two by the hyperfine interaction. In fact, such splittings have been observed in the microwave spectrum of CF_3 [2.42].

When the molecule is in a vibronically nondegenerate state, the three nuclei are equivalent, and the $\boldsymbol{T}_1$ tensor may be expressed as

$$\boldsymbol{T}_1 = \begin{bmatrix} T_{xx} & 0 & T_{xz} \\ 0 & T_{yy} & 0 \\ T_{xz} & 0 & T_{zz} \end{bmatrix},$$

where the z axis represents the symmetry axis of the molecule, and the nucleus 1 is assumed to be in the xz plane. The diagonal sum $T_{xx} + T_{yy} + T_{zz}$ is zero. The $\boldsymbol{T}_2$ and $\boldsymbol{T}_3$ tensors are easily obtained by rotating $\boldsymbol{T}_1$ by $2\pi/3$ and $4\pi/3$, respectively, about the z axis. Substitution of these results in (2.3.89) leads to the results

$$\boldsymbol{T}_0 = \begin{bmatrix} -T_{zz}/2 & 0 & 0 \\ 0 & -T_{zz}/2 & 0 \\ 0 & 0 & T_{zz} \end{bmatrix}, \quad \text{and} \tag{2.3.94 a}$$

$$\boldsymbol{T}_\pm = \begin{bmatrix} (T_{xx} - T_{yy})/4 & \mp \mathrm{i}(T_{xx} - T_{yy})/4 & T_{xz}/2 \\ \mp \mathrm{i}(T_{xx} - T_{yy})/4 & -(T_{xx} - T_{yy})/4 & \pm \mathrm{i}\,T_{xz}/2 \\ T_{xz}/2 & \pm \mathrm{i}\,T_{xz}/2 & 0 \end{bmatrix}. \tag{2.3.94 b}$$

These tensors may further be expressed in the spherical tensor form as follows:

$$g_S g_N \beta \beta_N T_0^2(C_0) = T_{zz}/2, \tag{2.3.95 a}$$

$$g_S g_N \beta \beta_N T_{\pm 1}^2(C_\mp) = \mp (1/\sqrt{6})\, T_{xz}, \quad \text{and} \tag{2.3.95 b}$$

$$g_S g_N \beta \beta_N T_{\pm 2}^2(C_\pm) = (1/\sqrt{24})\,(T_{xx} - T_{yy}). \tag{2.3.95 c}$$

These show that only T_{zz} contributes to the diagonal terms. Equation (2.3.95 c) indicates that the $K = 1$ splitting allows us to determine $T_{xx} - T_{yy}$, whereas T_{xz}

may be difficult to obtain, because it concerns only the $\Delta K = \pm 1$ elements; some accidental degeneracy between two levels of $\Delta K = \pm 1$ would be necessary to determine T_{xz}.

The above discussion needs to be modified, if the state under consideration is vibronically degenerate. An example is the CH_3O radical in the ^2E ground vibronic state. The vibronic wave function may be given as $|\Lambda\rangle$, with $\Lambda = \pm 1$ representing the degeneracy. The coupling constants are now written as $\langle \Lambda' | a_{F\alpha} | \Lambda \rangle$ and $\langle \Lambda' | T_q^2 (C_\alpha) | \Lambda \rangle$. The symmetry consideration again leads to the selection rule $-\Delta\Lambda + \alpha = 0, \pm 3$ for the Fermi coupling constant and $-\Delta\Lambda + \alpha + q = 0, \pm 3$ for the dipole-dipole interaction constant, where $\Delta\Lambda = \Lambda' - \Lambda$. Since the selection rule is thus relaxed from the nondegenerate vibronic case, more coupling constants exist, namely $a_{F\pm}$, $T^2_{\pm 2}(C_\mp)$, $T^2_{\pm 1}(C_0)$, $T^2_0(C_\pm)$, $T^2_{\pm 1}(C_\pm)$, and $T^2_{\mp 2}(C_0)$. This fact is primarily ascribed to a "deformation" of the molecule; the vibronic wave function made slightly nonaxially symmetric by the Jahn-Teller effect causes the three nuclei to be nonequivalent. Because Λ is not zero, Γ is A_1 and $E_\pm$ when $K - \Lambda = 3n$ and $\neq 3n$, respectively.

The CH_3O radical approximates the case $(a)_\beta$ more closely than the case $(b)_{\beta J}$. The matrix elements are thus derived using the case $(a)_\beta$ function $|\Lambda, S\Sigma, JP, I_0 F M_F, \Gamma\rangle$ as a basis as follows:

$$\begin{aligned}
&\langle \Lambda', S\Sigma', J'P', I_0' F' M_F'; \Gamma' | H_{\mathrm{HF}} | \Lambda, S\Sigma, JP, I_0 F M_F; \Gamma\rangle \\
&= \delta_{\Lambda\Lambda'}\,\delta_{PP'}\,\delta_{FF'}\,\delta_{M_F M_F'} (-1)^{J+I_0'+F} \langle I_0'; \Gamma' \| T^1(\boldsymbol{I}_0) \| I_0; \Gamma\rangle \\
&\quad \times [(2J'+1)(2J+1)]^{1/2} \begin{Bmatrix} I_0' & J' & F \\ J & I_0 & 1 \end{Bmatrix} (-1)^{J'-P'} \begin{pmatrix} J' & 1 & J \\ -P' & 0 & P \end{pmatrix} \\
&\quad \times \langle \Lambda' \| a_L \| \Lambda \rangle \zeta_e d \\
&\quad + \delta_{FF'}\,\delta_{M_F M_F'} \sum_{\alpha=0,\pm} (-1)^{J+I_0'+F} \begin{Bmatrix} J' & I_0' & F \\ I_0 & J & 1 \end{Bmatrix} \\
&\quad \times \langle I_0; \Gamma' \| T^1(\boldsymbol{I}_{-\alpha}) \| I_0; \Gamma\rangle \\
&\quad \times \sum_q (-1)^{J'-P'} [(2J'+1)(2J+1)]^{1/2} \begin{pmatrix} J' & 1 & J \\ -P' & q & P \end{pmatrix} \\
&\quad \times (-1)^{S-\Sigma'} [S(S+1)(2S+1)]^{1/2} \begin{pmatrix} S & 1 & S \\ -\Sigma' & q & \Sigma \end{pmatrix} \langle \Lambda' \| a_{F\alpha} \| \Lambda \rangle \\
&\quad + \sqrt{30}\,\delta_{FF'}\,\delta_{M_F M_F'}\, g_S g_N \beta \beta_N \sum_{\alpha=0,\pm} (-1)^{J+I_0'+F} \begin{Bmatrix} J' & I_0' & F \\ I_0 & J & 1 \end{Bmatrix} \\
&\quad \times \langle I_0'; \Gamma' \| T^1(\boldsymbol{I}_{-\alpha}) \| I_0; \Gamma\rangle \\
&\quad \times \sum_{q q_1 q_2} (-1)^{J'-P'} [(2J'+1)(2J+1)]^{1/2} \begin{pmatrix} J' & 1 & J \\ -P' & q & P \end{pmatrix}
\end{aligned}$$

$$\times (-1)^q \begin{pmatrix} 1 & 2 & 1 \\ q_1 & q_2 & -q \end{pmatrix}$$

$$\times (-1)^{S-\Sigma'} [S(S+1)(2S+1)]^{1/2} \begin{pmatrix} S & 1 & S \\ -\Sigma' & q_1 & \Sigma \end{pmatrix}$$

$$\times \langle \Lambda' \| T^2_{q_2}(C_\alpha) \| \Lambda \rangle . \qquad (2.3.96)$$

Because the orbital angular momentum is not quenched completely in CH_3O, the $\boldsymbol{H}_{IL}$ term is retained in (2.3.96) (the first term). When the symmetrized basis function is used, it is easily seen that levels with $P = \pm 0.5$ will be split in the first order. Since the splitting is caused by $\Delta\Lambda = \pm 2$ matrix elements, the parameters obtainable from doublet separations are different from those in nondegenerate cases.

2.4 Vibronic Interaction Including the Renner-Teller Effect

2.4.1 Born-Oppenheimer Approximation and Vibronic Interaction

As is well known, the Born-Oppenheimer separation of the electronic, vibrational, and rotational motions of a molecule is primarily based upon the fact that the electron mass m_e is about 1800 times smaller than the proton mass M_p. In fact, *Born* and *Oppenheimer* [2.43, 44] have shown than the vibrational and rotational energies are smaller than the electronic energy by an order of κ^2 and κ^4, respectively, namely

$$E_V \sim \kappa^2 E_E \quad \text{and} \quad E_R \sim \kappa^4 E_{E'} \qquad (2.4.1)$$

where

$$\kappa = (m_e/M_p)^{1/4} .$$

The separation ceases to be valid, however, when two electronic and two vibrational states approach and their energy differences become comparable with the vibrational and rotational energies, respectively. The latter case is referred to as vibration-rotation interaction, Sect. 2.2. The former is called the vibronic interaction, which is classified into three cases.

1) *Renner-Teller Effect* (or Renner Effect). In a degenerate electronic state of a linear molecule, the vibrational angular momentum induced by a degenerate mode couples with the electronic angular momentum, resulting in further splittings of vibrational levels [2.45].

2) *Jahn-Teller Effect.* A nonlinear polyatomic molecule in a degenerate electronic state will be unstable with respect to a coordinate not totally symmetric, and accordingly its symmetry will be lowered [2.46]. An excited state of a degenerate vibration will be split, a phenomenon which is referred to as the dynamical Jahn-Teller effect [2.33, 47].

3) *Herzberg-Teller Effect*. A pair of nearly degenerate electronic states of a polyatomic molecule may be mixed by excitation of a vibrational mode, provided that the direct product of the representations of the two electronic states involves the representation of the vibrational mode. This effect often distorts the vibrational potential function and causes intensity borrowing of the electronic transitions [2.48].

Both Jahn-Teller and Herzberg-Teller effects may occur through a term linear in vibrational coordinate Q, whereas the Renner effect is caused by a quadratic term. Many theoretical and experimental studies of the Renner effect have been reported, following the original report by *Renner* [2.45], as recently reviewed by *Jungen* and *Merer* [2.49]. This section is primarily devoted to the Renner effect in a Π linear polyatomic molecule, paying special attention to the rotational energy levels.

2.4.2 Renner Effect on a Linear Polyatomic Molecule in a Π Electronic State

When a linear polyatomic molecule has an unquenched orbital angular momentum component Λ along the molecular axis, the electronic state is doubly degenerate because of cylindrical symmetry, in other words, the states with $\pm\Lambda$ are degenerate. When a bending vibration is excited, the cylindrical symmetry is no longer retained, dipole and higher-pole fields being created which are different in the plane and out of the plane of the molecule bent by the bending vibration. Thus the degeneracy is removed; in other words, the vibrational potential function is split into two parts which join with each other at a linear configuration.

We consider the electrostatic interaction between a nucleus at cylindrical coordinates r_k, χ_k, ϕ_k and an electron at r_i, χ_i, θ_i, which is given by

$$V_i = \sum_k Z_k e^2/|\boldsymbol{r}_{ik}|, \tag{2.4.2}$$

then the total Coulomb potential is given by $V = \sum_i V_i$. In most cases, one electron is responsible for the degeneracy, so we simplify the discussion by omitting the summation over i. The coordinates of that electron are hereafter designated by r_e, χ_e, θ_e. When the kth nucleus is displaced by δr_k, the electrostatic potential is changed by

$$\delta V = \sum_k Z_k e^2/|\boldsymbol{r}_{ik} - \delta \boldsymbol{r}_k| - \sum_k Z_k e^2/|\boldsymbol{r}_{ik}|, \tag{2.4.3}$$

where

$$|\boldsymbol{r}_{ik} - \boldsymbol{r}_k|^{-1} = \sum_{l=0}^{\infty} [(\delta r_k)^l / r_{ik}^{l+1}] P_l(\cos x) \tag{2.4.4}$$

with x denoting the angle between $\boldsymbol{r}_{ik}$ and $\delta \boldsymbol{r}_k$. By using the addition theorem of spherical harmonics

$$P_l(\cos x) = [4\pi/(2l+1)] \sum_{q=-l}^{l} (-1)^q Y_l^{-q}(\chi_k, \phi_k)\, Y_l^q(\chi_e, \theta_e), \tag{2.4.5}$$

δV is rewritten as

$$\begin{aligned} \delta V = & \sum_k \sum_{l=1}^{\infty} Z_k e^2 [(\delta r_k)^l / r_{ik}^{l+1}]\, [4\pi/(2l+1)] \\ & \times \sum_{q=-l}^{l} (-1)^q Y_l^{-q}(\chi_k, \phi_k)\, Y_l^q(\chi_e, \theta_e) \\ = & \sum_k Z_k e^2 (\delta r_k)/r_{ik}^2 (4\pi/3) \sum_{q=-1}^{1} (-1)^q Y_1^{-q}(\chi_k, \phi_k)\, Y_1^q(\chi_e, \theta_e) \\ & + \sum_k Z_k e^2 (\delta r_k)^2/r_{ik}^3 (4\pi/5) \sum_{q=-2}^{2} (-1)^q Y_2^{-q}(\chi_k, \phi_k)\, Y_2^q(\chi_e, \theta_e) + \ldots . \end{aligned} \tag{2.4.6}$$

Because all the nuclei move perpendicularly to the molecular axis in a bending mode of infinitesimal amplitude, we set all $\chi_k = \pi/2$. Furthermore, we may choose ϕ common to all nuclei, which specifies the molecular plane when the bending is excited. The Renner-Teller Hamiltonian $\boldsymbol{H}_{\mathrm{RT}}$ then becomes

$$\boldsymbol{H}_{\mathrm{RT}} = [V'_{11}(\delta r)\cos(\theta - \phi) + V'_{22}(\delta r)^2 \cos 2(\theta - \phi) + \ldots]/hc, \tag{2.4.7}$$

where

$$V'_{11} \propto -\sum_k Z_k e^2 \sin \chi_e / r_{ik}^2,$$

$$V'_{22} \propto \sum_k Z_k e^2\, 3 \sin^2 \chi_e / r_{ik}^3,$$

and use is made of the fact that all δr_k are proportional to the bending vibration amplitude δr. The suffix e of the electron coordinate θ_e is dropped for simplicity. Note that since the V'_{nn} term is proportional to $1/r^{n+1}$, the first and second terms of (2.4.7) are smaller than the electrostatic energy by the factors of $\delta r/r$ and $(\delta r/r)^2$, respectively, which are of the orders of κ and κ^2.

Equation (2.4.7) may be rewritten, using $Q_\pm = \sqrt{\mu_2}(\delta r)\exp(\pm \mathrm{i}\phi)$ (μ_2 denoting the effective mass for the bending mode), as

$$\boldsymbol{H}_{\mathrm{RT}} = [V_{11}(Q_+ \mathrm{e}^{-\mathrm{i}\theta} + Q_- \mathrm{e}^{\mathrm{i}\theta}) + V_{22}(Q_+^2 \mathrm{e}^{-2\mathrm{i}\theta} + Q_-^2 \mathrm{e}^{2\mathrm{i}\theta}) + \ldots]/hc, \tag{2.4.8}$$

where $V_{nn} = V'_{nn}(\mu_2)^{n/2}/2$.

The vibronic basis function we chose is a product of the electronic and vibrational wave functions as follows:

$$|n, \Lambda, v_2, l\rangle = |n, \Lambda\rangle\, |v_2, l\rangle, \quad \text{where} \tag{2.4.9}$$

$$|n, \Lambda\rangle = \sum_L F_L Y_L^\Lambda(\chi, \theta) = \sum_L F_L (2\pi)^{-1/2} e^{i\Lambda\theta} \Theta_{L,\Lambda}(\chi) \quad \text{and} \tag{2.4.10}$$

$$|v_2, l\rangle = (2\pi)^{-1/2} e^{il\phi} \Psi_{v,l}(\delta r). \tag{2.4.11}$$

The parameter n specifies all properties of the electronic state considered besides the electron orbital angular momentum component given by Λ. The function $\Psi_{v,l}(\delta r)$ is, as defined by *Di Lauro* and *Mills* [2.50], independent of ϕ and satisfies $\Psi_{v,l} = \Psi_{v,-l}$. The part of the spherical harmonics which depends on χ, $\Theta_{L,\Lambda}(\chi)$, fulfills [2.51]

$$\Theta_{L,-|\Lambda|}(\chi) = (-1)^\Lambda \Theta_{L,|\Lambda|}(\chi). \tag{2.4.12}$$

For a Π electronic state ($\Lambda = \pm 1$) the second term of (2.4.8) gives first-order contributions, whereas the first term may be reduced to a form similar to the second term by a second-order perturbation treatment in which Σ and Δ electronic states act as intermediate states. The first term may thus be regarded as causing a Herzberg-Teller type interaction. First, we consider the second term or Renner interaction term only by paying attention to how it perturbs vibrational levels, while the first term (or Herzberg-Teller type interaction term) is discussed below.

For a two-dimensional harmonic oscillator the nonvanishing matrix elements of the vibrational coordinates are given by

$$\langle v_2 \pm 1, l+1 | Q_+ | v_2, l\rangle = (2\gamma_2)^{-1/2}[(v_2+1) \pm (l+1)]^{1/2}, \tag{2.4.13 a}$$

$$\langle v_2 \pm 1, l-1 | Q_- | v_2, l\rangle = (2\gamma_2)^{-1/2}[(v_2+1) \mp (l-1)]^{1/2}, \tag{2.4.13 b}$$

where

$$\gamma_2 = 2\pi c \omega_2/\hbar. \tag{2.4.14}$$

The nonvanishing matrix elements of $Q_\pm^2$ are derived from (2.4.13) as follows:

$$\langle v_2+2, l \pm 2 | Q_\pm^2 | v_2, l\rangle = (1/2\gamma_2)[(v_2 \pm l + 2)(v_2 \pm l + 4)]^{1/2}, \tag{2.4.15 a}$$

$$\langle v_2, l \pm 2 | Q_\pm^2 | v_2, l\rangle = (1/2\gamma_2)[(v_2 \mp l)(v_2 \pm l + 2)]^{1/2}, \tag{2.4.15 b}$$

$$\langle v_2-2, l \pm 2 | Q_\pm^2 | v_2, l\rangle = (1/2\gamma_2)[(v_2 \mp l)(v_2 \mp l - 2)]^{1/2}. \tag{2.4.15 c}$$

Since most Π molecules so far investigated have electron spin, we must consider the spin-orbit interaction, which is assumed to take the form

$$\boldsymbol{H}_{\mathrm{SO}} = A \boldsymbol{L} \cdot \boldsymbol{S}. \tag{2.4.16}$$

(Since only linear molecules are discussed in this section, the spin-orbit coupling constant A_{SO} is simply written as A.) The rotational Hamiltonian may be written as

$$\boldsymbol{H}_{\mathrm{ROT}} = B(\boldsymbol{J} - \boldsymbol{W}) \cdot (\boldsymbol{J} - \boldsymbol{W}) + \gamma \boldsymbol{N} \cdot \boldsymbol{S}, \quad \text{where} \tag{2.4.17}$$

$$\boldsymbol{W} = \boldsymbol{L} + \boldsymbol{S} + \boldsymbol{G},$$

$\boldsymbol{G}$ denoting the vibrational angular momentum. The rotational part, which is a Hund's case (a) function, is added to the vibronic base function

$$|n\Lambda, v_2, l, \Sigma; J, P\rangle = |n, \Lambda\rangle\, |v_2, l\rangle\, |S, \Sigma\rangle\, |J, P\rangle,$$

where $P = K \pm \Sigma$ and $K = |\Lambda + l|$.

The parity of the base function is determined in the following way. The inversion operator E^* is defined as

$$E^* f(X, Y, Z) = f(-X, -Y, -Z), \tag{2.4.18 a}$$

where f is a function of the space-fixed coordinates X, Y, Z. The Euler angles α, β, and γ are accordingly transformed as

$$E^* f(\alpha, \beta, \gamma) = f(\pi + \alpha, \pi - \beta, \pi - \gamma). \tag{2.4.18 b}$$

The effect on the molecule-fixed coordinates x, y, and z is thus

$$E^*(x, y, z) = f(x, -y, z). \tag{2.4.18 c}$$

Therefore, E^* is completely equivalent to σ_{xz}, Sect. 2.3.4, as far as the molecule-fixed coordinates are concerned, and the results (2.3.18–20) may readily be applied to the present problem. They are reproduced here with the addition of $|v_2, l\rangle$:

$$\sigma_{xz}\, |n, \Lambda\rangle = (-1)^{\Lambda + s}\, |n, -\Lambda\rangle, \tag{2.4.19}$$

$$\sigma_{xz}\, |v_2, l\rangle = |v_2, -l\rangle, \tag{2.4.20}$$

$$\sigma_{xz}\, |S, \Sigma\rangle = (-1)^{S - \Sigma}\, |S, -\Sigma\rangle, \tag{2.4.21}$$

$$\sigma_{xz}\, |J, P\rangle = (-1)^{J - P}\, |J, -P\rangle. \tag{2.4.22}$$

The total basis function is thus transformed as

$$\sigma_{xz}\, |n, \Lambda, v_2, l, S, \Sigma; J, P\rangle$$
$$= (-1)^{J - S - l + s}\, |n, -\Lambda, v_2, -l, S, -\Sigma; J, -P\rangle \tag{2.4.23}$$

and the eigenfunctions of σ_{xz} with the eigenvalues ± 1 are given by

$$|n, \Lambda, v_2, l, S, \Sigma; J, P; \pm\rangle = (1/\sqrt{2})\, [|n, \Lambda, v_2, l, S, \Sigma; J, P\rangle$$
$$\pm (-1)^{J - S - l + s}\, |n, -\Lambda, v_2, -l, S, -\Sigma; J, -P\rangle], \tag{2.4.24}$$

which is an extension of (2.3.22).

Next, we evaluate the matrix elements of the Renner effect [the second term of (2.4.8)] and the spin-orbit interaction equation (2.4.16) for a doublet state ($S = 1/2$), while neglecting the rotation for a moment, Sect. 2.4.3. We have then a two-by-two matrix of the following form:

$$|\Lambda = 1, v_2, l = K-1, \Sigma = 1/2; \pm\rangle \quad |\Lambda = -1, v_2, l = K+1, \Sigma = 1/2; \pm\rangle$$

$$\begin{bmatrix} (v_2+1)\,\omega_2 + A_{\text{eff}}\,\Sigma - (\varepsilon\,\omega_2)^2 (v_2+1)(K+1)/8\,\omega_2 & (1/2)\,\varepsilon\,\omega_2 [(v_2+1)^2 - K^2]^{1/2} \\ & (v_2+1)\,\omega_2 - A_{\text{eff}}\,\Sigma + (\varepsilon\,\omega_2)^2 (v_2+1)(K-1)/8\,\omega_2 \end{bmatrix}, \tag{2.4.25}$$

where the $\Delta v_2 = \pm 2$ matrix elements of the Renner effect within the electronic state have been reduced by second-order perturbation and thus the A constant is modified to [2.52]

$$A_{\text{eff}} = A[1 - (\varepsilon^2/8)\,K(K+1)]. \tag{2.4.26}$$

The parameter $\varepsilon\,\omega_2$ in (2.4.23) receives contributions from both the first- and second-order terms of (2.4.8). The term $\varepsilon^{(1)}\omega_2$ derived from V_{22} by the first-order perturbation is given by

$$\varepsilon^{(1)}\omega_2 = \langle\xi|\,V_{22}\,|\xi\rangle/hc\gamma_2, \tag{2.4.27}$$

where $|\xi\rangle$ denotes the electronic wave function of the Π state considered, i.e., $|n, \Lambda\rangle$ except for the factor $(2\pi)^{-1/2}\,e^{i\Lambda\theta}$. Strictly speaking, $|\xi\rangle$ on the left-hand side corresponds to $\Lambda = +1$ and $|\xi\rangle$ on the right to $\Lambda = -1$. The other term $\varepsilon^{(2)}\omega_2$ from V_{11} is given below. Because both Λ and l are not well defined under a strong vibronic interaction, we replace them by $K = |\Lambda + l|$ to designate each vibronic level as Σ, Π, Δ, ... according to $K = 0, 1, 2, \ldots$.

The solutions of (2.4.25) are given by

$$E^{\kappa}_{v_2,K,\Sigma} = \omega_2(v_2+1) + r - [\varepsilon^2\omega_2(v_2+1)/8](1 + A_{\text{eff}}\,K\,\Sigma/r), \tag{2.4.28 a}$$

$$E^{\mu}_{v_2,K,\Sigma} = \omega_2(v_2+1) - r - [\varepsilon^2\omega_2(v_2+1)/8](1 - A_{\text{eff}}\,K\,\Sigma/r), \tag{2.4.28 b}$$

where

$$r = (1/2)\,\{A^2_{\text{eff}} + \varepsilon^2\omega_2^2[(v_2+1)^2 - K^2]\}^{1/2}. \tag{2.4.29}$$

The eigenfunctions are accordingly expressed as

$$\psi^{\kappa} = \cos\beta\,|1, v_2, l = K-1, \Sigma = 1/2; \pm\rangle + \sin\beta\,|-1, v_2, l = K+1, \Sigma = 1/2; \pm\rangle, \tag{2.4.30 a}$$

$$\psi^{\mu} = -\sin\beta\,|1, v_2, l = K-1, \Sigma = 1/2; \pm\rangle + \cos\beta\,|-1, v_2, l = K+1, \Sigma = 1/2; \pm\rangle, \tag{2.4.30 b}$$

where

$$r \sin \beta = (\varepsilon \omega_2/2) [(v_2 + 1)^2 - K^2]^{1/2}, \tag{2.4.31 a}$$

$$r \cos \beta = A_{\text{eff}}/2, \tag{2.4.31 b}$$

and $\Delta v_2 \neq 0$ matrix elements are all neglected.

Substituting $\Sigma = \pm 1/2$ in (2.4.28) yields four energy levels for each pair of v_2 and K, provided that K is equal to neither zero nor $v_2 + 1$. Following *Hougen* [2.52], we use κ and μ to designate the upper and lower levels with a particular K value in a vibrational state.

Figure 2.1 shows an example of the energy level diagram consisting of a series of excited bending (v_2) states, drawn to scale using the following parameters of BO_2: $\omega_2 = 464\ \text{cm}^{-1}$, $\varepsilon\omega_2 = -92\ \text{cm}^{-1}$, and $A = -150\ \text{cm}^{-1}$. It should be noted that vibronic interaction is most conspicuous for levels with $K \neq v_2 + 1$, as indicated by the large splittings of these levels. On the other hand, $K = v_2 + 1$ levels are split primarily by the spin-orbit interaction, which is slightly modified by the Renner effect, as discussed below.

It is worthwhile to examine the case where the electron spin (i.e., the spin-orbit interaction) may be neglected. Then each vibronic level is either symmetric or antisymmetric with respect to reflection on the molecular plane. For $K = 0$ levels in a $^1\Pi$ state, for example, (2.4.28) becomes

$$E^{\pm}_{v, K=0} = \omega_2 (1 \pm \varepsilon/2) (v_2 + 1) - \varepsilon^2 \omega_2 (v_2 + 1)/8 \tag{2.4.32}$$

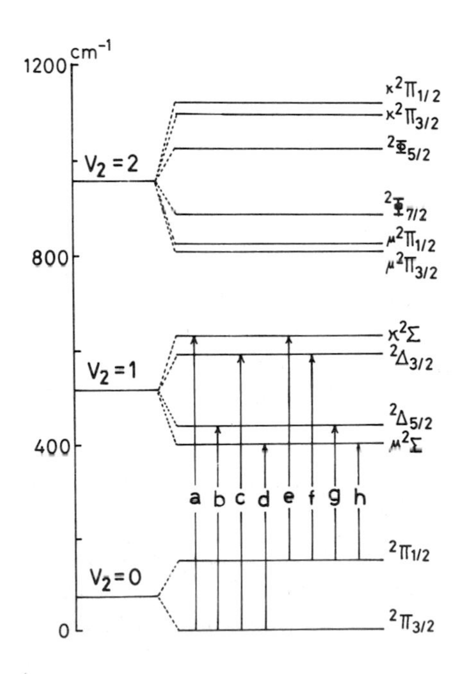

Fig. 2.1. Lowest bending states of BO_2 radical in the $\tilde{X}^2\Pi_g$ round electronic state

and the corresponding eigenfunctions are given by

$$|v_2, K = 0, \pm\rangle = (1/\sqrt{2})\,[|1, v_2, l = -1\rangle \pm |-1, v_2, l = +1\rangle], \qquad (2.4.33)$$

where higher-order mixing with other vibronic states is ignored. The + and − signs correspond to the vibronic states which are symmetric and antisymmetric with respect to the molecular plane.

Now, consider the effect of the first term of (2.4.8), which is referred to as the Herzberg-Teller interaction. The second-order perturbation correction to the vibronic energy level is calculated as follows [2.53]:

$$E^{(2)} = \Delta T/hc - (v_2 + 1)\,\Delta\omega_2^{(2)} + g_K K \Lambda + \dots, \qquad (2.4.34)$$

where

$$\Delta T = (\hbar^2/4) \sum_{\xi' \subset \Delta} |\langle \xi, \Lambda = 1|\, V_{11}\, |\xi', \Lambda = 2\rangle|^2/(\Delta E)^2, \qquad (2.4.35)$$

$$\Delta\omega_2^{(2)} = (1/4hc\gamma_2) \sum_{\xi' \subset \Sigma, \Delta} |\langle \xi|\, V_{11}\, |\xi'\rangle|^2/\Delta E, \qquad (2.4.36)$$

$$g_K = (\omega_2/4\gamma_2) \sum_{\xi' \subset \Sigma, \Delta} (-1)^p\, |\langle \xi|\, V_{11}\, |\xi'\rangle|^2/(\Delta E)^2. \qquad (2.4.37)$$

In these equations $\Delta E = E^0(\xi') - E^0(\xi)$ and p is an even and an odd integer when ξ' is an Σ and a Δ state, respectively. The second correction term in $\Delta\omega_2^{(2)}$ simply reduces the bending frequency, whereas the third correction term in g_K contributes to changing the energy differences between vibronic levels of different K values.

The Renner parameter also receives second-order correction from the V_{11} term:

$$\varepsilon^{(2)}\omega_2 = -(1/2hc\gamma_2) \sum_{\xi' \subset \Sigma} (-1)^s\, |\langle \xi|\, V_{11} e^{i\theta}\, |\xi'\rangle|^2 \times [1 + (hc\omega_2/\Delta E)^2]/\Delta E, \qquad (2.4.38)$$

where s is even and odd for Σ^+ and Σ^-, respectively.

The A_{eff} constant equation (2.4.26) has added to it a third-order correction term arising from interactions with other vibronic states within the $^2\Pi$ ground electronic state and also from those with other electronic states through the first term of (2.4.8), given below.

The third-order perturbation correction to the nth eigenvalue is given by

$$E_n^{(3)} = \sum_k{}' \sum_m{}' V_{nm} V_{mk} V_{kn}/(\omega_{mn}\omega_{kn}) - V_{nn} \sum_m{}' |V_{nm}|^2/\omega_{mn}^2, \qquad (2.4.39)$$

where V denotes the perturbation operator, ω_{mn} the difference between the zero-order eigenvalues, i.e., $\omega_{mn} = [E_n^{(0)} - E_m^{(0)}]/hc$, and the prime on the summation symbols means the terms with $k = n$ or $m = n$ are to be omitted. The

Renner operator (2.4.8) has matrix elements of $\Delta v_2 = \pm 2$ within the electronic state to which the vibronic state considered belongs and of $\Delta v_2 = \pm 1$ between different electronic states. On the other hand, the spin-orbit interaction matrix elements obey the $\Delta v_2 = 0$ selection rule, since the operator equation (2.4.16) does not contain $Q_\pm$. The third-order perturbation correction (2.4.39) then slightly modifies *Hougen*'s expression for $A_{\text{eff}} = A_{\text{true}}\,[1 - (\varepsilon^2/8)\,K(K+1)]$, (2.4.26), as follows:

$$A_{\text{eff}} = A_{\text{true}}[1 - (\varepsilon^2/8)\,K(K+1) + K\eta] \tag{2.4.40}$$

for $K = v_2 + 1$, but for $K = 0$ Σ vibronic states, A_{eff} is given by

$$A_{\text{eff}} = A_{\text{true}}[1 - (\varepsilon^2/32)\,(v_2+1)\,(v_2+3) + (v_2+1)\,\eta], \tag{2.4.41}$$

where η represents the third-order correction term caused by the interactions with other electronic states and is expressed by

$$\eta = \sum_{\xi' \subset \Sigma, \Delta} |\langle \xi |\, V_{11}\, | \xi' \rangle|^2 / \Delta E^2 (1/4\gamma_2)\,\{[1 - (-1)^p]\,A_{\xi'}/A_\xi - 1\} \tag{2.4.42}$$

with p equal to an even or an odd integer according to whether ξ' is a Σ or a Δ state and with $A_{\xi'}$ and A_ξ representing the spin-orbit coupling constants in the ξ' and ξ states, respectively. It should be noted that (2.4.41) does not take into account the ordinary vibrational effect on the A constant. Therefore, the η constant obtained by fitting (2.4.41) to the experimental A values should be interpreted with care. The third-order perturbation also contributes to the g factor, as discussed in Sect. 2.5.

The sign of the Renner parameter ε may be inferred with the aid of the *Walsh* diagram [2.54], as shown by an example of an XY_2-type molecule illustrated in Fig. 2.2. Suppose a molecule to have three electrons in a π_g orbital

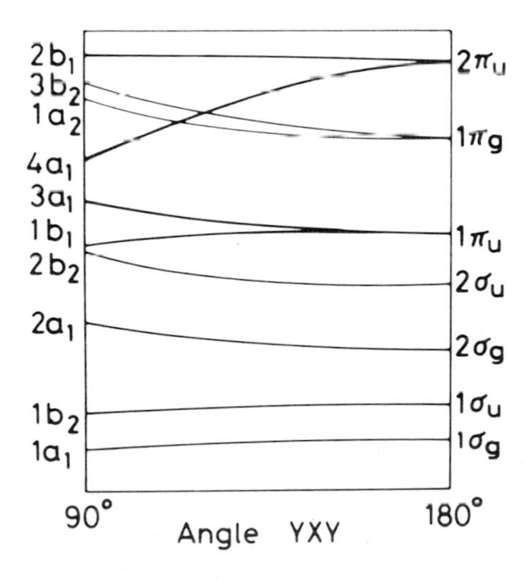

Fig. 2.2. Walsh diagram for a symmetric XY_2-type molecule

in the limit of linearity, like BO_2 and CO_2^+. When the bending mode is excited, this π_g orbital is split into one b_2 and one a_2 orbital, generating a $^2B_2$ state with the electron configuration $(a_2)^2(b_2)$ and a $^2A_2$ state with $(a_2)(b_2)^2$ from the original $^2\Pi_g$ state. Since the a_2 orbital is likely to be lower than the b_2 orbital, which is symmetric with respect to reflection on the molecular plane, the Renner parameter will be negative. When a molecule has only one electron in the π_g orbital like CNC, the $^2A_2$ state with the $(a_2)^1$ configuration is lower than the $^2B_2$ state with $(b_2)^1$, making ε positive.

The $\tilde{A}^2\Pi_u$ states of BO_2 and CO_2^+ are obtained by $(1\pi_u)^4(1\pi_g)^3 \rightarrow (1\pi_u)^3(1\pi_g)^4$ excitation, and a similar consideration for the π_u orbital predicts ε to be negative also in the $\tilde{A}$ state. However, the a_1 and b_1 orbitals derived from the π_u orbital do not seem to differ much in energy, as long as the bending of the molecule is not appreciable, and thus the magnitude of ε would be smaller in the $\tilde{A}$ state than in the $\tilde{X}$ state. This is really the case; ε in $\tilde{A}$ and $\tilde{X}$ are -0.026 and -0.194, respectively, for BO_2 [2.55] and -0.093 and -0.189, respectively, for CO_2^+ [2.56]. For XYZ-type molecules, the a_1 and b_2 orbitals correlate with a' and a_2 and b_1 with a'', and a similar argument predicts ε to be positive and negative for CCN and NCO, respectively.

The NO_2 molecule may be considered to have an extremely large Renner effect; the "original" linear molecule would have the configuration $(1\pi_u)^4(1\pi_g)^4(2\pi_u)^1$ and ε would be much less than -1. A similar consideration may be applied to other "bent" molecules like NH_2 and PH_2 in the ground electronic states and CS_2 in the $\tilde{A}$ state.

Finally, let us consider the relative intensity of the ν_2 vibrational band. Equations (2.4.30) may be written for $v_2 = 1$ and 0 as follows:

$$\psi(\kappa^2\Sigma) = \cos\beta\,|1, v_2 = 1, -1, 1/2; \pm\rangle + \sin\beta\,|-1, v_2 = 1, 1, 1/2; \pm\rangle, \tag{2.4.43 a}$$

$$\psi(^2\Delta_{3/2}) = |-1, v_2 = 1, -1, 1/2; \pm\rangle, \tag{2.4.43 b}$$

$$\psi(^2\Delta_{5/2}) = |1, v_2 = 1, 1, 1/2; \pm\rangle, \tag{2.4.43 c}$$

$$\begin{aligned}\psi(\mu^2\Sigma) = &-\sin\beta\,|1, v_2 = 1, -1, 1/2; \pm\rangle \\ &+\cos\beta\,|-1, v_2 = 1, 1, 1/2; \pm\rangle,\end{aligned} \tag{2.4.43 d}$$

$$\psi(^2\Pi_{1/2}) = |-1, v_2 = 0, 0, 1/2; \pm\rangle, \tag{2.4.43 e}$$

$$\psi(^2\Pi_{3/2}) = |1, v_2 = 0, 0, 1/2; \pm\rangle. \tag{2.4.43 f}$$

Then the relative intensities of the a through h transitions indicated in Fig. 2.1 are given by

$$a:b:c = h:f:e = \cos^2\beta : 1 : \sin^2\beta, \tag{2.4.44}$$

whereas the d- and g-type transitions are allowed by either the magnetic dipole moment or an electric dipole moment induced by rotational mixing between

$P = K + 1/2$ and $P = K - 1/2$ states and are thus expected to be very weak, especially near Hund's case (a) limit.

2.4.3 Rotational Energy Levels of a Linear Triatomic Molecule Affected by Vibronic Interactions

Vibronic interactions perturb rotational energy levels of $K \neq v_2 + 1$ vibronic levels in characteristic ways, while those in $K = v_2 + 1$ states remain apparently unchanged and are quite similar to those of a $^2\Pi$ diatomic molecule. A close inspection shows, however, that the vibration-rotation constant α of the bending mode includes contributions from vibronic interactions like the Renner effect, in addition to those from the ordinary vibration-rotation interactions, and thus takes slightly different values for different vibronic states within an electronic state.

Equations (2.4.16, 17) may be expanded in terms of the molecule-fixed components of the angular momenta involved:

$$\begin{aligned} \boldsymbol{H}_{\mathrm{SO}} + \boldsymbol{H}_{\mathrm{ROT}} = {} & A\Lambda\Sigma + B(\boldsymbol{J}^2 - J_z^2) + (B-\gamma)(\boldsymbol{S}^2 - S_z^2) + B(L_+ L_- + G_+ G_-) \\ & - B(L_+ J_- + L_- J_+) - (B - \gamma/2)(S_+ J_- + S_- J_+) \\ & - B(G_+ J_- + G_- J_+) + (A/2 - \gamma/2 + B)(L_+ S_- + L_- S_+) \\ & - (B - \gamma/2)(G_+ S_- + G_- S_+) - B(L_+ G_- + L_- G_+) . \end{aligned} \tag{2.4.45}$$

First we consider the $K = v_2 + 1$ vibronic state. By using (2.4.24) as a base function, the parity-independent matrix elements of $\boldsymbol{H}_{\mathrm{SO}} + \boldsymbol{H}_{\mathrm{ROT}}$ are derived as follows:

$$\begin{aligned} & \langle K; J, P = K - 1/2; \pm | H_{\mathrm{SO}} + H_{\mathrm{ROT}} | K; J, P = K - 1/2; + \rangle \\ & \quad = - A_{\mathrm{eff}}/2 + (B + q^*/2 - A_J)(X + K) - D[(X + K)^2 + X], \end{aligned} \tag{2.4.46 a}$$

$$\begin{aligned} & \langle K; J, P = K + 1/2; \pm | H_{\mathrm{SO}} + H_{\mathrm{ROT}} | K; J, P = K + 1/2; \pm \rangle \\ & \quad = A_{\mathrm{eff}}/2 + (B + q^*/2 + A_J)(X - K) - D[(X - K)^2 + X], \end{aligned} \tag{2.4.46 b}$$

$$\begin{aligned} & \langle K; J, P = K + 1/2; \pm | H_{\mathrm{SO}} + H_{\mathrm{ROT}} | K; J, P = K - 1/2; \pm \rangle \\ & \quad = - [B + q^*/2 - \gamma/2 + p^*/4 - 2DX]\sqrt{X}, \end{aligned} \tag{2.4.46 c}$$

where $X = (J + 1/2)^2 - K^2$ and A_J denotes the centrifugal correction term for the spin-rotation interaction. Note that A_J corresponds to $A_D/2$ of Sect. 2.3.

The parity-dependent terms, which lead to Λ-type doubling, are given for a $^2\Pi$ state by

$$\langle K = 1; J, P = 1/2; \pm | H_{SO} + H_{ROT} | K = 1; J, P = 1/2; \pm \rangle$$
$$= \mp (-1)^{J-1/2} (q + p/2)(J + 1/2), \tag{2.4.47 a}$$

$$\langle K = 1; J, P = 3/2; \pm | H_{SO} + H_{ROT} | K = 1; J, P = 1/2; \pm \rangle$$
$$= \pm (-1)^{J-1/2} (q/2)(J + 1/2)\sqrt{X}. \tag{2.4.47 b}$$

When the bending mode is excited, the parity-dependent terms assume expressions that are slightly different for different vibronic states, such as $\mu\,^2\Sigma$, $\kappa\,^2\Sigma$, and $^2\Delta$ in $v_2 = 1$, as shown by *Bolman* et al. [2.57].

The vibrational dependence of the B rotational constant may be expressed (with $i = 1$ and 3 denoting two stretching modes) as

$$B_{v_1, v_2, v_3} = B_e - \alpha_1 (v_1 + 1/2) - \alpha_2 (v_2 + 1) - \alpha_3 (v_3 + 1/2) \tag{2.4.48}$$

just as in the case of an ordinary $^1\Sigma$ triatomic molecule, for which the α_i constant may be divided into three parts:

$$\alpha_i = \alpha_i^{\text{harm}} + \alpha_i^{\text{Cor}} + \alpha_i^{\text{anharm}}. \tag{2.4.49}$$

Explicit expressions for α_i may be found, for example, in [2.58]. The Renner effect will primarily modify the α_2 constant, while leaving α_1 and α_3 relatively unaffected. The Coriolis and anharmonic parts of α_2 are given by

$$\alpha_2^{\text{Cor}} = (B_e^2/\omega_2)\,[1 + 4\zeta_{21}^2\,\omega_2^2/(\omega_1^2 - \omega_2^2) + 4\zeta_{23}^2\,\omega_2^2/(\omega_3^2 - \omega_2^2)], \tag{2.4.50}$$

$$\alpha_2^{\text{anharm}} = -(2B_e)^{3/2}\,[\zeta_{23}\,k_{122}/(2\omega_1^{3/2}) - \zeta_{21}\,k_{322}/(2\omega_3^{3/2})] \tag{2.4.51}$$

for a $^1\Sigma$ molecule, where two Coriolis coupling constants defined by (2.14) satisfy $\zeta_{21}^2 + \zeta_{23}^2 = 1$.

The rovibronic interaction modifies the B rotational constant (and thus the α_2 constant) in two ways, one through the change in the v_2 frequency and the other due to the different anharmonic potential [2.59]. *Hougen* [2.60] has introduced a cubic anharmonic potential of the following form:

$$\boldsymbol{H}_{\text{anharm}} = \boldsymbol{H}_{F1} + \boldsymbol{H}_{F2}, \tag{2.4.52}$$

$$\boldsymbol{H}_{F1} = (1/2) \sum_{i=1,3} (f'_{i22} + f''_{i22})\,Q_i\,\mu_2\,(\delta r)^2, \tag{2.4.53}$$

$$\boldsymbol{H}_{F2} = (1/2) \sum_{i=1,3} (f'_{i22} - f''_{i22})\,Q_i\,\mu_2\,(\delta r)^2\, 2\cos 2(\theta - \phi), \tag{2.4.54}$$

where the single and double primes refer to electronic states that are symmetric and antisymmetric with respect to the reflection on the plane of the bent molecule, respectively. Using (2.4.13) the nonvanishing matrix elements of the anharmonic potential may be evaluated. Hougen has employed two parameters $W_{1,i}$ and $W_{2,i}$, instead of f'_{i22} and f''_{i22}, defined as

$$W_{1,i} = (1/2)\,(f'_{i22} + f''_{i22})\,(\hbar/4\pi c\,\omega_2)\,(\hbar/4\pi c\,\omega_i)^{1/2}, \quad (2.4.55)$$

$$W_{2,i} = (1/2)\,(f'_{i22} - f''_{i22})\,(\hbar/4\pi c\,\omega_2)\,(\hbar/4\pi c\,\omega_i)^{1/2}. \quad (2.4.56)$$

For a symmetrical XY_2-type molecule, the $i = 3$ terms are absent, and $W_{1,1}$ and $W_{2,1}$ are simply written as W_1 and W_2, respectively. The first-order terms in the expansion of the moment of inertia in terms of the vibrational coordinate Q_i give the following off-diagonal matrix elements between two vibrational states:

$$\langle v_1, v_2, v_3; \Lambda, l, \Sigma; J|\, H_{\mathrm{ROT}}\, |v_1 + 1, v_2, v_3; \Lambda, l, \Sigma; J\rangle$$
$$= -\,\zeta_{23}(2B_e^{3/2}/\omega_1^{1/2})\,(v_1 + 1)^{1/2}\,[(J + 1/2)^2 - K^2], \quad (2.4.57)$$

$$\langle v_1, v_2, v_3; \Lambda, l, \Sigma; J|\, H_{\mathrm{ROT}}\, |v_1, v_2, v_3 + 1; \Lambda, l, \Sigma; J\rangle$$
$$= +\,\zeta_{21}\,(2B_e^{3/2}/\omega_3^{1/2})\,(v_3 + 1)^{1/2}\,[(J + 1/2)^2 - K^2]. \quad (2.4.58)$$

As discussed in Sect. 2.2, these terms are combined with the cubic anharmonic potential to generate the $\alpha_i^{\mathrm{anharm}}$ term. In fact, a second-order perturbation calculation based upon (2.4.55, 57, 58) leads to (2.4.51). When the treatment is extended to third order, using (2.4.39), the following correction term is obtained for B:

$$B^{(3)}(K = v_2 + 1) = \{\;\; 2\varepsilon Z_2 + (\varepsilon^2/16)\,[B(v_2 + 1) - B(v_2)]\}\,(v_2 + 1)\,(v_2 + 2), \quad (2.4.59)$$

where

$$Z_2 = \zeta_{23}(B_e/\omega_1)^{3/2}\,\overline{W}_{2,1} - \zeta_{21}\,(B_e/\omega_3)^{3/2}\,\overline{W}_{2,3}\,. \quad (2.4.60)$$

For a $K = 0\ \Sigma$ vibronic state, we use (2.4.30) as the vibronic base functions which diagonalize $\boldsymbol{H}_{\mathrm{SO}}$ and $\boldsymbol{H}_{\mathrm{RT}}$, shown above, to evaluate matrix elements of the rotational Hamiltonian $\boldsymbol{H}_{\mathrm{ROT}}$ shown in (2.4.45). The results are

$$\langle \kappa\, {}^2\Sigma, J; \pm|\, H_{\mathrm{ROT}}\, |\kappa\, {}^2\Sigma, J; \pm\rangle$$
$$= r + (B_v + s)\,(J + 1/2)^2 - D[(J + 1/2)^4 + (J + 1/2)^2]$$
$$\pm (-1)^{J+1/2}\sin 2\beta (J + 1/2)\,[B_v - \gamma/2 - 2D(J + 1/2)^2]$$
$$\pm (\;\;\, 1)^{J+1/2}\,\gamma_{\mathrm{vib}}(J + 1/2), \quad (2.4.61\,\mathrm{a})$$

$$\langle \mu\, {}^2\Sigma, J; \pm|\, H_{\mathrm{ROT}}\, |\mu\, {}^2\Sigma, J; \pm\rangle$$
$$= -r + (B_v - s)\,(J + 1/2)^2 - D[(J + 1/2)^4 + (J + 1/2)^2]$$
$$\mp (-1)^{J+1/2}\sin 2\beta (J + 1/2)\,[B_v - \gamma/2 - 2D(J + 1/2)^2]$$
$$\pm (-1)^{J+1/2}\,\gamma_{\mathrm{vib}}(J + 1/2), \quad \text{and} \quad (2.4.61\,\mathrm{b})$$

$$\langle \kappa\,^2\Sigma, J; \pm |\, H_{\mathrm{ROT}}\, | \mu\,^2\Sigma, J; \pm \rangle$$
$$= \pm(-1)^{J+1/2}[B_v - \gamma/2 - 2D(J+1/2)^2]$$
$$\times (J+1/2)\cos 2\beta - t(J+1/2)^2, \tag{2.4.61 c}$$

where J-dependent contributions arising from second-order and third-order perturbations by other vibronic states are included in s, γ_{vib}, and t, which are expressed as

$$s = Z\sin 2\beta + A_J\cos 2\beta + (\text{Coriolis term}), \tag{2.4.62}$$

$$\gamma_{\mathrm{vib}} = Z + A_J \sin 2\beta\cos 2\beta, \tag{2.4.63}$$

$$t = -Z\cos 2\beta, \quad \text{where} \tag{2.4.64}$$

$$Z = \zeta_{2,3}(B_e/\omega_1)^{3/2}(v_2+1)[8W_{2,i} - (v_2+3)\varepsilon W_{1,i}/2]$$
$$-\zeta_{2,1}(B_e/\omega_3)^{3/2}(v_2+1)[8\bar{W}_{2,3} - (v_2+3)\varepsilon\bar{W}_{1,3}/2], \tag{2.4.65}$$

assuming that $v_1 = v_3 = 0$. The small Coriolis term in (2.4.62) arises from the difference in the effective ω_2 frequency between the $\kappa\,^2\Sigma$ and $\mu\,^2\Sigma$ states (2.4.50). The B_v constant in (2.4.61) contains higher-order vibronic contributions, given by

$$B^{(3)}(K=0) = \{-\varepsilon Z_2(v_2+3)/2 + (\varepsilon^2/16)$$
$$\times [B(v_2+1) - B(v_2)]\}(v_2+1). \tag{2.4.66}$$

This expression is similar but not identical to (2.4.59) for the $K = v_2 + 1$ state, reflecting that the Renner interaction affects the $K = 0\,\Sigma$ and $K = v_2 + 1$ states differently.

When $r \gg B(J+1/2)$ is satisfied, (2.4.61) may be expanded to obtain the following expressions for the rotational energy levels, which are similar in form to those for a $^2\Sigma$ Hund's case (b) diatomic molecule:

$$F_1(\kappa\,^2\Sigma, N) = C + r + \gamma^\kappa/2 + B^\kappa_{\mathrm{eff}}N(N+1) + (1/2)\gamma^\kappa N$$
$$-[D^\kappa_{\mathrm{eff}}N^2(N+1)^2 + 2\gamma^\kappa_D(N+1)^3], \tag{2.4.67}$$

$$F_2(\kappa\,^2\Sigma, N) = C + r + \gamma^\kappa/2 + B^\kappa_{\mathrm{eff}}N(N+1) - (1/2)\gamma^\kappa(N+1)$$
$$-[D^\kappa_{\mathrm{eff}}N^2(N+1)^2 - 2\gamma^\kappa_D N^3], \tag{2.4.68}$$

$$F_1(\mu\,^2\Sigma, N) = C - r + \gamma^\mu/2 + B^\mu_{\mathrm{eff}}N(N+1) + (1/2)\gamma^\mu N$$
$$-[D^\mu_{\mathrm{eff}}N^2(N+1)^2 + 2\gamma^\mu_D(N+1)^3], \tag{2.4.69}$$

$$F_2(\mu\,^2\Sigma, N) = C - r + \gamma^\mu/2 + B^\mu_{\mathrm{eff}}N(N+1) - (1/2)\gamma^\mu(N+1)$$
$$-[D^\mu_{\mathrm{eff}}N^2(N+1)^2 - 2\gamma^\mu_D N^3], \tag{2.4.70}$$

where

$$B^{\kappa,\mu}_{\mathrm{eff}} = B_v[1 \pm (B_v/2r)\cos^2 2\beta] \pm s, \tag{2.4.71}$$

$$\gamma^{\kappa,\mu}_{\mathrm{eff}} = 2[B^{\kappa,\mu}_{\mathrm{eff}} - (B_v - \gamma/2)|\sin 2\beta| \pm \gamma_{\mathrm{vib}}], \tag{2.4.72}$$

$$\gamma_D^{\kappa,\mu} = D_{\text{eff}}^{\kappa,\mu} - |D \sin 2\beta + (B_v^3/4r^2) \sin 2\beta \cos^2 2\beta \pm (B_v t/2r) \cos 2\beta|, \tag{2.4.73}$$

$$D_{\text{eff}}^{\kappa,\mu} = D[1 \pm (2B_v/r) \cos^2 2\beta]. \tag{2.4.74}$$

The + and − signs in (2.4.71 – 74) apply to the $\kappa\,^2\Sigma$ and $\mu\,^2\Sigma$ states, respectively. The above results are identical in form to those derived by *Hougen* [2.52], but include contributions of rovibronic interactions, which are by no means negligible in most cases. For example, if the s term is absent from B_{eff}, as Hougen found, the B_{eff} constant must always be larger in $\kappa\,^2\Sigma$ than in $\mu\,^2\Sigma$, contrary to the observations made for a number of molecules. Hougen also derived expressions for the rotational energy levels in the states for which K is equal to neither $v_2 + 1$ nor 0, but again higher-order rovibronic contributions were neglected.

2.4.4 Hyperfine Structure in a $^2\Pi$ Polyatomic Molecule Affected by Rovibronic Interactions

The hyperfine Hamiltonian (2.3.52) derived for a $^2\Pi$ diatomic molecule which includes both the magnetic and electric nuclear quadrupole interactions may be applied to a $^2\Pi$ linear polyatomic molecule in the $K = v_2 + 1$ state, provided that Ω is replaced by P. The parity-dependent terms such as those in d and eQq_2 must be retained for $K = 1$ states, as for $^2\Pi$ diatomic molecules, but for $K > 1$ such terms have matrix elements only between different vibronic states and thus contribute little to the energy levels.

An argument which leads to the v_2 dependence of the A constant [(2.4.40, 41) for $K = v_2 + 1$ and $K = 0$, respectively] may be applied to the hyperfine constant a to account for its v_2 dependence:

$$a_{\text{eff}} = a_{\text{true}}[1 - \varepsilon^2 K(K + 1)/8 + K\eta], \tag{2.4.75}$$

for $K = v_2 + 1$ and

$$a_{\text{eff}} = a_{\text{true}}[1 - \varepsilon^2 (v_2 + 1)(v_2 + 3)/32 + (v_2 + 1)\eta] \tag{2.4.76}$$

for $K = 0$. The ordinary vibrational dependence is likely to be less important for a than for A, and thus the η term obtained from (2.4.75 or 76) may be a good estimate of the V_{11} term, i.e., the Herzberg-Teller interaction term.

The hyperfine structure in the $K = 0\,^2\Sigma$ vibronic state may be derived using (2.4.30) as base functions and (2.3.52) as the hyperfine Hamiltonian as follows:

$$\begin{aligned}\langle \kappa\,^2\Sigma, J; \pm | H_{\text{HF}} | \kappa\,^2\Sigma, J; \pm \rangle \\ = [-a \cos 2\beta - (b + c)/2 \mp (-1)^{J+1/2} b \sin 2\beta (J + 1/2)] R/[4J(J + 1)] \\ + eQq_1 [3R(R + 1) - 8J(J + 1)][(3/4) - J(J + 1)]/ \\ \times [8(2J + 3)(2J - 1) J(J + 1)],\end{aligned} \tag{2.4.77 a}$$

$$\begin{aligned}\langle \mu\, ^2\Sigma, J; \pm |\, H_{\mathrm{HF}}\, | \mu\, ^2\Sigma, J; \pm \rangle \\ &= [a \cos 2\beta - (b + c)/2 \pm (-1)^{J+1/2} b \sin 2\beta (J + 1/2)]\, R/[4J(J+1)] \\ &\quad + eQq_1 [3R(R+1) - 8J(J+1)]\,[(3/4) - J(J+1)] \\ &\quad /[8(2J+3)(2J-1)\,J(J+1)], \qquad (2.4.77\,\mathrm{b})\end{aligned}$$

$$\begin{aligned}\langle \kappa\, ^2\Sigma, J; \pm |\, H_{\mathrm{HF}}\, | \mu\, ^2\Sigma, J; \pm \rangle \\ &= [a \sin 2\beta \mp (-1)^{J+1/2} b \cos 2\beta (J + 1/2)]\, R/[4J(J+1)], \qquad (2.4.77\,\mathrm{c})\end{aligned}$$

where $R = J(J+1) + I(I+1) - F(F+1)$. Terms with $\Delta J \neq 0$ have been neglected in deriving (2.4.77).

2.5 Zeeman and Stark Effects

The Zeeman and Stark effects have been widely employed in spectroscopy. They are particularly useful in high-resolution spectroscopy, because currently observable Zeeman and Stark energies are of the order of a few cm^{-1} at most, requiring moderately high resolution to study such small effects. The Stark effect primarily provides the electric dipole moment which is one of the most important molecular constants. On the other hand, the magnetic moment of a paramagnetic molecule (free radical) is fairly well understood, because its main part is ascribed to the unpaired electron spin and/or the electron orbital angular momentum which is not quenched. Still we expect to obtain some information on the electronic structure of molecules through fine details of the magnetic moment or g factors. This is particularly so in excited electronic states; we often observe large Zemman effects even for singlet states.

The Zeeman and Stark effects allow us to observe signals from atoms and molecules, because they underlie the interactions of electromagnetic radiation with matter, i.e., the basic processes in spectroscopy. We may exploit them in detecting faint signals from atoms and molecules; various techniques such as Zeeman and Stark modulation and polarization spectroscopy have been developed and widely applied as most sensitive and yet versatile techniques of spectroscopy (Chap. 3). Two such extremes would be laser magnetic resonance (LMR) and laser Stark spectroscopy, in which the Zeeman and Stark effects play the most essential roles of scanning spectra.

In this section we focus attention primarily on Zeeman effects and to a lesser extent on Stark effects affecting an open shell molecule (i.e., free radical). The magnetic moment of a molecule consists of two terms proportional to the unpaired electron spin angular momentum $\boldsymbol{S}$ and the electron orbital angular momentum $\boldsymbol{L}$ as mentioned above, with some higher-order terms. The leading term of the Zeeman energy is thus given by

$$\boldsymbol{H}_{\mathrm{Z}}^{\mathrm{e}} = g_S \beta \boldsymbol{S} \cdot \boldsymbol{B} + g_L \beta \boldsymbol{L} \cdot \boldsymbol{B}, \qquad (2.5.1)$$

where $g_S = 2.00232$ and $g_L = 1.0$ denote the spin and orbital g factors of a free electron, respectively, and β stands for the Bohr magneton, which is equal to 1.3996 MHz/G or 0.046686 cm^{-1}/kG. The nuclei which form the molecular framework also interact with the external magnetic field through the rotation of the framework in space and also through their nuclear magnetic moments, namely

$$\boldsymbol{H}_{\mathrm{Z}}^{\mathrm{n}} = -\, g_{\mathrm{r}}^{\mathrm{N}} \beta \boldsymbol{R} \cdot \boldsymbol{B} - g_N \beta_N \boldsymbol{I} \cdot \boldsymbol{B}, \tag{2.5.2}$$

where β_N stands for the nuclear magneton, which is equal to $5.45 \times 10^{-4}\,\beta$, and $\boldsymbol{R}$ and $\boldsymbol{I}$ represent the overall rotational angular momentum of the nuclear framework and the nuclear spin angular momentum, respectively. The nuclear rotational g factor $g_{\mathrm{r}}^{\mathrm{N}}$ in (2.5.2) is about 10^{-4} times the electronic g factors g_S and g_L, because the nuclei weigh more than the electron. (Care should be taken since $g_{\mathrm{r}}^{\mathrm{N}}$ is often given in units of β_N, rather than β.) The expression for the nuclear rotational Zeeman effect, the first term of (2.5.2), applies only to a diatomic and a linear polyatomic molecule, while for a nonlinear polyatomic molecule the scalar $g_{\mathrm{r}}^{\mathrm{N}}$ factor must be replaced by a second-rank tensor. By substituting $\boldsymbol{R} = \boldsymbol{J} - \boldsymbol{L} - \boldsymbol{S} = \boldsymbol{N} - \boldsymbol{L}$ in (2.5.1, 2), the first-order Zeeman Hamiltonian is derived as follows:

$$\boldsymbol{H}_{\mathrm{Z}} = g_S \beta \boldsymbol{S} \cdot \boldsymbol{B} + (g_L + g_{\mathrm{r}}^{\mathrm{N}})\, \beta \boldsymbol{L} \cdot \boldsymbol{B} - g_{\mathrm{r}}^{\mathrm{N}} \beta \boldsymbol{N} \cdot \boldsymbol{B} - g_N \beta_N \boldsymbol{I} \cdot \boldsymbol{B}. \tag{2.5.3}$$

The angular momentum $\boldsymbol{L}$ will mix different electronic states, resulting in second-order Zeeman energy. For example, the second term of (2.5.3), which is proportional to $\boldsymbol{L} \cdot \boldsymbol{B}$, may be coupled with the spin-orbit interaction operator $A_{\mathrm{SO}} \boldsymbol{L} \cdot \boldsymbol{S}$ (2.4.16) to give the following second-order term:

$$E^{(2)} = g_l \beta (S_x B_x + S_y B_y), \quad \text{where} \tag{2.5.4}$$

$$g_l = -\, 2(g_L + g_{\mathrm{r}}^{\mathrm{N}}) \sum_n \langle 0 |\, A_{\mathrm{SO}} L_x \,| n \rangle \, \langle n |\, L_x \,| 0 \rangle / (E_n - E_0). \tag{2.5.5}$$

Similarly, when combined with the L-uncoupling term $-\,B(N_+ L_- + N_- L_+)$, (2.4.45), the electron contribution to the rotational g factor is derived as follows:

$$g_{\mathrm{r}}^{\mathrm{e}} = 4(g_L + g_{\mathrm{r}}^{\mathrm{N}}) \sum_n \langle 0 |\, L_x \,| n \rangle \, \langle n |\, B L_x \,| 0 \rangle / (E_n - E_0). \tag{2.5.6}$$

The total Zeeman Hamiltonian is thus given by

$$\begin{aligned} \boldsymbol{H}_{\mathrm{Z}} = {} & g_S \beta \boldsymbol{S} \cdot \boldsymbol{B} + (g_L' + g_{\mathrm{r}})\, \beta \boldsymbol{L} \cdot \boldsymbol{B} \\ & - g_{\mathrm{r}} \beta \boldsymbol{N} \cdot \boldsymbol{B} + g_l \beta (S_x B_x + S_y B_y) - g_N \beta_N \boldsymbol{I} \cdot \boldsymbol{B} \\ & - g_{\mathrm{r}}^{\mathrm{e}\prime} \beta (\mathrm{e}^{-2\mathrm{i}\phi} N_+ B_+ + \mathrm{e}^{2\mathrm{i}\phi} N_- B_-) \\ & + g_l' \beta (\mathrm{e}^{-2\mathrm{i}\phi} S_+ B_+ + \mathrm{e}^{2\mathrm{i}\phi} S_- B_-), \end{aligned} \tag{2.5.7}$$

where g_L' denotes g_L corrected for some second-order terms and

$$g_\mathrm{r} = g_\mathrm{r}^\mathrm{N} - g_\mathrm{r}^\mathrm{e}. \tag{2.5.8}$$

The last two terms of (2.5.7) need to be retained only for molecules in Π states which exhibit resolvable Λ-type doubling; $g_\mathrm{r}^{\mathrm{e}\prime}$ and g_l' are similar to those defined by (2.5.6, 5), respectively, except that the matrix elements of L_x are between the $\Lambda = \pm 1$ states.

2.5.1 Zeeman Effect of a Molecule Without Orbital Angular Momentum

The electronic orbital angular momentum is almost quenched in a Hund's case (b) or (c) diatomic molecule, a doublet asymmetric molecule, and a symmetric top molecule in a nondegenerate state. For a nonlinear polyatomic molecule among such molecules, for which the second and the last two terms of (2.5.7) may be neglected, the Zeeman Hamiltonian is reduced to

$$\begin{aligned} \boldsymbol{H}_\mathrm{Z} = {} & g_S \beta B_0 T_0^1(\boldsymbol{S}) - \beta B_0 \sum_k (-1)^k [(2k+1)/3]^{1/2} T_0^1(g^k, \boldsymbol{S}) \\ & + \beta B_0 \sum_k (-1)^k [(2k+1)/3]^{1/2} T_0^1(g_\mathrm{r}^k, \boldsymbol{N}) - g_N \beta_N B_0 T_0^1(\boldsymbol{I}), \end{aligned} \tag{2.5.9}$$

where the first-rank tensors are

$$T_0^1(g^k, \boldsymbol{S}) = (-1)^{-k+1} \sqrt{3} \sum_p T_p^k(g_l)\, T_{-p}^1(\boldsymbol{S}) \begin{pmatrix} k & 1 & 1 \\ p & -p & 0 \end{pmatrix}, \tag{2.5.10}$$

$$\begin{aligned} T_0^1(g_\mathrm{r}^k, \boldsymbol{N}) = {} & (-1)^{k+1} \sqrt{3} \sum_p (1/2) \\ & \times [T_q^k(g_\mathrm{r})\, T_{-p}^1(\boldsymbol{N}) + (-1)^k T_p^1(\boldsymbol{N})\, T_{-p}^k(g_\mathrm{r})] \begin{pmatrix} k & 1 & 1 \\ p & -p & 0 \end{pmatrix}. \end{aligned} \tag{2.5.11}$$

The components of the $T_p^k(g_l)$ and $T_p^k(g_\mathrm{r})$ tensors are related to those in the molecule-fixed coordinate system, and usually the three shown below may be determined experimentally:

$$T_0^0(g_\mathrm{r}) = -(g_\mathrm{r}^{aa} + g_\mathrm{r}^{bb} + g_\mathrm{r}^{cc})/\sqrt{3}, \tag{2.5.12 a}$$

$$T_0^2(g_\mathrm{r}) = (2g_\mathrm{r}^{aa} - g_\mathrm{r}^{bb} - g_\mathrm{r}^{cc})/\sqrt{6}, \tag{2.5.12 b}$$

$$T_{\pm 2}^2(g_\mathrm{r}) = (g_\mathrm{r}^{bb} - g_\mathrm{r}^{cc})/2. \tag{2.5.12 c}$$

Similar expressions hold for $T_p^k(g_l)$.

The matrix elements of (2.5.9) may be evaluated using the following angular momentum coupling scheme

$$\boldsymbol{N} + \boldsymbol{S} = \boldsymbol{J} \tag{2.5.13}$$

and a symmetric top wave function $|N, K, S, J\rangle$ as a basis. The tensor relations in Table 2.4 are used in calculating the matrix elements of (2.5.9).

The matrix elements of the first term of (2.5.9) are readily evaluated using the Wigner-Eckart theorem

$$\langle N', K', S, J', M_J'|\; T_0^1(\mathbf{S})\; |N, K, S, J, M_J\rangle$$
$$= \delta_{M_J, M_J'}(-1)^{J'-M_J'}\begin{pmatrix} J' & 1 & J \\ -M_J' & 0 & M_J \end{pmatrix}\langle J'||T^1(\mathbf{S})||J\rangle, \qquad (2.5.14)$$

of which the last factor is calculated with the fourth equation in Table 2.4:

$$\langle J'||T^1(\mathbf{S})||J\rangle = \delta_{N',N}\delta_{K',K}(-1)^{N+S+J'+1}$$
$$\times [(2J'+1)(2J+1)S(S+1)(2S+1)]^{1/2}\begin{Bmatrix} S & J' & N \\ J & S & 1 \end{Bmatrix}. \qquad (2.5.15)$$

Other terms of (2.5.9) are calculated similarly, giving the matrix elements as

$$\langle N', K', S, J', M_J'|\; H_Z\; |N, K, S, J, M_J\rangle$$
$$= \delta_{M_J', M_J}\beta B_0(-1)^{J'-M_J}\begin{pmatrix} J' & 1 & J \\ -M_J & 0 & M_J \end{pmatrix}$$
$$\times \{\delta_{N',N}\delta_{K',K}g_S(-1)^{N+S+J'+1}$$
$$\times [(2J'+1)(2J+1)S(S+1)(2S+1)]^{1/2}\begin{Bmatrix} S & J' & N \\ J & S & 1 \end{Bmatrix}$$
$$- \sum_{k=0}^{2}(-1)^k[(2k+1)/3]^{1/2}\langle N', K', S, J'||T^1(g_l^k, \mathbf{S})||N, K, S, J\rangle$$
$$+ (-1)^{N'+S+J+1}[(2J'+1)(2J+1)]^{1/2}\begin{Bmatrix} N' & J' & S \\ J & N & 1 \end{Bmatrix}$$
$$\times \sum_k(-1)^k[(2k+1)/3]^{1/2}\langle N', K'||T^1(g_r^k, \mathbf{N})||N, K\rangle\}. \qquad (2.5.16)$$

The two reduced matrix elements are given by

$$\langle N', K', S, J'||T^1(g^k, \mathbf{S})||N, K, S, J\rangle$$
$$= [(2J'+1)(2J+1)/3]^{1/2}\begin{Bmatrix} N' & N & k \\ S & S & 1 \\ J' & J & 1 \end{Bmatrix}\sum_q T_q^k(g_l)(-1)^{N'-K'}$$
$$\times \begin{pmatrix} N' & k & N \\ -K' & q & K \end{pmatrix}[(2N'+1)(2N+1)S(S+1)(2S+1)]^{1/2} \qquad (2.5.17)$$

and

$$\langle N', K' | |T^1(g_r^k, N)| | N, K\rangle$$
$$= -(1/\sqrt{3}) \sum_q T_q^k(g_r)(-1)^{N-K'} \begin{pmatrix} N' & k & N \\ -K' & q & K \end{pmatrix}$$
$$\times (1/2) \left\{ \begin{Bmatrix} N' & N & 1 \\ 1 & k & N \end{Bmatrix} (2N+1)[N(N+1)(2N'+1)]^{1/2} \right.$$
$$\left. + (-1)^k \begin{Bmatrix} N & N' & 1 \\ 1 & k & N' \end{Bmatrix} (2N'+1)[N(N+1)(2N+1)]^{1/2} \right\}. \quad (2.5.18)$$

When hyperfine structures are resolved, the $3-j$ symbol in (2.5.16) $\begin{pmatrix} J' & 1 & J \\ -M_J & 0 & M_J \end{pmatrix}$ must be replaced by

$$(-1)^{F'-M_F} \begin{pmatrix} F' & 1 & F \\ -M_F & 0 & M_F \end{pmatrix} (-1)^{J'+I+F+1}$$
$$\times [(2F'+1)(2F+1)]^{1/2} \begin{Bmatrix} J' & F' & I \\ F & J & 1 \end{Bmatrix}, \quad (2.5.19)$$

where the coupling scheme $\boldsymbol{J} + \boldsymbol{I} = \boldsymbol{F}$ is assumed.

As mentioned earlier, the term in the anisotropic g-tensor element $T_q^k(g_l)$ arises from a second-order perturbation treatment of the $\boldsymbol{L} \cdot \boldsymbol{B}$ term in the Zeeman energy combined with the spin-orbit interaction. This sort of mechanism may be compared with that which causes the spin-rotation interaction; the $\boldsymbol{L} \cdot \boldsymbol{B}$ Zeeman term is here replaced by $-2B\boldsymbol{L} \cdot \boldsymbol{N}$, the L-uncoupling (or electronic Coriolis interaction) term, provided that the direct interaction between the electron spin and small magnetic field induced by the molecular rotation is much less important than the second-order perturbation term mentioned above. In fact, *Curl* [2.61] pointed out that a relation exists between the $g_l^{\alpha\alpha}$ factor and the spin-rotation coupling constant $\varepsilon_{\alpha\alpha}$:

$$g_l^{\alpha\alpha} = -\varepsilon_{\alpha\alpha}/2B_{\alpha\alpha} \quad (\alpha = a, b, c), \quad (2.5.20)$$

where $B_{\alpha\alpha}$ denotes the rotational constant associated with the α axis.

As shown in (2.5.8), the rotational g factor g_r consists of two terms of opposite signs, the magnitude of each being gauge-dependent. An expression for the electron contribution (2.5.6) which holds for a linear molecule may be extended to a nonlinear molecule as follows:

$$g_r^{\alpha\alpha}(\text{el}) = 4g_L \sum_n{}' \langle 0| L_\alpha |n\rangle \langle n| B_{\alpha\alpha} L_\alpha |0\rangle/(E_0 - E_n)$$
$$= 4g_L B_{\alpha\alpha} \sum_n{}' |\langle 0| L_\alpha |n\rangle|^2/(E_0 - E_n). \quad (2.5.21)$$

Following the argument advanced by *Curl*, *Barnes* et al. [2.62] derived the relation

$$g_r^{\alpha\alpha}(\mathrm{el}) = -\,\varepsilon_{\alpha\alpha}/A_{\mathrm{SO}}, \tag{2.5.22}$$

using the spin-orbit interaction constant A_{SO}, instead of the rotational constant. The two relations (2.5.20, 22) can explain the experimental data obtained by electron paramagnetic resonance and LMR spectroscopy well.

For a *doublet state* ($S = 1/2$), (2.5.16) may be rewritten more explicitly:

$$\begin{aligned}
&\langle N, K, S, J, M_J|\, H_Z\, |N, K, S, J, M_J\rangle \\
&\quad = B_0 M_J/[J(J+1)]\,\{[g_S - T_0^0(g_l)/\sqrt{3}]\,\Gamma(SJN) \\
&\qquad - T_0^2(g_l)\,\Gamma(SNJ)\,[3K^2 - N(N+1)]/[\sqrt{6}N(N+1)] \\
&\qquad + T_0^0(g_r)\,\Gamma(JNS)/\sqrt{3} \\
&\qquad - T_0^2(g_r)\,\Gamma(JNS)\,[3K^2 - N(N+1)]/[\sqrt{6}N(N+1)] \\
&\qquad \mp T_2^2(g_l)\,\Gamma(SNJ)/2 \mp T_2^2(g_r)\,\Gamma(JNS)/2\},
\end{aligned} \tag{2.5.23}$$

$$\begin{aligned}
&\langle N, K, S, J-1, M_J|\, H_Z\, |N, K, S, J, M_J\rangle \\
&\quad = -\,\beta B_0 (J^2 - M_J^2)^{1/2}/(2J) \\
&\qquad\quad \times \{[g_S - T_0^0(g_l)/\sqrt{3} - T_2^2(g_l)/\sqrt{6}] \\
&\qquad\quad \times [3K^2 - N(N+1)]/[12N(N+1)] \\
&\qquad - T_0^0(g_r)/\sqrt{3} + T_0^2(g_r)\,[3K^2 - N(N+1)]/[\sqrt{6}N(N+1)] \\
&\qquad \mp T_2^2(g_l)/4 \mp T_2^2(g_r)/2\},
\end{aligned} \tag{2.5.24}$$

where $\Gamma(abc) = a(a+1) + b(b+1) - c(c+1)$. Only those matrix elements diagonal in both N and K are given, and the upper and lower signs apply only to $K = 1$ upper and lower levels, respectively.

Equation (2.5.23) shows that the most dominant term in $g_S - T_0^0(g_l)/\sqrt{3}$ is positive for the F_1 ($J - N + 1/2$) and negative for the F_2 ($J - N - 1/2$) spin components, whereas the $T^k(g_r)$ terms have the same signs for both components and increase rapidly with the N value. The energy matrix which includes the most dominant Zeeman term (g_S) may be explicitly given as

$$\begin{array}{cc}
|N, J = N + 1/2, M_J\rangle & |N, J = N - 1/2, M_J\rangle \\
\left[\begin{array}{l} E_{\mathrm{VR}} + \gamma N/2 + g_S \beta B_0 M_J/(2N+1) \\ \end{array}\right. & \left.\begin{array}{r} -\,g_S\beta B_0[(N+1/2)^2 - M_J^2]^{1/2}/(2N+1) \\ E_{\mathrm{VR}} - \gamma(N+1)/2 - g_S\beta B_0 M_J/(2N+1) \end{array}\right]
\end{array} \tag{2.5.25}$$

where γ denotes the spin-rotation coupling constant, which depends on the K value for a nonlinear molecule, and E_{VR}, the vibration-rotation energy. The $M_J = \pm(N + 1/2)$ levels are exceptional; their energy is simply given by the upper left corner element of the above matrix. The solution of (2.5.25) is given by

$$E = E_{VR} - \gamma/4 \pm (1/2)\,[\gamma^2(N + 1/2)^2 + 2\gamma g_S \beta B_0 M_J + (g_S \beta B_0)^2]^{1/2} \qquad (2.5.26)$$

for $|M_J| < (N + 1/2)$ and

$$E = E_{VR} + \gamma N/2 \pm g_S \beta B_0/2 \qquad (2.5.27)$$

for $M_J = \pm (N + 1/2)$. Figure 2.3 shows an example of the Zeeman splitting when γ is positive.

Because the γ constant is of the order of 1 cm^{-1} or less, the Zeeman energy becomes comparable with γ when the magnetic field is as large as a few kG. Then the Zeeman effect deviates from the linearity, and finally the electron spin is decoupled from the molecular framework, a phenomenon referred to as the Paschen-Back effect. In this limit the electron spin couples to the external magnetic field, and all Zeeman components are well specified by M_S and again move linearly with the applied magnetic field. The Paschen-Back effect limits the versatility of LMR, because the electric dipole transitions between rotational or vibrational-rotational levels cannot be tuned anymore, unless crossing, normally avoided, occurs. Therefore, the spin-rotation interaction constant is of particular importance for LMR spectroscopy; it determines the tuning range we may achieve with an external magnetic field.

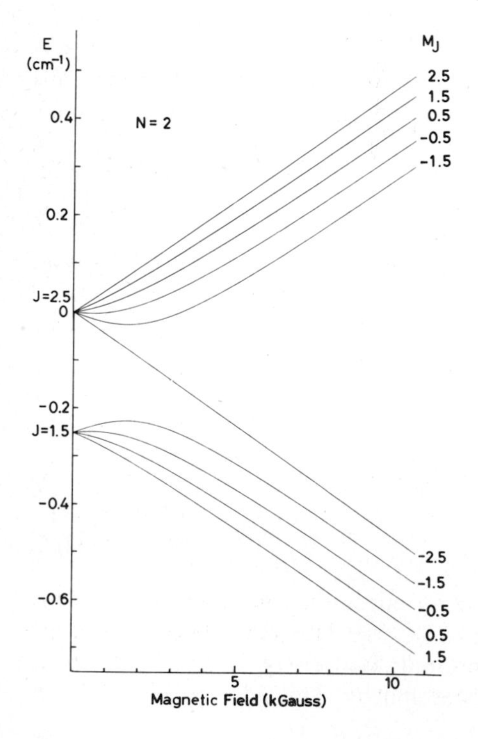

Fig. 2.3. Zeeman effect of a rotational level of $N = 2$

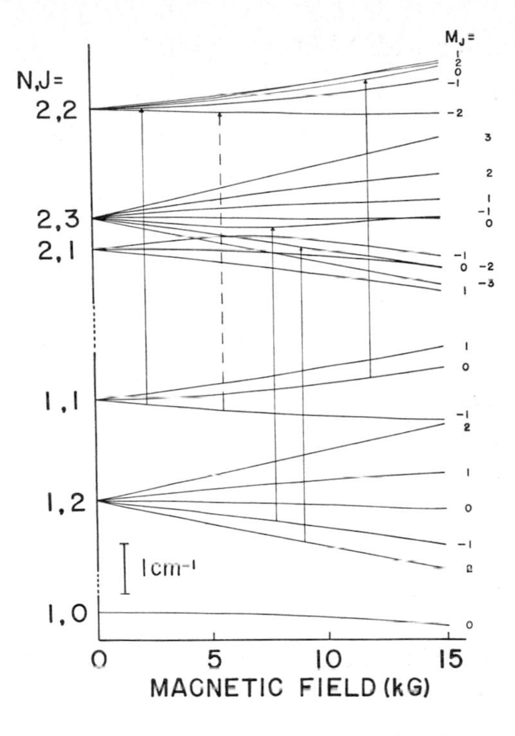

Fig. 2.4. Zeeman effects of the six low-J levels of the PD radical in the $X^3\Sigma^-$ ground vibronic state [2.63]

A magnetic dipole transition is allowed between Zeeman components and also between split levels even at the Paschen-Back limit. However, the intensity is a few orders of magnitude smaller than that of the electric-dipole transition.

For a *triplet state* ($S = 1$), the diagonal matrix elements become

$$E_Z(J = N) = g_S \beta B_0 M_J/[N(N+1)], \tag{2.5.28 a}$$

$$E_Z(J = N - 1) = -g_S \beta B_0 M_J/N, \tag{2.5.28 b}$$

$$E_Z(J = N + 1) = g_S \beta B_0 M_J/(N+1), \tag{2.5.28 c}$$

where Hund's case (b) is assumed. The matrix elements off-diagonal in J are given by

$$\begin{aligned} &\langle N, J-1, M_J | H_Z | N, J, M_J \rangle \\ &\quad = -g_S \beta B_0/(2J) \, [(J^2 - M_J^2)(S + J + N + 2)(-S + J + N) \\ &\qquad \times (S - J + N + 2)(S + J - N)/(4J^2 - 1)]^{1/2} . \end{aligned} \tag{2.5.29}$$

It is easily seen that the Zeeman effect is smaller for F_2 ($J = N$) than for F_1 ($J = N + 1$) and F_3 ($J = N - 1$) by a factor of N when N is large. In the absence of an external magnetic field, the F_1 and F_3 levels are usually very close. Therefore, they will repel each other when the magnetic field is on, as (2.5.29) indicates. Figure 2.4 illustrates the Zeeman effect of the PD radical, which closely approximates Hund's case (b) [2.63].

2.5.2 Zeeman Effect of a Molecule with Electron Orbital Angular Momentum

We shall mainly consider a $^2\Pi$ molecule for which Hund's case (a) is most appropriate. The matrix elements of (2.5.7) evaluated using a Hund's case (a) basis set are given by

$$\begin{aligned}&\langle \Pi_{3/2}, J, M_J, \pm |\, H_Z \,| \Pi_{3/2}, J, M_J, \pm \rangle \\ &\quad = \beta B_0 M_J \{3(g'_L + g_S/2) - g_r[2(J+1/2)^2 - 4]\}/[2J(J+1)],\end{aligned} \tag{2.5.30 a}$$

$$\begin{aligned}&\langle \Pi_{1/2}, J, M_J, \pm |\, H_Z \,| \Pi_{1/2}, J, M_J, \pm \rangle \\ &\quad = \beta B_0 M_J \{(g'_L - g_S/2) - 2 g_r (J+1/2)^2 \\ &\qquad \pm (-1)^{J-1/2} (g'_l - g_r^{e\prime})(J+1/2)\}/[2J(J+1)],\end{aligned} \tag{2.5.30 b}$$

$$\begin{aligned}&\langle \Pi_{3/2}, J, M_J, \pm |\, H_Z \,| \Pi_{1/2}, J, M_J, \pm \rangle \\ &\quad = \beta B_0 M_J [(g_S + g_l + g_r) \mp (-1)^{J-1/2} g_r^{e\prime}(J+1/2)] \\ &\qquad \times [(J-1/2)(J+3/2)]^{1/2}/[2J(J+1)],\end{aligned} \tag{2.5.30 c}$$

where only $\Delta J = 0$ terms are retained and the Zeeman effects associated with the nuclear spins are neglected. The upper and lower signs refer to the e and f levels of the Λ-type doublet, respectively. In the Hund's case (a) limit, the magnetic dipole moment is almost zero in the $^2\Pi_{1/2}$ state because the spin and orbital magnetic moments cancel each other. In an intermediate case, the $^2\Pi_{1/2}$ state acquires magnetic moment through the off-diagonal term (2.5.30 c) combined with a spin-uncoupling term $-B[(J-1/2)(J+3/2)]^{1/2}$ (Sect. 2.3.3).

In a $^2\Pi$ linear molecule, the Herzberg-Teller effect generates a correction term for the g_L factor, given by

$$\Delta g_L = -\hbar \sum_{\xi'} (-1)^p \,|\langle \xi |\, V_{11} \,| \xi' \rangle|^2 / [8\pi c \omega_2 (\Delta E)^2] \tag{2.5.31}$$

(for notations, see Sect. 2.4.2). Therefore, Δg_L, when experimentally determined, allows us to determine V_{11} and V_{22} separately, by combining it with other data.

The Zeeman effect in the excited bending states of a $^2\Pi$ triatomic molecule is similar to (2.5.30) when $K = v_2 + 1$. For a $^2\Sigma$ vibronic state ($K = 0$), the effective g_J factor in the weak field limit is calculated using (2.4.30) as a basis to be

$$g_J(\kappa\, ^2\Sigma) = [\cos^2\beta \pm (J+1/2)\sin 2\beta]/[J(J+1)], \tag{2.5.32 a}$$

$$g_J(\mu\, ^2\Sigma) = [-\sin^2\beta \mp (J+1/2)\sin 2\beta]/[J(J+1)], \tag{2.5.32 b}$$

where the upper and lower signs correspond to the F_1 ($J = N + 1/2$) and F_2 ($J = N - 1/2$) spin components, respectively, and $\sin 2\beta$ is defined by (2.4.31 a).

Equations (2.5.32) show that the Zeeman effect in the Σ vibronic state differs much from that of the vibronic Π state, being rather closer to that in a $^2\Sigma$ state.

A symmetric top molecule in a degenerate electronic state will show Zeeman effects similar to those given by (2.5.31) provided that appropriate changes are made for the quantum numbers of the angular momentum components along the symmetry axis and g_r and g_l are replaced by second-rank tensors.

The relative intensities of the Zeeman components are discussed below, along with those of Stark components.

2.5.3 Stark Effect

The interaction of a molecular dipole moment with an applied electric field (i.e., the Stark effect) is expressed as

$$\boldsymbol{H}_S = -T^1(\boldsymbol{\mu}) \cdot T^1(\boldsymbol{E}). \tag{2.5.33}$$

We evaluate the matrix elements of this Hamiltonian using the symmetric top wave function $|J, K, M_J\rangle$ as a base:

$$\begin{aligned}\langle J', K', M'_J | H_S | J, K, M_J\rangle &= \sum_p (-1)^p T_p^1(\boldsymbol{E}) (-1)^{J'-M'_J} (2J'+1)(2J+1) \begin{pmatrix} J' & 1 & J \\ -M'_J & -p & M_J \end{pmatrix} \\ &\times \sum_q T_q^1(\boldsymbol{\mu}) (-1)^{J'-K'} \begin{pmatrix} J' & 1 & J \\ -K' & q & K \end{pmatrix},\end{aligned} \tag{2.5.34}$$

where all interactions involving the electron and nuclear spins are neglected. When they are included, the matrix elements become

$$\begin{aligned}&\langle N', K', S, J, I, F', M'_F | H_S | N, K, S, J, I, F, M_F\rangle \\ &= -\sum_p (-1)^p T_p^1(\boldsymbol{E}) (-1)^{F'-M'_F} \begin{pmatrix} F' & 1 & F \\ -M'_F & p & M_F \end{pmatrix} \\ &\times (-1)^{J'+I+F+1} [(2F'+1)(2F+1)]^{1/2} \begin{Bmatrix} J' & F' & I \\ F & J & 1 \end{Bmatrix} \\ &\times (-1)^{N'+S+J+1} [(2J'+1)(2J+1)]^{1/2} \begin{Bmatrix} N' & J' & S \\ J & N & 1 \end{Bmatrix} \\ &\times \sum_q T_q^1(\boldsymbol{\mu}) (-1)^{N'-K'} [(2N'+1)(2N+1)]^{1/2} \begin{pmatrix} N' & 1 & N \\ -K' & q & K \end{pmatrix},\end{aligned} \tag{2.5.35}$$

where coupled representation is used. When the molecular dipole moment is located along the a axis, q in (2.5.34) is equal to 0 and the diagonal matrix element reduces to the well-known expression

$$\langle J, K, M_J | H_S | J, K, M_J\rangle = -\mu_a E K M_J / [J(J+1)], \tag{2.5.36}$$

and the off-diagonal element is given by

$$\langle J-1, K, M_J | H_S | J, K, M_J \rangle = (\mu_a E/J)\,[(J^2-K^2)(J^2-M_J^2)/(4J^2-1)]^{1/2}. \tag{2.5.37}$$

When the b component of the dipole moment is present, the nonvanishing matrix elements are given by

$$\langle J, K-1, M_J | H_S | J, K, M_J \rangle = \mu_b E M_J\,[(J-K+1)(J+K)]^{1/2}/[2J(J+1)], \tag{2.5.38 a}$$

$$\langle J-1, K-1, M_J | H_S | J, K, M_J \rangle = \mu_b E M_J\,[(J^2-M_J^2)(J+K-1)(J+K)/(4J^2-1)]^{1/2}/2J, \tag{2.5.38 b}$$

$$\langle J, K-1, M_J | H_S | J-1, K, M_J \rangle = -\mu_b E M_J\,[(J^2-M_J^2)(J-K)(J-K+1)/(4J^2-1)]^{1/2}/2J. \tag{2.5.38 c}$$

Because the dipole moment has odd parity, its average is finite only when the state considered consists of degenerate pair levels with opposite parity. The $+K$ and $-K$ levels of a symmetric top are degenerate and form such pair levels, resulting in the first-order Stark effect equation (2.5.36). This is in sharp contrast with the magnetic moment, which has even parity. Since the $+M_J$ and $-M_J$ components have the same parity and are degenerate at zero magnetic field, the Zeeman effect is always first order as long as the external magnetic field is so small that off-diagonal matrix elements of the Zeeman Hamiltonian can be neglected.

The Stark effect can become large when two levels of opposite parity form a closely spaced pair. Examples include an inversion doublet (like that of NH_3), l-type doublet, Λ-type doublet, and K-type doublet of a near symmetric top which has a dipole moment component along the a (or c) axis. When the Stark energy is larger than the doublet spacing, the Stark shifts become approximately linear. These large first-order or pseudo-first-order Stark effects are especially important in laser Stark spectroscopy since a wide range tuning can be achieved only by exploiting such large Stark shifts.

As mentioned earlier, the electron spin is easily decoupled from the molecular framework when the spin-rotation coupling constant is not large in comparison with the Zeeman energy (the Paschen-Back effect). A similar phenomenon arises for the Stark effect, but does not interfere with the tuning range of laser Stark spectroscopy because different parity imposes different selection rules.

The intensity of the electric dipole transition is easily calculated by taking absolute squares of the Stark effect expressions (2.5.34, 35), because we only need to replace the static field by the oscillating field of radiation. The $p = 0$

and $p = \pm 1$ terms in (2.5.34) correspond to the $\Delta M_J = 0$ and $\Delta M_J = \pm 1$ transitions, respectively. The relative intensities of the $M_J \leftarrow M_J$ Stark or Zeeman components are thus given by the square of the $3-j$ symbol

$$\begin{pmatrix} J' & 1 & J \\ -M_J & -p & M_J \end{pmatrix}, \quad \text{namely}$$

$$I(R \text{ branch}, J+1 \leftarrow J) \propto (J+1)^2 - M_J^2, \tag{2.5.39 a}$$

$$I(Q \text{ branch}, J \leftarrow J) \propto M_J^2, \tag{2.5.39 b}$$

$$I(P \text{ branch}, J-1 \leftarrow J) \propto J^2 - M_J^2. \tag{2.5.39 c}$$

Similarly, those of the $M_J \pm 1 \leftarrow M_J$ components are

$$I(R \text{ branch}, J+1 \leftarrow J) \propto (J \pm M_J + 1)(J \pm M_J + 2)/4, \tag{2.5.40 a}$$

$$I(Q \text{ branch}, J \leftarrow J) \propto (J \mp M_J)(J \pm M_J + 1)/4, \tag{2.5.40 b}$$

$$I(P \text{ branch}, J-1 \leftarrow J) \propto (J \mp M_J - 1)(J \mp M_J)/4. \tag{2.5.40 c}$$

The relative intesities of the fine and hyperfine components may be obtained from (2.5.35); $6-j$ symbols involved there lead to the result that the transitions satisfying $\Delta J = \Delta N$ and $\Delta F = \Delta J$ are stronger than others when the three quantum numbers are much larger than 1.

3. Experimental Details

This chapter describes spectroscopic systems for observing the spectra of transient molecules, by placing main emphasis on the spectrometers which my group has set up and employed to observe high-resolution spectra of transient species. The high reactivity and thus the short lifetime of such molecules require that the spectroscopic methods are highly sensitive. Introduction of lasers as sources has increased the sensitivity, in particular in the IR region, where the light sources hitherto available were of low output. Improvement of detectors has also contributed much to the increase of sensitivity. In conventional optical and IR spectroscopy the resolution has been incompatible with the sensitivity, because to attain high resolution one had to reduce the slit width and thus lose the power entering the detector. This difficulty has been eliminated to a great extent also by using lasers as sources. One might think that high resolution may not be indispensable for the study of transient molecules, rather it may make the spectroscopic method cumbersome to use and may limit its applicability range. That this is not the case is indicated by noting that transient molecules exist only with many other molecules that are either reactants or products of the reaction generating them. High resolution enables us to observe spectra of transient species nearly or completely isolated from much stronger spectra of much more abundant, chemically stable molecules. Obviously, high resolution brings about very precise data on molecular constants including the fine and hyperfine coupling constants characteristic of free radicals. These data are central in understanding the molecular structure of free radicals.

Another important phase of the free-radical study concerns the efficient generation of transient molecules. Here there is a close interplay between spectroscopy and reaction kinetics. The results obtained in one field will contribute much to the progress in the other field, which will in turn facilitate research in the former area. Section 3.5 is entirely devoted to the chemical reactions by which our group has generated short-lived molecules.

3.1 Microwave Spectrometer

3.1.1 Requirements to be Met by a Microwave Spectrometer for the Study of Transient Molecules

The lifetime of transient molecules we wish to study typically ranges from a few seconds to a fraction of a ms, and their concentration from a few percent to one ppm. Since the total pressure of a sample in a microwave absorption cell is normally maintained at about 10 mTorr, one ppm concentration corresponds to 3.6×10^8 molecules/cm^3. Suppose we have generated a linear molecule with a dipole moment of 1 D to a concentration of 0.01 % in a cell. Its absorption line at 100 GHz will then have a peak absorption coefficient of 5.5×10^{-7} cm^{-1}, provided that the half linewidth pressure broadening parameter is assumed to be 10 MHz/Torr [3.1]. We may easily observe this line on a CRO using a conventional Stark modulated spectrometer; the signal-to-noise ratio would be several to ten. However, for a transient species, this sensitivity may not be sufficient, because our knowledge of such species is usually very limited; we often have to scan the spectrum as widely as a few thousand MHz, while maintaining the spectrometer at best performance conditions.

Furthermore, the chemistry leading to the production of the transient molecule poses more serious problems; we do not know much about the production conditions unless we observe a spectral line of the species. The chemical conditions we choose at an initial stage of searching for the spectra can be far from the best. The spectrometer must be at least one or two orders of magnitude more sensitive than a conventional Stark modulated spectrometer whose sensitivity is normally of the order of 10^{-8} cm^{-1}. The 10^{-10} cm^{-1} sensitivity may be attained in the cm wave region, if one slowly scans the spectrum with a long time constant for the phase-sensitive detector combined with a long path cell, but such procedure is very difficult to apply to observing the spectra of transient molecules [3.2].

It is advantageous to conduct spectroscopic studies at much higher frequencies, because the rotational transition increases the peak absorption in proportion to the square and cube of the transition frequency, respectively, for the asymmetric top and the linear/symmetric top molecules. It must be admitted, however, that several experimental difficulties must be overcome for high-frequency spectroscopy to be practical. Transmission of a microwave through a Stark cell such as a parallel-plate cell becomes extremely poor when the microwave frequency exceeds 100 GHz, so necessitating the cell length to be cut down. The pumping speed is another problem for the parallel-plate Stark cell, linear flow rate of 10 m/s being a typical value. When the lifetime of a transient molecule is of the order of 10 ms, it survives only in a length of 10 cm from the place where it is synthesized. For a parallel-plate Stark cell, we may generate transient species at the inlet of the cell, but an in situ production in the cell would be almost impossible. These two factors, poor transmission of the microwave and slow pumping rate of the sample gas, limit the reasonable cell length to 40–100 cm. *Saito* [3.2] has shown that the minimum detectable absorption coefficient is about $3 \times 10^{-9}\ \mathrm{cm}^{-1}$ at 120 GHz for a Stark spectrometer with a 40 cm long absorption cell. This value is by no means satisfactory, because in the case of Stark modulation, we have to worry about deterioration of transient molecules at the surface of the Stark electrodes; even when the surface is coated with gold, it certainly shortens the lifetime of many reactive species.

The difficulties inherent in a parallel-plate Stark cell when it is applied to free radical studies make a free space cell attractive. It certainly guarantees good transmission of microwaves, especially of millimeter and submillimeter waves; the cell length may be increased up to 3–5 m. Since simple glass tubing may be used as the cell, its inside wall is rather inactive to many transient molecules and may even be coated with suitable material, if necessary. The most serious disadvantage in using a free space cell lies in abandoning Stark modulation which makes microwave spectroscopy highly sensitive. It is, however, to be noted that recent progress in computers has remedied the shortcoming of source frequency modulation which has to be employed with a free space cell.

The microwave source to be employed is required to deliver an output of a few mW at least, free of excessive noise, and must be combined with detectors of high efficiency and also of low noise level.

3.1.2 Historical Survey of Microwave Studies of Transient Molecules

The microwave study of transient molecules was initiated by *Dousmanis* et al. [3.3] in 1955, who observed several Λ-type doubling transitions of OH in the region 7–37 GHz using a Zeeman modulated microwave spectrometer with a 1.5 m long cell consisting of a split circular wave guide. The sensitivity of this spectrometer was estimated to be $5 \times 10^{-9}\ \mathrm{cm}^{-1}$. The OH radical was generated by a discharge in water vapor outside the cell and pumped through the cell continuously. The lifetime of the OH radical was estimated to be 1/3 s and its concentration reached 10% or so in the cell.

About ten years elapsed before the second microwave detection of a transient molecule was reported. *Powell* and *Lide* [3.4] and *Winnewisser* et al. [3.5] independently observed the microwave spectrum of SO in 1964. *Powell* and *Lide* used Stark modulation with a parallel-plate absorption cell 60 cm in length and attained a sensitivity of about $10^{-8}\ \mathrm{cm}^{-1}$. On the other hand, *Winnewisser* et al. employed a video detection scheme combined with a 40 cm long free space absorption cell. Although their spectrometer was low in sensitivity, the minimum detectable absorption coefficient being 10^{-5}–$10^{-6}\ \mathrm{cm}^{-1}$, it was well adapted to spectroscopy in the millimeter and submillimeter wavelength region [3.6]. Both groups generated a SO radical outside the cell by the reactions of discharged oxygen with a sulfur-containing compound such as

OCS and H_2S, or with elemental sulfur (solid). *Powell* and *Lide* thus observed five transitions of SO between 13 and 66 GHz, whereas *Winnewisser* et al. measured six transitions in 86–172 GHz. The SO molecule has a long life, perhaps about 1 s, and thus the concentration was as high as several percent in both experiments.

Because the two free radicals OH and SO are so long lived, they provided a good footing for the start of microwave free radical studies. However, in spite of many trials, further detections of other species have proved to be difficult. In the meantime, gas-phase electron paramagnetic resonance (EPR) spectroscopy has had an impact. The EPR has several advantages over zero-field microwave spectroscopy. Since it employs a resonant microwave cavity as an absorption cell, transient species need to be generated in a small limited space within the cavity, while the effective path length is maintained quite long; it is rather easy to attain a high concentration of such species. Scanning the spectrum can easily be accomplished by changing the magnetic field. *Carrington* and co-workers [3.7] set up a Stark or Zeeman modulated gas-phase EPR spectrometer with a cavity cell whose inside was coated with gold. They observed the spectra of diatomic and triatomic free radicals, including SO ($^1\Delta$), ClO, BrO, IO, NF, SF, SeF, CF, NCO, NCS, and HCO [3.7]. These results were of great use in subsequent zero-field microwave observations of these radicals, because EPR studies supplied quite accurate molecular constants including the rotational constants and also information on chemistry generating these free radicals. The following are the transient species on which microwave spectra have successfully been observed in the period 1969–1975 mostly using parallel-plate absorption cells: ClO [3.8], BrO [3.9], NS [3.10], SO($^1\Delta$) [3.11], NCO [3.12], HNO [3.13], HCO [3.14], IO [3.15], SF [3.16], NF_2 [3.17], and HO_2 [3.18].

These studies have, however, been rather difficult to extend to other important species of shorter lifetimes, primarily because of shortcomings inherent in the parallel-plate cell. In 1973 *Woods* [3.19] presented a new free space absorption cell 300 cm long and 15 cm in diameter, primarily to generate molecular ions by a DC discharge in the cell and to observe their microwave spectra in absorption. He has shown that discharge plasma does not interfere much with microwave transmission and thus with the observation of absorption spectra. With this cell he and his associates have observed several molecular ions, some of which had been reported to exist in interstellar space. The CO^+ ion thus detected [3.20] is the first example of a molecular ion observed by microwave spectroscopy. It was then followed by other ions and neutrals: HCO^+ [3.21], HNC [3.22], N_2H^+ [3.23], CN [3.24], HCS^+ [3.25], and HOC^+ [3.26].

3.1.3 High Sensitivity Millimeter-Wave Spectrometer to Study Transient Molecules

Figure 3.1 shows a block diagram of a source frequency modulation millimeter-wave (mm-wave) spectrometer equipped with a free-space absorption cell which has been set up at the Institute for Molecular Science to

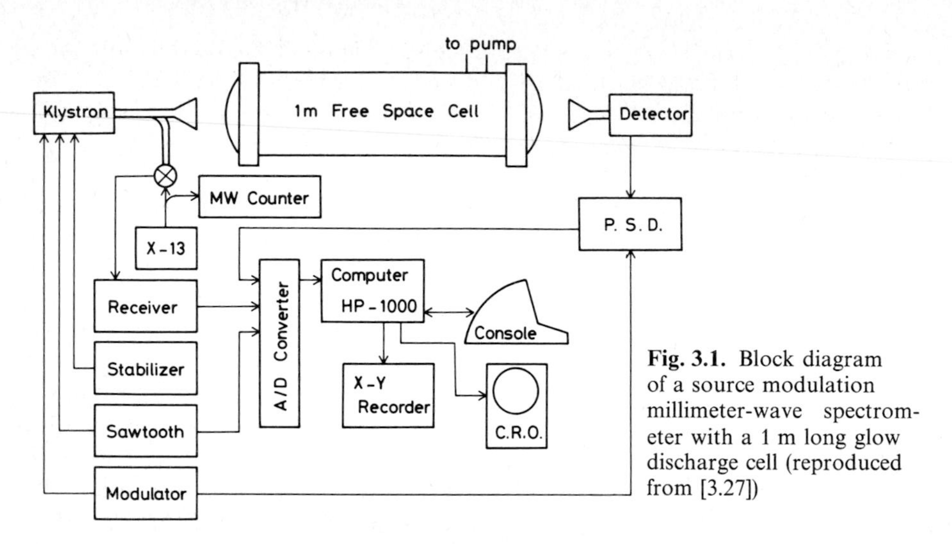

Fig. 3.1. Block diagram of a source modulation millimeter-wave spectrometer with a 1 m long glow discharge cell (reproduced from [3.27])

investigate transient molecules [3.27]. As described below, the minimum detectable absorption coefficient of this spectrometer is as low as 1.1×10^{-9} cm^{-1} at 122 GHz, and it may be used satisfactorily up to 400 GHz or higher. In place of the 3.5 m long free space absorption cell 15 cm in diameter, a shorter one may be employed, 1 m in length and 10 cm in diameter, shown in Fig. 3.1; both are made of Pyrex glass tubes and designed following *Woods* [3.19]. They are equipped with a cylindrical stainless stell electrode at each end, and may be cooled down to any temperature between room temperature and liquid nitrogen temperature by adjusting the amount of liquid nitrogen flowing through copper tubes soldered onto copper sheets surrounding the cell; the entire cell is installed in a polystyrene box for thermal shielding. Two Helmholtz coils are wound around the cell to compensate the Earth's magnetic field, one generating a vertical field and the other a horizontal field. The latter coil is also employed to distinguish paramagnetic lines from diamagnetic lines. The cells are sealed with a Teflon lens at each end which collimates and focuses microwaves, and are pumped by a mechanical booster pump (3300 l/min) followed by two liquid nitrogen traps and a rotary pump. The 1.1 m cell has two inlets for gases, which enables microwave discharge products to be mixed with a second gas inside the cell. When a large dc discharge current is required, one of the cylindrical electrodes in the 1.1 m cell is replaced by a hollow cathode 60 cm long [3.28]; the discharge current may be increased up to about 400 mA. In some experiments the free space cells are replaced by a 40 cm long parallel-place cell, and accordingly the modulation scheme is switched from source frequency to Stark. This cell is described in detail in [3.2, 12].

The microwave source consists of a series of OKI klystrons up to 200 GHz, and beyond this limit a Si-W point contact harmonic generator is used up to 400 GHz [3.29]. The modulation signal consists of two identical square waves of 50 kHz with a phase difference of $\frac{\pi}{2}$ [3.30]. The microwave thus modulated is fed to the absorption cell through a horn and the Teflon lens, and, after passing through the cell, is focused onto a detector again by another Teflon lens and a horn. The detectors employed are GaAs Schottky diodes, Hitachi T3420A and W3420A, respectively, above and below 110 GHz, and an InSb photoconductive detector manufactured by QMC, which is operated at 4.2 K. The absorption signal from the detector is preamplified and phase-sensitive detected at 100 kHz.

The frequency of mm-wave klystrons is measured using a Varian X-13 klystron as a secondary frequency standard, which is phase-locked to an RF synthesizer HP3335A by a Microwave Systems MOS-5. The beat notes between the output of a mm-wave klystron and a harmonic of the X-13 output generated by a 1N26 or 1N78 mixer diode are observed by an all-wave receiver, JRC NRD-10, which receives signals in the range 0.1 – 30 MHz. The spectrometer is interfaced to a YHP 2170A minicomputer which can signal average, correct baseline, display signals on a CRT, and has a frequency read out of absorption peaks. In observing an absorption line the source klystron is repetitively swept by a 5 Hz sawtooth over a few MHz region, which encompasses both plus and minus beat notes obtained from the receiver. As Fig. 3.2 indicates, a gate opens with a fixed time delay after the start of the sawtooth, and the minicomputer initiates storing the signal delivered from the phase-sensitive detector and also the shaped output of the beat notes through an eight-channel A/D converter. Both the signal and the beat notes are digitized at uniform time intervals during one cycle of the scan and are alternately allocated to 1000 data points. The computer finds the center frequency of each beat note which is broadened by source modulation, calculates the frequency of each data point

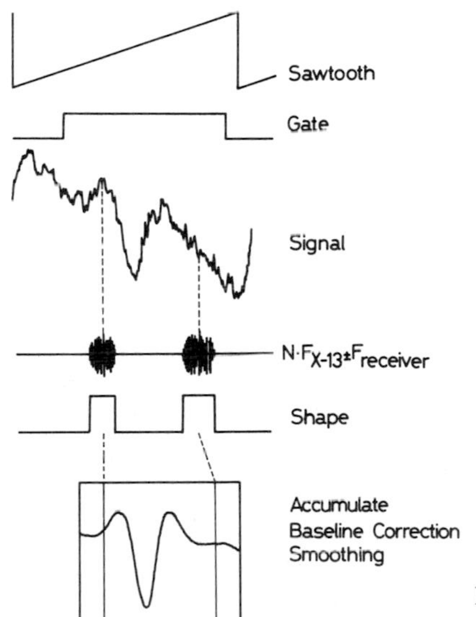

Fig. 3.2. Timing for data acquisition by a minicomputer (reproduced from [3.27])

by interpolation or extrapolation, and accumulates the data on each channel in the memory. An array of 500 words is allocated for signals, of which the 50th and 450th words are assigned to the minus and plus beat center frequencies, respectively. This process is repeated for every scan so that the signal accumulation is not influenced by slow drift in the source klystron frequency. A feedback loop keeps the klystron frequency within the region specified by the two markers (beat notes). This system enables us to accumulate the signal for longer than 30 min without broadening it. The accumulated signals are stored on a disc, and, if necessary, baseline distortions are corrected for by fitting them to a polynomial function. The signals thus processed may be displayed on a CRO or on an *X-Y* recorder. A program is written to read out the center frequency of the absorption line.

The sensitivity of this spectrometer has been examined carefully. It depends on the frequency stability and noise of the source klystron. A low-noise OKI 120V10 of good stability was operated at 122 GHz to observe the $J = 11 \leftarrow 10$ transition of $^{18}O^{13}C^{34}S$ in natural abundance of 0.96 ppm, using the 3.5 m long cell. As Fig. 3.3 shows, it was recorded with a signal-to-noise (S/N) ratio of about 10 when integrated for 50 s. Since the peak absorption coefficient of this line is calculated to be 5.24×10^{-9} cm^{-1} [3.31], the minimum detectable absorption coefficient of the present spectrometer is about 1.1×10^{-9} cm^{-1} at 122 GHz, where the S/N ratio of 2 is assumed to be the detection limit. This minimum absorption corresponds to 10^{12} molecules in total, or 1.8×10^7 molecules/cm^3.

Figure 3.4 shows an example of the spectrum above 200 GHz, recorded using the output of a harmonic generator as a source. The transition observed is the $J = 7/2 - 5/2$, $F = 3 - 2$ transition of the CF radical in the $^2\Pi_{3/2}$ state [3.29], which appears at 301 GHz. This spectrum may be compared with a trace

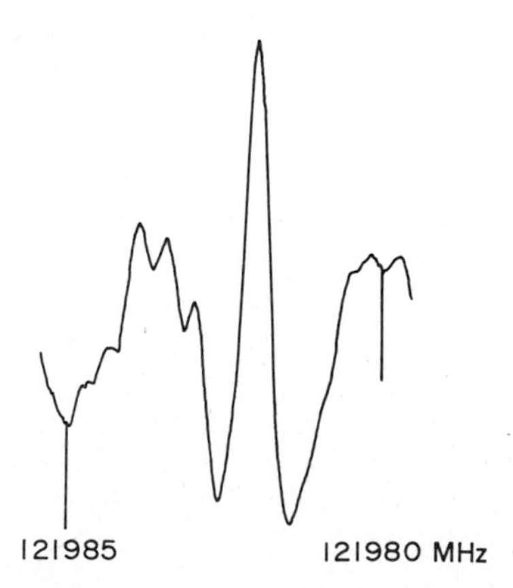

Fig. 3.3. The $J = 11 \leftarrow 10$ transition of $^{18}O^{13}C^{34}S$ in natural abundance (0.96 ppm) observed using a source modulation spectrometer

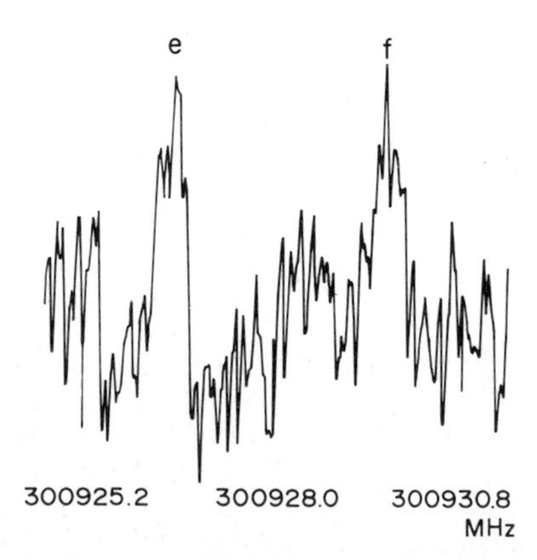

Fig. 3.4. The $J = 7/2 - 5/2$, $F = 3 - 2$ transition of the CF radical, recorded using a harmonic generator as a source (reproduced from [3.29])

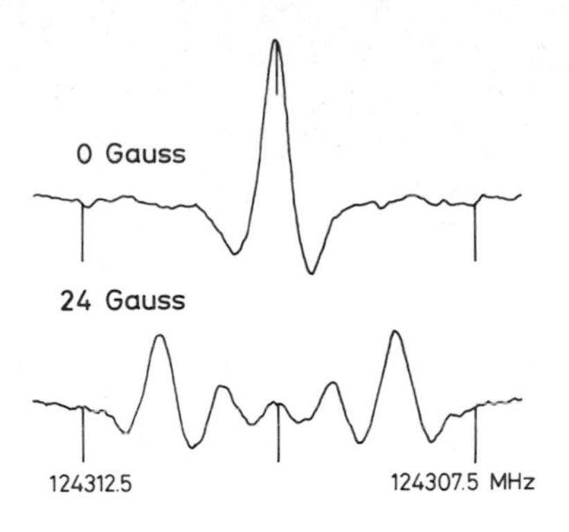

Fig. 3.5. The $J = 3/2 - 1/2$, $F = 2 - 1$ transition of the CF radical. An OKI 120V10 klystron was employed as a source (reproduced from [3.29])

Table 3.1. Transient molecules investigated by microwave spectroscopy at the IMS

Diatomic molecule:	$^2\Pi_r$	CF, CCl, SiF, SiCl, GeF, PO
	$^2\Pi_i$	SF
	$^3\Sigma^-$	NCl, PF, PCl
	$^2\Sigma^+$	SiN
	$^5\Delta$	FeO
Triatomic molecule:	linear	NCO, CCO
	bent	PH_2, PO_2, PF_2, SF_2, H_2D^+
		HO_2, HSO, FSO, ClSO, HPO
Polyatomic molecule:	linear	HCCN
	symmetric top	CF_3, CH_3O, CH_3S
	asymmetric top	CH_2F, CH_2Cl, CH_2CHO

of a line reproduced in Fig. 3.5, which corresponds to the $J = 3/1 - 1/2, f$, $F = 2 - 1$ transition at 124 GHz of CF in the $^2\Pi_{1/2}$ state. Since the $^2\Pi_{3/2}$ state is higher than the $^2\Pi_{1/2}$ state by 77.11 cm^{-1}, the former should be six times stronger than the latter. The poor S/N ratio of the former spectrum thus indicates that the sensitivity in the 200 400 GHz region is 80 times less than that below 200 GHz. This large difference in sensitivity is entirely due to the low microwave power available in the region above 200 GHz.

Table 3.1 summarizes the molecules which have so far been investigated using the spectrometer described here. Section 3.5 describes how these molecules were generated in the three types of absorption cells mentioned above.

3.2 Infrared Laser Spectrometers

Infrared spectroscopy has been behind in both resolution and sensitivity compared with spectroscopy in the visible and uv regions, for example. This drawback is primarily ascribed to the low output power of the IR light source. The introduction of IR lasers has had great impact on spectroscopy in this region. In an earlier stage of IR laser spectroscopy, only fixed-frequency lasers were available. The CO_2, N_2O, and CO lasers are three of the most familiar such lasers in the IR region. Since they have high performance, they are still frequently employed as sources of molecular spectroscopy. Because they are fixed in frequency, either an electric or magnetic field is applied to samples to bring molecular transitions into resonance with the laser lines. The electric field case is referred to as laser Stark spectroscopy, whereas the magnetic field tuning technique is called laser magnetic resonance, or LMR. The latter has been extensively used to detect, identify, characterize, and monitor paramagnetic species of short lifetimes. Section 3.2.2 discusses IR LMR in detail.

For the far-infrared region, the H_2O/D_2O and HCN/DCN lasers were invented in an early stage of laser development, but they have been gradually taken over by CO_2-laser pumped lasers. Since all of these lasers are fixed in frequency, they have been employed as sources of FIR LMR, as described in Sect. 3.2.3.

Remarkable progress has recently been made in developing tunable laser sources. The most typical example would be the IR diode laser, which is now available for nearly the entire IR region. Section 3.2.1 describes IR spectroscopy with diode lasers as sources and its application to the study of transient molecules. Two other examples of a tunable IR laser source are the color-center laser and the difference-frequency laser, the latter being briefly discussed in Sect. 3.2.4.

3.2.1 Diode Laser Spectrometer

Because of its narrow linewidth, high spectral brightness, ease of tuning the oscillation frequency, and wide coverage of the IR region, the IR diode laser has recently attracted much attention in molecular spectroscopy and also in many other related fields. Although the quality of the diode was rather low in late 1970s and early 1980s, the improvement has been spectacular in these few years. This section describes the details of a diode laser spectrometer constructed at the Institute for Molecular Science which is specially designed for free-radical studies [3.32].

The laser diode consists of a p-type and an n-type semiconductor, each being a ternary alloy which includes lead as one of the constituents; examples are $Pb_{1-x}Sn_xSe$ (8–30 μm), $Pb_{1-x}Sn_xTe$ (7–30 μm), $PbS_{1-x}Se_x$ (4–8 μm), and $Pb_{1-x}Cd_xS$ (2.8–4 μm), which are commercially available. The factor x specifies the region of laser oscillation, and one diode typically covers a region of $50-100\ \mathrm{cm}^{-1}$. When the forward current flows through the $p-n$ junction, laser action is induced by the stimulated carrier-hole recombination emission, with two cleaved facets at the opposite ends of the diode as laser mirrors. The stripe-type double heterostructure has been employed to reduce the threshold current and thus to lower the heat dissipated in the crystal. The diode may be tuned in oscillation frequency by either changing the band gap or the index of refraction of the material, both of which are functions of temperature. Therefore, the oscillation wavelength may be chosen by adjusting the temperature of the diode or by varying the current fed to the diode. Although the current is eventually transformed into heat by ohmic loss, it is much more convenient in fine tuning and modulating the oscillation frequency. The oscillation frequency increases with temperature and current. It is not possible to cover the wavelength region continuously; the output consists of modes, each 0.5–1 cm^{-1} long, separated by gaps of a few cm^{-1}. The output power is typically of the order of 0.1 mW per mode. The diode oscillates in the temperature range 15–70 K, in contrast with the near IR diode which operates at room temperature. Therefore, a cryogenic refrigerator is indispensable together with a temperature controller required to maintain the temperature of the diode within a few mK. Frequency jitter is of the order of 1–10 MHz.

Figure 3.6 schematizes an IR diode laser spectroscopic system. Since the laser beam comes out from a limited area on a facet of the diode, the beam divergence is as large as 30° because of diffraction. The output is collimated by a lens of small f number, and then focused onto an entrance slit of a 25-cm monochromator to detect a single mode from among others; multimode oscillation is quite common. The beam coming out from the monochromator is collimated by a concave mirror, fed into an absorption cell, and then detected by a semiconductor detector. As summarized in Fig. 3.7, there are a few kinds of detectors designed for different IR wavelength regions.

In high-resolution spectroscopy using a tunable laser as a source, the wavelength needs to be known precisely. The easiest way is to refer to the spectra of some appropriate reference gases which have already been precisely measured. Figure 3.8 indicates such reference samples and the wavelength ranges where they may be used. The spectral lines of a sample are measured from a reference line with the aid of an etalon. A "vacuum spaced" etalon such as the one shown in Fig. 3.9 is much better than that made of a germanium block, since the free spectral range (FSR) of the former is much stabler than that of

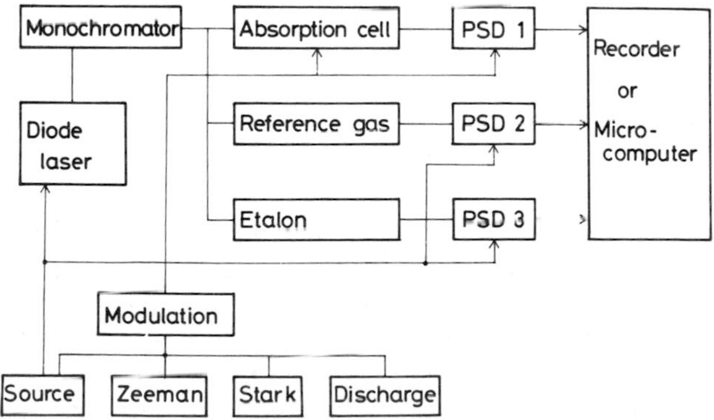

Fig. 3.6. Block diagram of a diode laser spectroscopic system

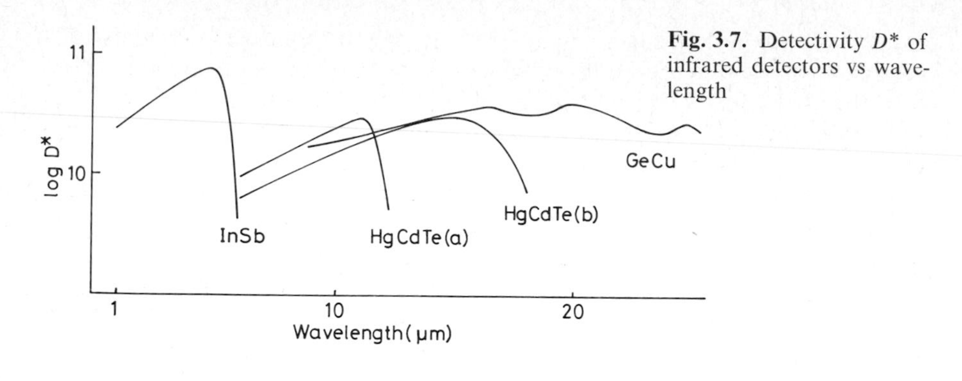

Fig. 3.7. Detectivity D^* of infrared detectors vs wavelength

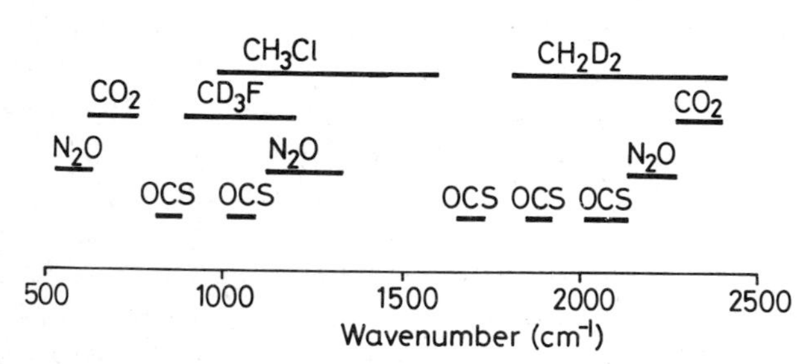

Fig. 3.8. Reference samples used as the wavelength standards

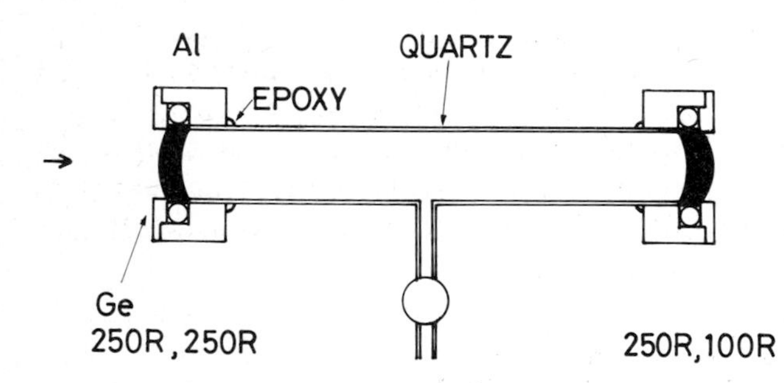

Fig. 3.9. Vacuum spaced etalon

the latter. When no appropriate references are available, the wavelength must be measured by a wavelength meter (λ meter). Figure 3.10 illustrates a block diagram of such a λ meter [3.33], which is an IR version of the λ meter *Hall* and *Lee* [3.34] designed for the cw visible laser. Because the refractive index in the IR region is not well known, the carriage carrying corner reflectors is installed in an evacuated box.

Several kinds of modulation may be introduced to make the measurement of absorption lines easier and the sensitivity higher. As in a conventional IR spectrometer, chopping the IR beam is one of the simplest schemes. This

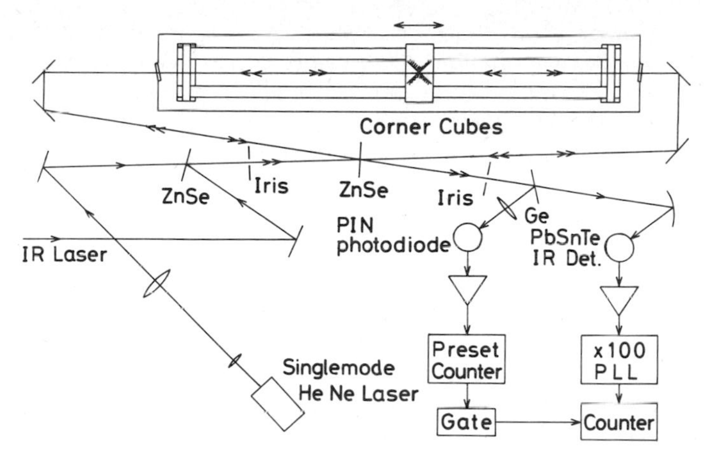

Fig. 3.10. Block diagram of a wavelength meter (reproduced from [3.33])

method is most suitable for the quantitative measurement of the absolute intensities and linewidths of absorption lines. It does not increase the sensitivity, however; the base line of the spectrum reflects the variation with wavelength of the output of the source laser and hence it is difficult to increase the gain. Source frequency modulation is most widely employed in diode laser spectroscopy. The diode is easily modulated by simply superimposing a small-amplitude sinusoidal wave on the injection current; the modulation frequency may be increased up to 10 kHz or so. The signal is demodulated by a phase-sensitive detector (PSD), resulting in an order of magnitude improvement in sensitivity. The noise spectrum of the source laser and the modulation efficiency suggest 5 kHz to be a best choice as the modulation frequency, although it is not critical. It is convenient to operate the PSD at 10 kHz, enabling the harmonic component of the modulated signal to be observed (referred to as $2f$ detection). This scheme makes the baseline flat and enables us to measure the peak wave number precisely. The sensitivity attained by source frequency modulation is limited by neither source noise nor detector noise, provided that the laser power exceeds a few ten μW, but by Fabry-Pérot type fringes caused by reflection of the laser beam by optical components in the path. The beam is often reflected back to the source diode, resulting in instability of the laser oscillation. The feedback may be eliminated by slightly detuning optical elements. In favorable cases, we may detect absorptions as small as 10^{-4} of the incident power. This limit corresponds to 2×10^9 CH_3 radicals/cm^3, for example, since 1.65×10^{13} CH_3 radicals/cm^3 absorb nearly 50% of the incident power, as judged from the ν_2 $Q(4,4)$ line transition moment of 0.28 D [3.35]. The minimum detectable number of molecules may even be smaller for diatomic and linear polyatomic molecules, because level densities are smaller. In fact, it has been estimated that 6×10^8 CF radicals/cm^3 may be detected, if the transition moment is as large as 0.2 D [3.36].

Fabry-Pérot type fringes which limit the sensitivity of source frequency modulation may be removed by molecular modulation such as Zeeman modulation, which allows us to pick out only molecular lines sensitive to modulation fields from among many other diamagnetic absorption lines and also fringes. For free-radical studies, Zeeman modulation is by far the most important. The magnetic field may be generated by winding a solenoid on an absorption cell. An example of such a solenoid is illustrated in Fig. 3.11. It consists of resin-coated copper wire with a rectangular cross section of 2×3 mm wound on a glass-fiber-reinforced epoxy tube of 80 mm diameter, in four layers for a total length of 500 mm, with one additional shim coil 45 mm long at each end to improve the magnetic field inhomogeneity to 1% over the central part (430 mm long), as shown in Fig. 3.11. The whole coil is immersed in an oil tank cooled by circulating water through a copper pipe dipped in the tank. The oil tank is made of nonmagnetic stainless steel with an epoxy glass tube passing through the central part of the tank, enabling an absorption cell to be installed. This glass tube also contributes in preventing secondary or eddy current from flowing through the wall of the tank.

A dc power supply delivers a current up to 50 A, which produces the magnetic field of 840 G. The Zeeman coil forms an LC circuit with a 10 μF capacitor, which is driven by an ac power source operated at about 1 kHz; the maximum ac field of 700 G is thus generated by a current of 15 A rms. The ac circuit is decoupled from the dc power supply by a choke coil which consists of 3 mm copper wire wound on a bobbin of about 30 cm diameter, dipped in another oil tank with water cooling. Usually the dc field is adjusted so that the total modulation field is nearly zero-based in every other half-cycle.

Figure 3.12 shows an example of a spectrum recorded with Zeeman modulation, which is assigned to the CH_3 radical generated by a glow discharge in di-*tert*-butylperoxide [3.37]. The spectrum in the same region recorded by source frequency modulation, reproduced in the lower part of Fig. 3.12 for comparison, exhibits many diamagnetic lines in addition to the paramagnetic

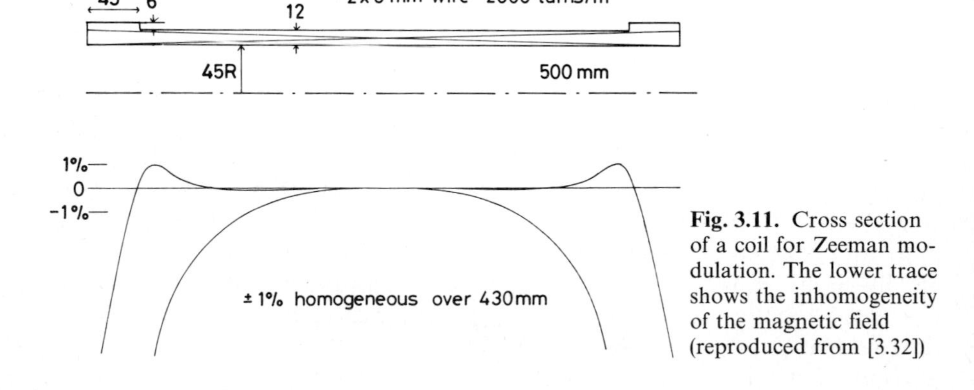

Fig. 3.11. Cross section of a coil for Zeeman modulation. The lower trace shows the inhomogeneity of the magnetic field (reproduced from [3.32])

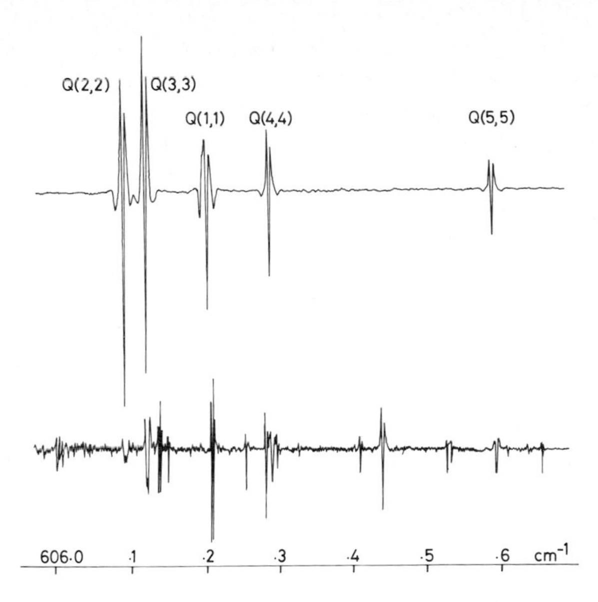

Fig. 3.12. *Q* branch transitions of the CH_3 ν_2 band. The upper trace was obtained using Zeeman modulation; the lower trace was obtained source frequency modulation (reproduced from [3.37])

lines of CH_3. These additional lines are perhaps due to ethane generated by the recombination reaction of the CH_3 radical.

Another modulation scheme which can be applied to transient molecules is discharge current modulation; here the abundance of molecules generated by the discharge is modulated by switching the discharge current on and off. *Endo* et al. [3.38] have successfully applied this technique to detect lines of the SF radical in $X^2\Pi_{1/2}$; SF in this spin substate is nearly diamagnetic because of cancellation of the electron spin and orbital magnetic moments. They inserted a triode in the discharge circuitry, and a low-frequency (a few hundred Hz) square wave is fed to its grid to modulate the discharge current. Because high-frequency current switching is difficult, discharge current modulation is combined with source frequency modulation to increase the sensitivity; the output of the first PSD for source frequency modulation is fed to the second PSD locked to the discharge current modulation frequency. Figure 3.13 shows a trace of a H_3^+ line observed by discharge current/source frequency double

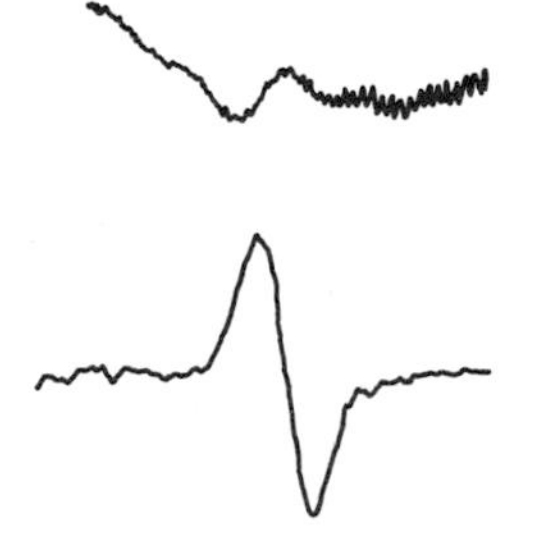

Fig. 3.13. $R(2)$ i.e. $J', k', l') = [(3, \pm 2, \pm 1), (3, 0, \mp 1)]_I \leftarrow (J'', k'') = (2, \pm 1)]$ transition of H_3^+. The upper trace was obtained using source frequency modulation, the lower trace by adding discharge current modulation

modulation [3.39]. The discharge current modulation frequency is limited by the finite bandwidth of the first PSD to be lower than 500 Hz. It must also be chosen with reference to the lifetime of the transient species; if the molecule lives longer than 20 ms, no modulation signals would be observed. In fact, *Yamada* et al. [3.40] have found that this method cannot be applied to HO_2 (although here difference frequency laser spectroscopy, rather than diode laser spectroscopy, was used).

Doppler-effect modulation or velocity modulation was invented by *Gudeman* et al. [3.41] to detect ionic species, mainly diamagnetic protonated ions, in discharge plasma. An alternating high voltage is applied along the discharge tube so that the transition frequency of the ion is modulated through the Doppler effect. *Hease* et al. [3.42] have shown that the Doppler shift of the frequency is comparable with the linewidth, ensuring efficiency of modulation. *Hease* and *Oka* [3.43] have recently applied this technique to observe the ν_2 band of H_3O^+ by diode laser spectroscopy.

A White-type multiple reflection cell with gold-coated mirrors has been most widely used in the diode laser studies of free radicals; it increases the sensitivity by an order of magnitude or more, compared with a single-pass cell. The laser beam is focused just outside the edge of the input mirror in order to observe well separated beam spots on the mirror so that the interference fringes are minimized and the path length is maximized. Twenty passes correspond to a 10 m path length which is under the Zeeman field. The path length could be made longer unless solid deposits on the surface of the mirrors due to chemical reactions deteriorate the optical throughput of the cell. The cell design may be modified in a number of ways, according to the type of reactions generating the transient species. When the discharge, either ac or dc, is induced in the cell, an electrode is inserted near each end. When the F atom is used to initiate a reaction, it is generated by a microwave discharge in CF_4 or F_2 placed in a side arm of the cell and is mixed with the second gas in the cell. The cell shown in Fig. 3.14 is designed so that the reaction takes place uniformly over the entire cell. This cell has been successfully applied to CH_2 and CD_2 [3.44], which were generated by the reaction of F with CH_2CO and CD_2CO, respectively; here gas *A* stands for ketene, which enters the cell through the series of holes drilled in the wall of the concentric jacket. A discharge in a mixture of CH_2CO and CF_4 in the cell gave no CH_2 signal.

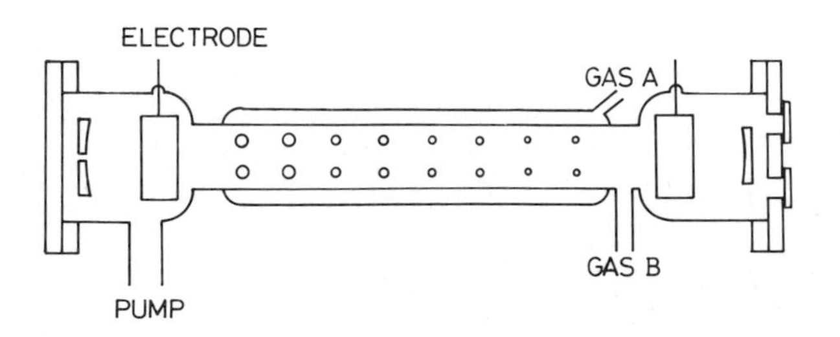

Fig. 3.14. A multiple-reflection discharge cell designed so as to give a uniform mixture of *A* and *B* [3.44] (see [4.8])

The cell is pumped by a mechanical booster (Roots) pump, followed by liquid nitrogen traps and a rotary pump, just as for a microwave spectrometer (Sect. 3.1.3). The only difference is the larger sample pressure ($\sim$ 1 Torr) in diode laser spectroscopy than in microwave spectroscopy. Since the 1 Torr pressure corresponds to the viscous flow region, the pumping speed is much higher, but in many cases such high pumping speed is unnecessary. An example is the F + O_2 reaction to generate FO_2 [3.150]. Because this reaction requires a third body (O_2 in the present case), the pumping speed needs to be much reduced to match the slow rate of the FO_2 formation.

It is quite tedious to measure the wave numbers for a large number of absorption lines with reference to the stardard spectra using etalon fringes as markers. These processes may be automated with the aid of a microcomputer; three kinds of signals, the sample and reference spectra and etalon fringes, are digitized by a 12 bit A/D converter and sent to the memory of a microcomputer SORD M343 which is divided into three parts, each consisting of 6000 words, to store the three types of data. The process is monitored by displaying the observed signals on a CRT scope in real time. The stored data are transferred to a floppy disk for later use. The computer is capable of averaging and smoothing the observed signals to improve the S/N ratio, of determining the peak wave numbers for the reference and sample spectral lines and also for fringes, and then of printing out the peak wave numbers thus determined with relative intensities. The spectrum may be displayed on an X-Y plotter, with the abscissa given in cm^{-1}. Figure 3.15 reproduces such a trace, a part of the PCl $v = 1 - 0$ transition observed by *Kanamori* et al. [3.45]. The computer program is written mostly in BASIC.

Diode lase spectroscopy may be applied to diagnosing chemical reaction systems in real time, by monitoring the spectra of transient species created or dissipated in the systems. A sensible application has recently been made to a few examples of excimer laser photolysis. This sort of study not only provides us with kinetic information on the systems, but also allows us to detect new molecular species that are difficult to generate by other means. Care must be taken to overlap the diode and excimer laser beams as closely as possible, to achieve good sensitivity. Since the excimer beam diameter is as large as 1.5–4 cm, it is easy to send the diode laser beam several times back and forth through the excimer laser beam. One conceivable arrangement is to introduce the excimer beam into a White-type multiple reflection cell through a hole just above the input mirror, while keeping the IR spots aligned just below the upper edge of the mirror.

The IR absorption may be recorded in either of the two modes, "spectroscopy" and "kinetics". In the spectroscopy mode, the computer receives signals for a certain period (gate width) before and after the laser shot and stores the difference between the two signals, while the diode laser is continuously swept as usual. In the kinetics mode, the diode laser sits on the top of an absorption line, and the computer stores the time variation of the signal intensity averaged for a number of excimer laser shots.

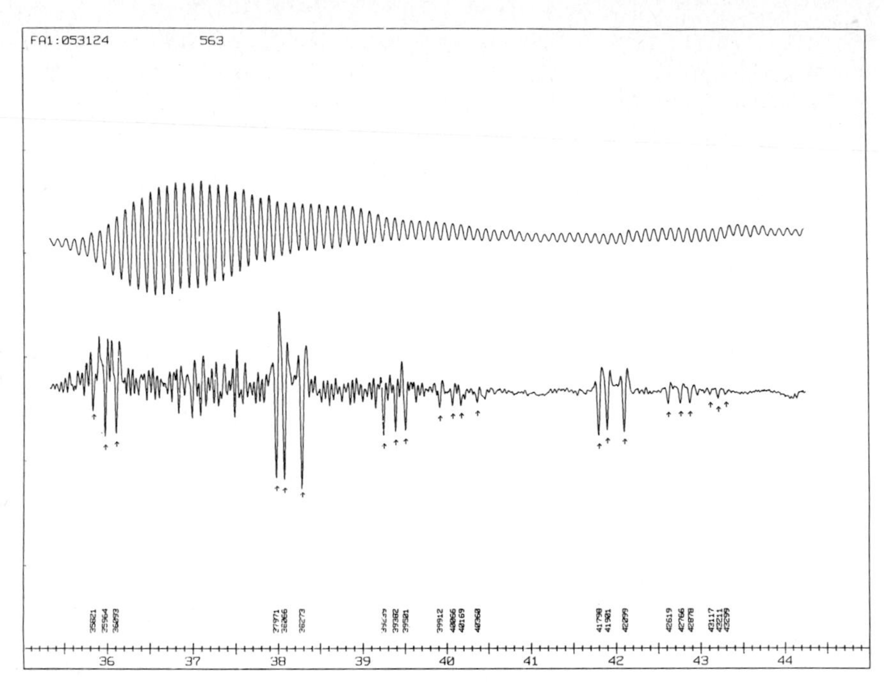

Fig. 3.15. Part of the PCl $v = 1 - 0$ band stored on a minicomputer; the line wave numbers were read out by the computer using etalon frings shown in the upper trace as markers

This method has been applied to the following transient molecules: CH_3 [CH_3I, $(CH_3)_2CO$], CS (CS_2), HCO (HCOOH), SO (SO_2), Cl (Cl_2, Cl_2SO), BH_2 or BH_3 (B_2H_6), and CCO (C_3O_2), where the starting materials are given in parentheses. The nascent SO molecule generated from SO_2 has been examined in detail, and interesting departures from the Boltzmann distributions have been observed for the vibrational, rotational, and fine-structure levels [3.46].

3.2.2 Infrared Laser Magnetic Resonance Spectrometer

Laser magnetic resonance (LMR) has proved to be one of the most sensitive ways of detecting paramagnetic species with high resolution in the gas phase. It was first invented for the FIR region and then extended to the mid-IR region where CO_2, N_2O, and CO lasers were employed as sources.

Figure 3.16 illustrates an energy level diagram with Zeeman effects which explains how LMR absorptions take place. Both upper and lower rotational levels are split into Zeeman components specified by the magnetic quantum number M_J, when the external magnetic field is applied. The laser frequency is fixed, but nearly coincides with the zero-field transition frequency, as indicated on the left of Fig. 3.16. When the magnetic field is applied, the transition is split into Zeeman components satisfying appropriate selection rules, and

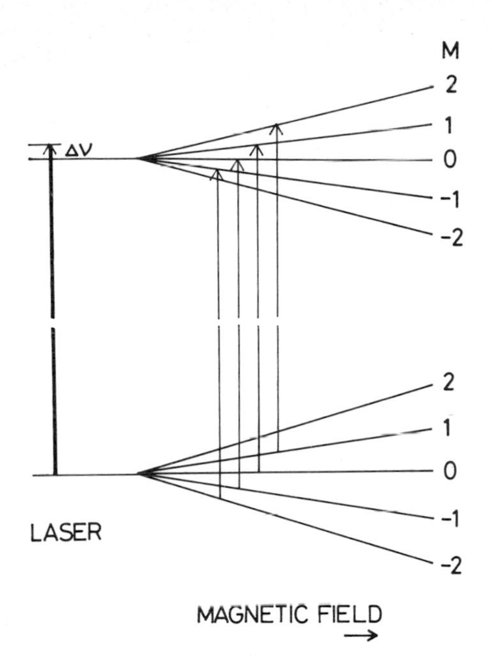

Fig. 3.16. Zeeman effects of two rotational levels. The laser frequency is off by $\Delta\nu$ from the molecular transition at zero magnetic field, but is brought into resonance with Zeeman components when an external magnetic field is applied

some of them are brought into resonance with the laser frequency as the magnetic field is swept, resulting in absorption lines. It is obvious that only paramagnetic species respond to LMR detection. Since the Bohr magneton is about 0.5 cm^{-1}/T, a region of up to 2 cm^{-1} may be scanned using a conventional electromagnet with maximum flux of 2 T (20 kG). In many cases, however, the Zeeman shift is much smaller and limits the applicability of LMR. As shown in Sect. 2.5, a first-order Zeeman effect is proportional to $[J(J+1)]^{-1}$, so that only low-J transitions can be detected by magnetic tuning.

The linewidth of a mid-IR LMR line is primarily determined by the Doppler effect, unless the sample pressure exceeds a few Torr. It is, however, quite routine to achieve sub-Doppler resolution by observing Lamb dips. The apparent linewidth is inversely proportional to the effective g factor defined by

$$g_{\mathrm{eff}} = d\nu/dB\,, \tag{3.2.1}$$

which is equal to the difference between the upper state and lower state gM_J factors:

$$g_{\mathrm{eff}} = g'M'_J - g''M''_J\,. \tag{3.2.2}$$

Lines with small Zeeman effects lead to LMR signals with large apparent linewidths.

The resonance field measured for a LMR line is affected by several factors such as the absolute field intensity and the laser stability. The magnetic field can be determined using NMR probes with an accuracy better than 0.1 mT

(1 G). The laser frequencies have been measured to 1 MHz or better, but the laser actually employed may oscillate with a frequency a few MHz off the reported value when it is locked to the peak of the Doppler gain profile. Therefore, the overall accuracy is about 10 MHz in most LMR measurements.

The sensitivity of a mid-IR LMR spectrometer is rather difficult to estimate, because the vibrational transition moment is not well known, especially for transient molecules. *Evenson* et al. [3.47] have estimated the minimum detectable number of molecules to be about 3×10^8 cm^{-3}, assuming the transition moment to be 0.03 D, which is an order of magnitude larger than the value the same authors gave for FIR LMR.

A typical arrangement for a mid-IR LMR spectrometer is illustrated in Fig. 3.17 [3.48]. It consists of a CO_2, N_2O, or CO laser with a gain tube and an absorption cell in the cavity. The absorption cell is placed between pole caps of an electromagnet, which delivers a magnetic field up to 2 T (20 kG). A 100 kHz modulation field of 2 mT peak-to-peak is generated by a pair of coils attached to the pole caps. The laser cavity is formed by a gold-coated concave mirror and a grating. The former is mounted on a PZT to lock the laser to the top of the gain profile, and the latter distinguishes a particular laser line. The zeroth-order reflection from the grating is used to monitor the laser power, and the output beam is detected by an IR detector, either Au:Ge in 5 μm or PbSnTe in 10 μm. The modulated signal is fed to a phase-sensitive detector with a passive filter of a high Q value.

The Brewster windows of the gain tube and the absorption cell determine the polarization of the laser with respect to the magnetic field. When the electric vector of the laser is perpendicular to the magnetic field, $\Delta M_J = \pm 1$ or σ electric dipole transitions are allowed, whereas $\Delta M_J = 0$ or π transitions appear with the parallel arrangement.

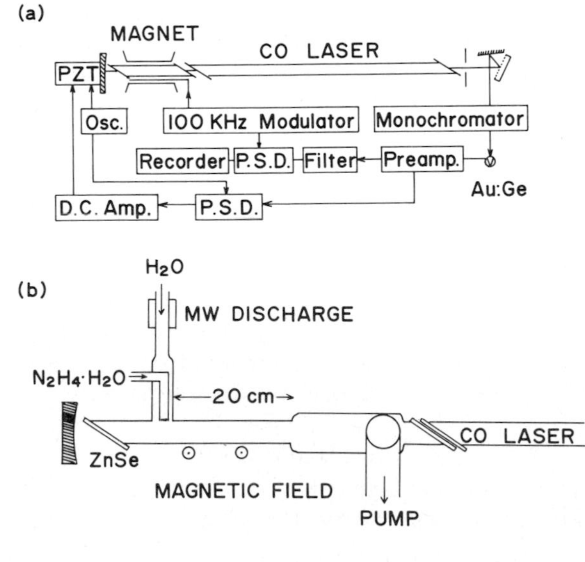

Fig. 3.17. (**a**) Block diagram of a mid-IR LMR spectrometer. (**b**) A sample cell for the observation of a transient molecule (reproduced from [3.48])

The CO_2 and N_2O lasers cover the wavelength range 875–1110 cm^{-1} when isotopic species containing ^{13}C, ^{18}O, or ^{15}N are employed. The lowest frequency lines are provided by the $^{14}C^{16}O_2$ laser. The separation between two adjacent lines is 2 cm^{-1} on the average. The frequency of each line has been measured accurately by heterodyne techniques [3.49,50]. The CO laser covers the region 1450–2000 cm^{-1}, which may be extended by cooling the laser with liquid nitrogen and/or by purging the laser cavity with dry nitrogen gas to eliminate water vapor. The spacing between two adjacent laser lines is about 4 cm^{-1}. The laser frequencies have also been measured [3.51,52].

The observed spectrum of mid-IR LMR provides us with information on both the upper and lower (ground) states and also on the band origin. This means, however, that more parameters are required in analyzing it than the FIR LMR spectrum. Furthermore, the resolution is lower in the mid-IR than in the FIR regions, because of the Doppler effect. Therefore, the mid-IR LMR spectrum is more difficult to analyze than the FIR LMR spectrum. One advantage of mid-IR LMR over FIR LMR is that low J transitions are easily observed.

3.2.3 Far-Infrared Laser Magnetic Resonance Spectrometer

The molecular spectra in the FIR region ($\lambda = 1000 - 30$ μm) correspond to the pure rotational transitions of relatively light molecules and also to the low-frequency large-amplitude vibrational transitions of molecules. Because the absorption coefficient of the pure rotational transition increases in proportion to the square and cube of the transition frequency for the nonlinear and linear molecules, respectively, rotational spectroscopy in the FIR region promises very high sensitivity. However, if blackbody radiation is used as a light source, the power available for the unit wavelength is too small to attain high sensitivity. In fact, no transient molecules had been detected by FIR spectroscopy until the introduction of the FIR laser magnetic resonance (FIR LMR) technique.

The LMR is a laser analog of electron paramagnetic (or spin) resonance (EPR or ESR); the microwave in EPR is simply replaced by a fixed-frequency laser in LMR. The FIR LMR is primarily designed to observe molecular rotational transitions and atomic fine-structure transitions. The first application of FIR LMR was to the O_2 molecule, where a HCN laser was employed as a light source [3.53], and since then quite a large number of transient molecules such as OH, CH, HO_2, and HCO were investigated by using either a HCN or a H_2O laser. However, the method had been seriously handicapped by the fact that these lasers provided only a very limited number of lines, which do not suffice for studying many other transient molecules. In 1970, *Chang* and *Bridges* [3.54] reported laser oscillation in the FIR of CH_3F when it was pumped by a Q-switched CO_2 laser. This technique has since been developed to such an extent that more than 1600 FIR laser lines generated from about 80 molecules had been reported by 1983. These laser lines have been compiled by

Knight [3.55]. The frequency has been measured for many FIR laser lines to an accuracy of 1 MHz by heterodyne mixing with microwave [3.56].

Radford and *Litvak* [3.57] have introduced an optically pumped FIR laser in their LMR spectrometer. To increase the sensitivity, they placed an absorption cell in the FIR laser cavity and separated it from the gain tube by a thin polypropylene sheet fixed at the Brewster angle. The pumping CO_2 laser beam is introduced coaxially with the FIR laser cavity through a small hole drilled at the center of the laser mirror. The pumping efficiency is fairly high for this arrangement, but, when the pumping CO_2 laser power is increased, it easily burns out the polypropylene membrane. *Evenson* et al. [3.58] introduced the CO_2 laser beam into the FIR cavity through a window on the side wall of a glass tube containing the lasing gas and the laser beam is reflected back and forth by two flat mirrors held parallel in the glass tube. *Evenson* [3.59] later replaced the parallel mirrors by a copper pipe or a gold-coated glass tube to increase the gain of the FIR laser in the long-wavelength region.

Because the absorption cell is placed in the laser cavity, the sensitivity of FIR LMR spectroscopy is very high; *Evenson* et al. [3.47] estimated that 1×10^6 OH radicals/cm^3 may be detected. This detection limit is nearly the same as that for laser-excited fluorescence, known to be one of the most sensitive methods.

Figure 3.18 shows a FIR LMR spectrometer constructed at the Institute for Molecular Science [3.60,61]. Pumping by a CO_2 laser may be accomplished either through the side wall of the laser tube or through one of the end mirrors, but the former is normally employed. The high-power CO_2 laser employed is an Apollo Model 560 A which has been modified so as to stabilize thermally

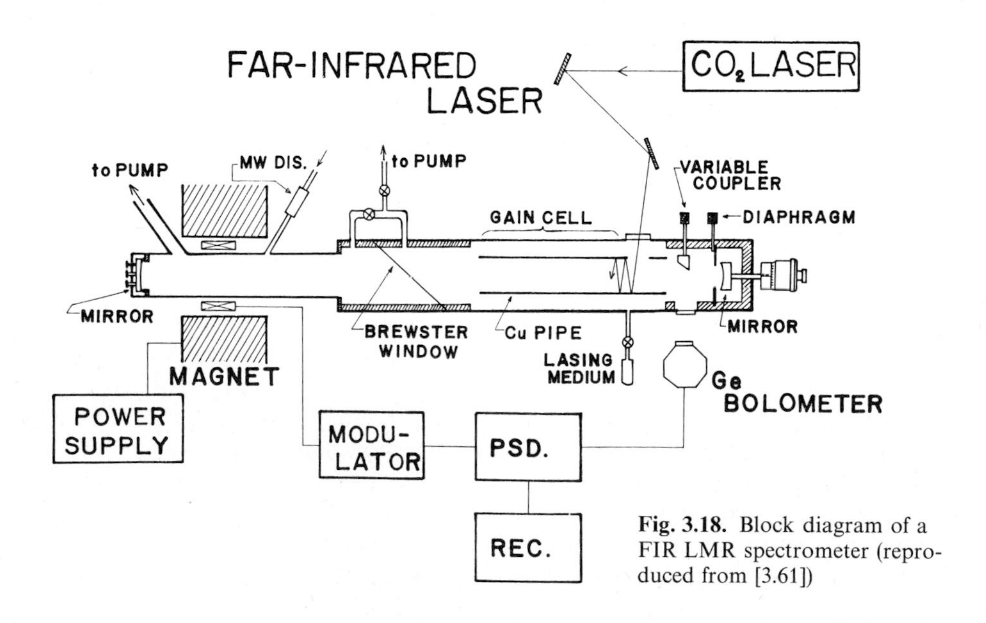

Fig. 3.18. Block diagram of a FIR LMR spectrometer (reproduced from [3.61])

its cavity length by three invar rods; it delivers an output power up to 55 W on the 10.6 μm P(22) line. A copper tube of 54 mm inner diameter is inserted in the FIR laser gain tube for CO_2 laser pumping. The FIR laser cavity is formed by a fixed concave mirror of 1.3 m curvature radius and a moving concave mirror of 2 m curvature radius, held 110 cm apart. The latter is mounted on a differential micrometer head and is placed so as to match the cavity length to one of the longitudinal modes of a FIR laser line. The beam waist is estimated to be 9 mm and 20 mm at a point 60 cm from an end mirror, for a laser wavelength of 100 μm and 500 μm respectively. The FIR laser cavity length is maintained fixed by three rods of "neoceram zero" which is a sort of ceramic with a low thermal expansion coefficient (-10^{-7} at 300 K). The stability of the FIR laser frequency is primarily determined by the pumping CO_2 laser; a feedback of the FIR laser output to the pump laser reduces the instability of the FIR laser.

The magnetic field is generated by a Varian 12″ electromagnet with pole cap separation of 76.2 mm which delivers a field up to 14 kG. The homogeneity of the magnetic field was measured by a Varian NMR gaussmeter to be 10^{-4} within a central area 5 cm in diameter, but is less along the magnetic flux, i.e., perpendicularly to the pole cap faces. During the course of an LMR experiment, the gaussmeter probe is inserted between an absorption cell and one of the pole caps, and the difference between the magnetic fields at the center of the cell and at the gaussmeter is corrected for after the spectrum is oberseved; the correction is 4 G or less.

The intracavity absorption cell is separated from the laser gain tube by a 12.5 μm thick polypropylene film fixed at the Brewster angle, which allows one to select the polarization of the laser radiation either parallel or perpendicular to the applied magnetic field. The modulation coil is attached to the absorption cell and delivers a field up to 20 G peak-to-peak.

The output of the FIR laser is coupled out by a small polished copper mirror 4 × 4 mm, which is placed at 45° to the axis of the laser cavity and may be moved perpendicularly to the cavity axis to adjust the degree of coupling. The laser beam reflected by the mirror passes through a polyethylene window and is detected by either a Ge bolometer or an InSb detector both operated at 4.2 K. The detected signal is fed into a lock-in amplifier and the absorption line is recorded as the first derivative on a strip chart recorder. The modulation frequency is 4.3 kHz and 100 kHz, respectively, for the Ge bolometer and the InSb detector. The Ge bolometer is of the so-called composite type, i.e., designed to respond as fast as possible; it is cooled down to 1.7 K when weak FIR laser lines are employed as sources. The InSb detector reaches maximum reponse at the 1 mm wavelength and is thus employed in the wavelength region longer than 300 μm. Because the laser noise decreases with frequency, it is desirable to choose the modulation frequency as high as possible.

The transient species is normally generated by the reaction of a stable molecule with discharge products of a second species, as described in Sect. 3.5; the discharge is placed well outside the magnetic field, because it is affected by the magnetic field. The reaction mixture is pumped rapidly through the cell.

3.2.4 Difference Frequency Laser Spectrometer

Another tunable coherent source is available in the IR region 2.2–4.2 μm. A few nonlinear optical elements may be employed to generate difference frequency radiation from the outputs of two visible lasers. Among them $LiNbO_3$ (lithium niobate) has been most extensively employed; *Pine* [3.62] was the first to develop a spectroscopic system using this crystal with a cw dye laser and an Ar^+ laser. Because the mixing is a nonlinear process, high peak power pulsed sources were initially employed to attain a high visible to IR conversion efficiency. However, it is hard to reduce the spectral width and the wavelength scatter of the emitted pulsed laser light, and, in fact, the resulting IR linewidth was of the order of 1 cm^{-1}. Because of its high conversion efficiency, the $LiNbO_3$ crystal has enabled a cw output up to 10 μW in the IR to be generated, which is by no means large, but is more than 10^5 times larger than the noise equivalent power of a good IR detector. The cw source has advantages over the pulsed system in that it is easy to stabilize and ready to scan continuously, which are of great significance for high-resolution spectroscopy.

Figure 3.19 is a schematic diagram of a difference frequency laser spectroscopic system developed at the Institute for Molecular Science, which follows the original design of *Pine*. The outputs from a single-mode Ar^+ laser and a tunable cw dye laser are combined collinearly by a dichroic mirror and are focused by a lens 25 cm in focal length to a 5 cm long a-cut $LiNbO_3$ crystal to generate the IR radiation with $\nu_{IR} = \nu_{Ar} - \nu_{Dye}$ frequency, where ν_{Ar} and ν_{Dye} denote the Ar^+ and dye laser frequencies, respectively. The IR power is maximized when the three laser beams are appropriately phase matched. The condition is satisfied by properly selecting the polarization of each laser beam and by maintaining the temperature of the crystal at a specified value using an

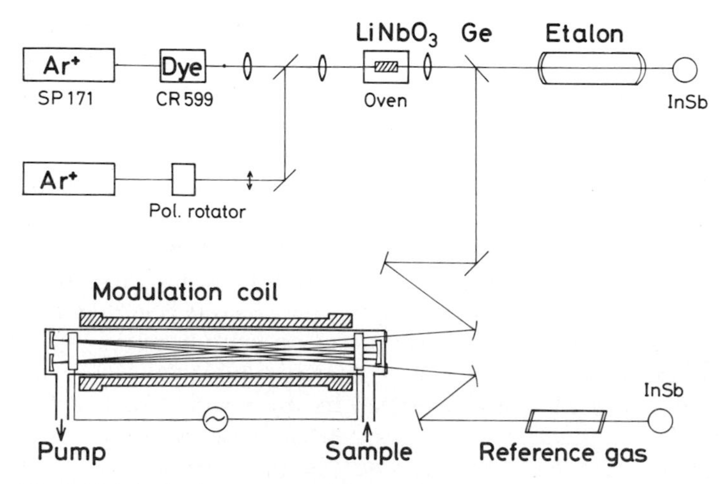

Fig. 3.19. Block diagram of a difference frequency laser spectroscopic system

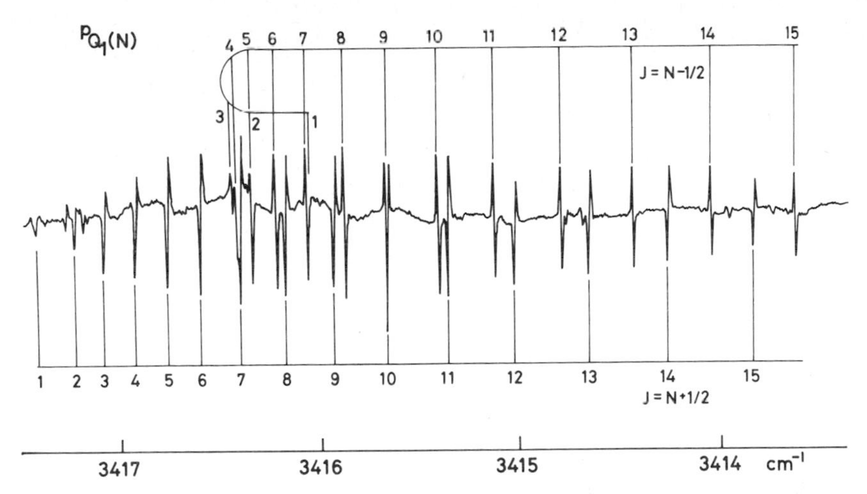

Fig. 3.20. A part of the HO_2 ν_1 band recorded with Zeeman modulation (reproduced from [3.40])

oven. The polarization of the Ar^+ laser beam is rotated by a polarization rotator so as to be orthogonal to that of the dye laser beam and parallel to the c axis of the $LiNbO_3$ crystal. A combination of the 514.5 nm or 488 nm Ar^+ laser line with the Rh 6 G dye laser beam gives rise to IR radiation in the range 2400–4400 cm^{-1}. The longest wavelength limit is imposed by the phonon absorption band of $LiNbO_3$ extending from 4.2–60 µm, and the shortest wavelength limit, by the damage of the crystal; the phase matching requires the temperature to be as low as 200 °C, when the IR wavelength becomes short. The linewidth of the IR radiation is mainly determined by the jitter of the Ar^+ ion laser, which is about 50 MHz.

Zeeman modulation has also been proved to be useful in sorting out the absorption spectra of paramagnetic transient species among others; there are a number of *M-H* stretching bands in the 2–4 µm region which are due to precursors and products of the reaction-generating transient molecules. Although the Doppler width may be as large as 100–150 MHz at 3 µm, no essential modifications of the spectroscopic techniques applied in the 10 µm region have been required. As an example of the spectrum, Fig. 3.20 shows a part of the ν_1 band of HO_2 observed by *Yamada* et al. [3.40]. The Zeeman effect is fairly large for this spectrum, because the spin-rotation splitting is large and the selection rule is b type, $\Delta K_a = \pm 1$, making the recorded line shape almost of first derivative. Care must be taken, however, that the true zero-field frequency does not necessarily coincide with either the maximum slope or the peak of the first derivative, because the Zeeman effect is not strictly linear, due to the repulsion between two spin states with the same N rotational quantum number. The observed spectrum must be carefully analyzed by simulation in order to determine the wave number of each absorption line precisely.

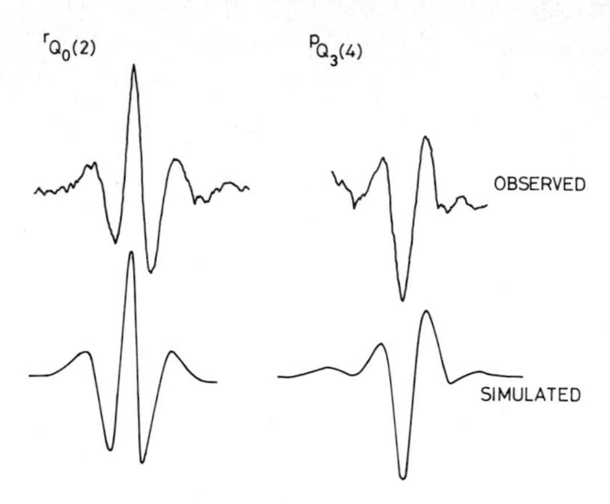

Fig. 3.21. Two rotational lines of the CH_3 ν_3 band recorded with Zeeman modulation (reproduced from [3.63])

An example of the other extreme is the CH_3 radical, for which the spin-rotation splitting is so small that the Zeeman effect is small and varies with the rotational transition, although the observed ν_3 band is of perpendicular type. In fact, as shown in Fig. 3.21, the phase is variable from line to line [3.63].

The advantage of the difference frequency laser is that it continuously covers a certain range of wavelength without any gaps. The source delivers an output which is low and often suffers from instability of the pump ion laser. Nevertheless, the noise inherent in the output beam seems to be smaller than that of the color-center laser, which often competes with the difference frequency laser.

3.3 Dye Laser Spectroscopic System

3.3.1 CW Dye Laser

A tunable cw dye laser pumped by an Ar^+ or a Kr^+ laser is one of the most suitable light sources for high-resolution spectroscopy in the visible region. The advantages of using a cw dye laser are summarized below.

1) Continuous single-mode coverage of the entire visible and near-IR (400–900 nm) regions using about ten kinds of dyes.
2) High spectral purity. The spectral resolution which can be achieved exceeds that of a conventional grating or Fourier transform spectrometer; it is not limited by the instrumental bandwidth, but by the linewidth of the observed spectrum.
3) High power density. This makes it possible to observe nonlinear phenomena which serve as bases for nonlinear spectroscopic techniques such as intermodulated fluorescence spectroscopy (Sect. 3.3.3) and double resonance spectroscopy (Sec. 3.4) of sub-Doppler resolution.

4) Small divergence of the output beam. This brings several advantages in setting up a spectroscopic system; a long path cell is easily designed for the observation of weak lines, and the background noise originating from scattering of the incident light by the window or wall of the cell is much more easily reduced than in the case of incoherent light.
5) Well-defined polarization. This makes polarization spectroscopy practical.

The disadvantage of the cw dye laser includes the narrow wavelength coverage and the low "peak" intensity, in comparison with the pulsed dye laser. The latter covers the wavelength region down to the vacuum ultraviolet, by means of frequency doubling and mixing techniques [3.64]. The low "peak" power of the cw dye laser makes it difficult to generate harmonics and sum-frequency light with good efficiency [3.65]. The multiphoton process is difficult to induce by the cw dye laser, also because of low "peak" power of the output. Another practical difficulty in using the cw dye laser stems from the pump laser; the ion laser is required to generate an output of 2–6 W for pumping in the visible or uv region, but such a pump laser is still relatively short lived (1–2 years) and the discharge tube is extremely expensive.

Recent progress in cw dye laser technology includes the development of the ring laser, which delivers a single-frequency output of more than one W. It has also been reported that more than 50 mW of stable single-frequency light was generated in the uv region, by placing a harmonic generator in the ring laser cavity [3.66] or in a passive enhancement cavity [3.67].

3.3.2 Doppler-Limited Excitation Spectroscopy Using a CW Dye Laser as a Source

We have found that laser excitation spectroscopy is one of the most efficient methods for investigating short-lived transient molecules. Provided that in the upper electronic state nonradiative processes do not play a dominant role, the excitation spectrum is a much more sensitive means of detecting transient species than that based on the absorption spectrum. If the spectral width of the source laser, either cw or pulsed, is narrower than the Doppler width, excitation spectroscopy enables the spectrum at the Doppler-limited resolution to be observed.

Figure 3.22 shows a typical arrangement of a laser excitation spectrometer which has been used at the Institute for Molecular Science. A single-frequency cw dye laser pumped by an Ar^+ laser is used as the light source. A part of the dye laser output is split from the main beam by a Pyrex glass beam splitter. The main part, after being mechanically chopped, enters a fluorescence cell. Fluorescence thus induced is collected through the wall of the cell with a lens system and is focused onto a photomultiplier. A few sheets of filters are placed in front of the photomultiplier to reduce scattered light and chemiluminescence. Phase-sensitive detection is of potential use in picking laser-excited fluorescence out from background chemiluminescence. The split part of the laser beam is further

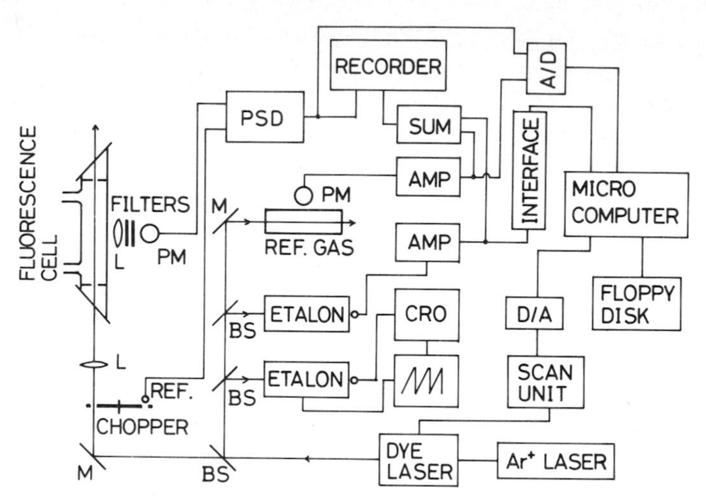

Fig. 3.22. Schematic diagram of a dye laser excitation spectrometer. Abbreviations used are BS for beam splitter, M for mirror, PM for photomultiplier, and L for lens system

divided into three parts: the first part monitors the dye laser mode using a spectrum analyzer, the second part generates wave-number markers from a temperature-stabilized etalon with the FSR (free spectral range) of 1.5 GHz, and the last part records the excitation or absorption spectrum of a reference molecule.

The wave-number of each observed spectral line is determined by recording the sample spectrum simultaneously with the reference spectrum and etalon fringes on a two-pen strip chart recorder. A typical example of the trace is shown in Fig. 3.23. The signals, after being appropriately digitized, are stored on a floppy disk and then transferred to a main computer in our Computer Center to search for the spectral position and to calculate the wave-number.

The absorption spectra of I_2 and $^{130}Te_2$ have been measured in the visible region with a precision of 10^{-3}–10^{-4} cm^{-1} by Fourier transform spectroscopy [3.68, 69]. The spectral lines of the two molecules serve as excellent wave-number standards in the regions 14,000–20,000 cm^{-1} and 18,500–23,800 cm^{-1}, respectively. For I_2 the absorption spectrum is well reproduced by excitation spectroscopy, whereas the Te_2 molecule shows an excitation spectrum that differs considerably from the absorption spectrum because of predissociation. Therefore, up to 20,000 cm^{-1} of the excitation spectrum of I_2 may be employed as a convenient reference, but beyond this limit the standard must be replaced by the absorption spectrum of $^{130}Te_2$.

The Doppler width is about 0.04 cm^{-1} at room temperature in the visible region for a molecule with molecular weight of 30–50. This resolution is sufficient for resolving rotational and fine structures in the electronic spectrum, and allows us to measure the position of an isolated line to an accuracy of 0.003 cm^{-1}, about one tenth of the Doppler width.

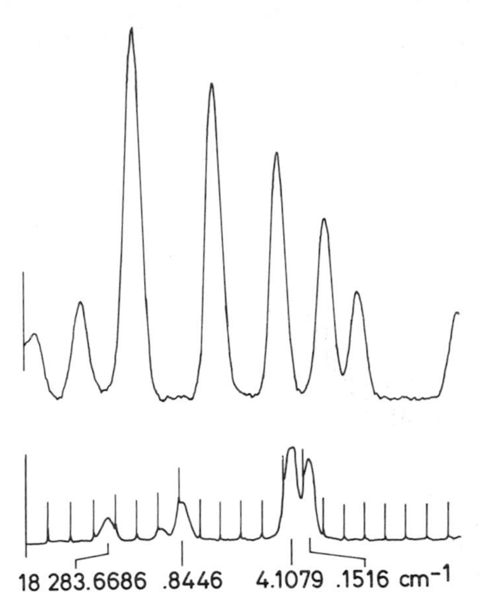

Fig. 3.23. A part of the HCF $\tilde{A}^1 A''(010) - \tilde{X}^1 A'(000)$ band. The lower trace shows the spectrum of I_2 used as a reference with 1.5 GHz markers generated by an etalon

The Doppler-limited laser excitation spectrum provides us with more precise and more reliable data on the molecular structure than the spectra obtained from conventional high-resolution spectroscopy. The results are useful in carrying out further detailed investigations such as those with double resonance spectroscopy (Sect. 3.4). The lower-state molecular parameters are often of great help in observing microwave and IR spectra in the ground electronic state manifold. As an example, Table 3.2 lists molecular constants of the HSO

Table 3.2. Molecular constants of HSO in the $\tilde{X}^2 A''$ (0 0 0) state (MHz)[a]

	Dye Laser [3.70]	Microwave [3.27]
A	299 478 (20)	299 484.63 (49)
B	20 504.4 (17)	20 504.56 (59)
C	19 135.6 (17)	19 133.93 (58)
Δ_N	0.031 8 (25)	0.030 70 (52)
Δ_{NK}	0.857 (58)	0.896 0 (26)
Δ_K	27.2 (20)	27.2[b]
δ_N	0.002 02 (56)	0.001 93 (22)
δ_K	0.45 (38)	0.89 (29)
ε_{aa}	−10 292 (64)	−10 365.99 (55)
ε_{bb}	−433 (32)	−426.656 (109)
ε_{cc}	−5 (32)	0.226 (161)

[a] Values in parentheses are 2.5 times standard errors and apply to the last digits
[b] Fixed

radical obtained from dye laser excitation spectroscopy [3.70] and those reported in a subsequent microwave study [3.27]. The excellent agreement between the two sets of the data demonstrates that the laser excitation spectroscopy described in this section is sufficiently reliable.

3.3.3 Intermodulated Fluorescence Spectroscopy

The hyperfine structure is an invaluable source of information on the electronic structure of the molecule, not only in the ground state, but also in excited electronic states, but is usually not well resolved in the spectrum of Doppler-limited resolution in the visible and uv regions. As mentioned earlier, the high-power density of the dye laser output permits us to conduct several types of nonlinear spectroscopy to overcome the barrier imposed by the Doppler effect. As an example of such spectroscopy, this section describes the details of the intermodulated fluorescence (IMF) method invented by *Sorem* and *Schawlow* [3.71]. This technique has already been applied with success to a number of transient molecules: BO_2 [3.72], NH_2 [3.73], PH_2 [3.74], and CaF [3.75].

Figure 3.24 shows a schematic diagram of a typical arrangement for observing IMF spectra. The source is again a single-mode cw dye laser, and its output is divided into two parts of nearly equal intensity, which are mechanically chopped at f_1 and f_2. The two modulation frequencies are chosen so as not to satisfy a simple relation $n_1 f_1 = n_2 f_2$, where n_1 and n_2 denote small integers. The two beams enter a fluorescence cell through windows at opposite ends and are focused nearly at the center of the cell. The two beams are almost collinear, but make a small angle with each other in order to prevent them from going back to the source laser. An optical isolator consisting of a 1/4-wavelength plate and a linear polarizer also suppresses beam feedback to the source.

Fluorescence is again focused onto a photomultiplier by a lens system including a few filters. The output signal of the photomultiplier is detected at the $f_1 + f_2$ sum frequency using a lock-in amplifier so that only the saturation signal is sorted out from the Doppler profile and noises. Figure 3.25 shows an example of the IMF spectrum, the 3_{30}, $J = 7/2 - 4_{40}$, $J = 9/2$ transition of PH_2 [3.74]; the hyperfine structures due to both P and H nuclei are well resolved (the triplets are due to the H nuclei, while the splitting between the two triplets is ascribed to the P nucleus). The linewidth of a single line is about 10 MHz, which is much larger than the pressure width and is mainly ascribed to the residual Doppler effects arising from the finite angle between the two beams and from the beam divergence.

For the IMF spectrum, the relative frequency, i.e., the splitting, is more important to measure than the absolute frequency, and a temperature-stabilized 50 cm long confocal etalon may be employed for such purposes which generates fringes at every 150 MHz, as shown in Fig. 3.25. The fringe separation may be calibrated against the hyperfine structure of a reference molecule. The NO_2 molecule would be the most suitable for the calibration,

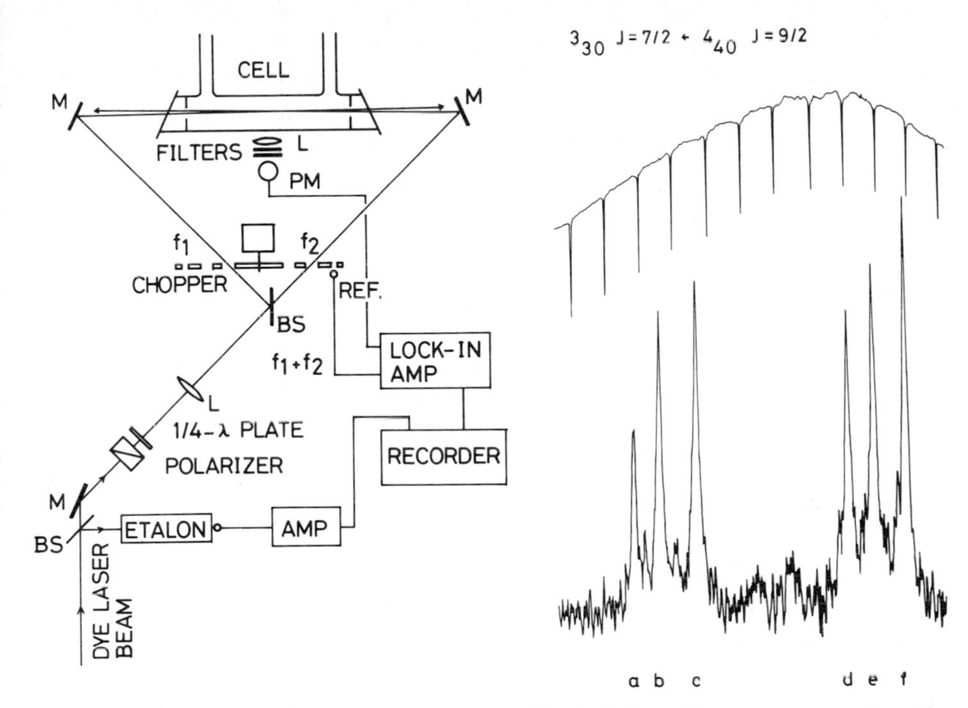

Fig. 3.24. Schematic diagram of an intermodulated fluorescense spectrometer. See Fig. 3.22 for the abbreviations employed

Fig. 3.25. Hyperfine structure in the $3_{30}-4_{40}$, $J=7/2-9/2$ transition of the PH_2 $\tilde{A}^2A_1(000)-\tilde{X}^2B_1(000)$ band, recorded using the intermodulated fluorescense technique. The upper trace shows the Doppler profile with 150 MHz markers by an etalon (reproduced from [3.74])

since the hyperfine structure has been precisely measured by MODR in the upper electronic state [3.76] and by microwave spectroscopy in the ground vibronic state [3.77].

3.4 Double Resonance Spectroscopy

The introduction of lasers has made it possible to perform various types of double resonance (DR) spectroscopy experiments in a number of wavelength regions and has opened a new field in molecular spectroscopy. This technique often allows us to observe energy levels that cannot be reached by a single-photon process. Double resonance spectroscopy may also be employed in making and/or confirming assignments. In many cases, the advantages (resolution, sensitivity, and so on) of spectroscopy in one wavelength region are transferred to spectroscopy in another wavelength region, without losing the advantages of the latter. In this section, two types of DR spectroscopy are discussed: one is microwave (or radio frequency) optical double resonance (MODR) and the other is IR-laser optical double resonance (IODR).

3.4.1 Microwave or Radio Frequency Optical Double Resonance

Because the Doppler width in the visible region is as large as several hundred to a few thousand MHz, straightforward laser excitation spectroscopy as described in Sect. 3.3 does not provide us width fine details of the spectra such as hyperfine structure and Stark and Zeeman effects. Some special techniques are required to break through the resolution limit imposed by the Doppler effect, i.e., the Doppler-limited resolution. Intermodulated fluorescence spectroscopy discussed in Sect. 3.3.3 is such an example. Microwave or radio-frequency optical double resonance, which is often referred to as MODR, is another method of achieving sub-Doppler resolution, and is described below.

Suppose we observe fluorescence from an excited electronic state by exciting an electronic transition from the ground state with a high-power laser. When we simultaneously irradiate the sample with microwave or radio frequency (rf), we may observe changes in intensity or polarization of fluorescence when microwave or rf resonates with a rotational transition which shares a level with the pumped electronic transition. Figure 3.26 shows such an example: the $\tilde{a}^3A_2$ $v_3 = 1$ and $\tilde{X}^1A_1$ $v = 0$ states of H_2CS. A dye laser pumps molecules from the 4_{14} level in the ground vibronic state to the 4_{14} F_3 level in the upper electronic state. Microwave pumping of the 4_{13}–4_{14} transition in the upper electronic state then causes fluorescence from 4_{14} to decrease and induces that from 4_{13}. Because the electronic transition is saturated, the net effect would be an increase in fluorescence intensitiy. The microwave transition is thus converted to the optical transition, without losing the high (sub-Doppler) resolution of microwave spectroscopy and the high sensitivity of electronic

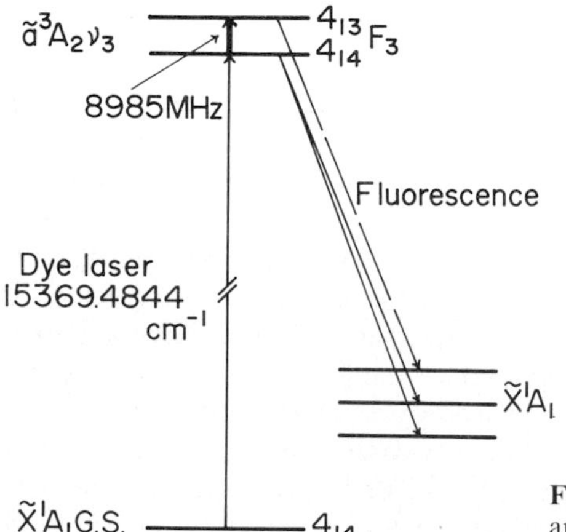

Fig, 3.26. Energy levels of H_2CS involved in an MODR experiment

spectroscopy. The method will find proper applications in the study of transient molecules, because the photomultiplier sorts out fluorescence emitted from a narrow region of space. As mentioned above, the method supports the assignment on which the DR experiment is based.

The MODR using a dye laser for pumping the electronic transition was initiated by *Field* and co-workers in the early 1970s [3.78–80]. They investigated BaO generated by the reaction of Ba metal vapor with O_2. The $A^1\Sigma - X^1\Sigma$ electronic transition was pumped by a Rh–6 G cw dye laser operated in a single mode with an output power of several ten mW. At the same time, microwaves were introduced via a horn in a region where fluorescence was emitted. In this way they observed 17 rotational transitions up to 76 GHz in various vibrational levels of both A and X states. Susequently they applied this technique to several other inorganic molecules [3.81–83].

Another important application should not be overlooked, namely the NH_2 radical investigated by *Curl* and collaborators [3.84–95]. They constructed a cylindrical resonant cavity as a DR cell. The radical was produced by the reaction of H_2O microwave discharge products with anhydrous hydrazine and was pumped through a quartz tube which passed coaxially through the cavity, namely through the region within the cavity where the magnetic field of microwave is strongest. They thus observed many magnetic dipole transitions between spin doublets in addition to a few electric-dipole-allowed rotational transitions in both the $\tilde{X}^2B_1$ ground electronic state and the $\tilde{A}^2A_1$ excited electronic state. They determined molecular constants precisely, including fine and hyperfine structure constants of both states from the observed DR spectra.

Brazier and *Brown* [3.96] recently observed the (0,0) band of the CH $A^2\Delta$-$X^2\Pi$ system at 432 nm and, by using MODR, have succeeded in detecting several transitions between Λ doublets of CH in the ground vibronic state which had been subjected to microwave spectroscopic examination for a period of ten years, because of their importance in astronomy.

3.4.2 An MODR Spectrometer to Study Transient Molecules

Figure 3.27 shows a block diagram of an MODR spectrometer which has been used for studying HNO [3.97] and H_2CS [3.98]. These two molecules are quite stable even in a metal container, with a lifetime of a few tens of seconds [3.13] to several minutes [3.99]. The MODR cell was thus made of a crossed metal wave guide of size appropriate for the frequency region to be scanned. The laser beam is introduced in the cell through a Brewster window and proceeds in an arm of the crossed wave guide, while the other arm is used to propagate the microwave; the microwave transmission is good up to 157 GHz. The sample is continuously pumped through the cell, and the fluorescence is observed through a slit cut at the center of the cross by a photomultiplier placed in front of the slit. Usually edge filters and/or band filters are inserted in front of the photomultiplier to reduce laser-light scattering. The millimeter wave is generated by a series of OKI klystrons, the output of which is frequency modulated

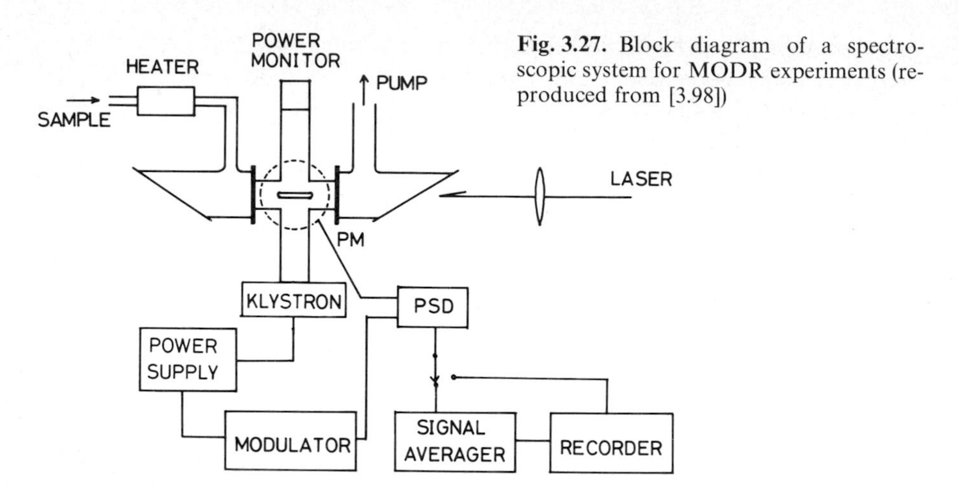

Fig. 3.27. Block diagram of a spectroscopic system for MODR experiments (reproduced from [3.98])

by a 10 kHz square wave to an amplitude of several MHz. In the centimeter wave region, a TWT amplifier driven by an X-13 klystron delivers an output up to 2 W, which is 100% amplitude modulated by 10 kHz square wave.

In the rf region a cell is employed which is made of a coaxial transmission line with inner and outer diameters of 9 and 21 mm, respectively, and with a characteristic impedance of about 50 Ω. Fluorescence is observed through a slit cut on the outer transmission line. The cell is terminated at one end by a 50 Ω resistance, and the rf power up to 10 W is fed to the other end of the cell; a transistor amplifier driven by a VHF oscillator is used as a power source. The VHF output is 100% amplitude modulated by a 10 kHz square wave. The microwave and rf power densities are of the order of several hundred mW to a few W per cm^2.

The optical transition is pumped by a dye laser, either a CR 599-21 standing-wave laser or a SP 380A ring laser, as described in Sect. 3.3. When the laser power is not large enough to saturate optical transitions, lenses are used to focus the laser beam at the place where the fluorescence is emitted. Since the beam waist is as small as 0.2 mm, the power density of the dye laser beam can reach a few hundred W per cm^2.

The dye laser frequency is normally set at a particular vibronic transition frequency, and the microwave or rf frequency is swept around the frequency expected for a rotational transition to be observed. The signal from the photomultiplier is demodulated by a phase-sensitive detector (PSD) operated at the modulation frequency, 10 kHz. A signal averager is often used to improve the signal-to-noise ratio; the output of the PSD is digitized and integrated by the averager, which is triggered by the beat note from an all-wave receiver used to measure the microwave frequency (Sect. 3.1). The system allows us to accumulate the double resonance signal for a period longer than 3 min.

The present spectrometer has been employed to observe the rotational transitions of HNO in the $\tilde{A}^1A''$ excited electronic state [3.97] and also hyper-

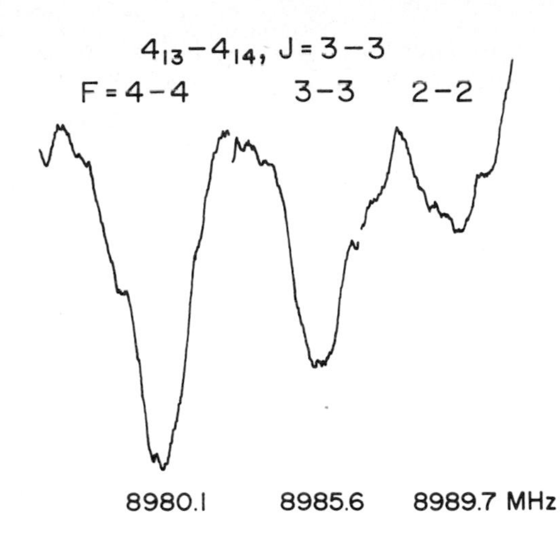

Fig. 3.28. The 4_{13}–4_{14}, $J = 3 - 3$ transition of H_2CS in the $\tilde{a}^3 A_2$ $v_3 = 1$ state observed by MODR

fine structures of the rotational transitions of H_2CS in the $\tilde{a}^3 A_2$ excited electronic state [3.98]. The HNO molecule was produced by the reaction of discharged oxygen with a propylene/NO mixture. A sample of H_2CS was prepared by flowing trimethylene sulfide through a quartz tube heated to about 850 °C. Figure 3.28 shows a typical example of the MODR spectrum, the 4_{13}–4_{14}, $J = 3 - 3$ transition observed for H_2CS in the $\tilde{a}^3 A_2$, $v_3 = 1$ state [3.98].

The MODR spectrum may be applied to measure the Stark effect, from which the dipole moment may be derived not only for the ground electronic state but also for excited electronic states. The Stark cell illustrated in Fig. 3.29 was used to determine the dipole moment of H_2CS in the $\tilde{A}^1 A_2$, $v = 0$ and $\tilde{a}^3 A_2$, $v_3 = 1$ states [3.100]. It consists of a piece of an X-band wave guide 15 cm long and has a stainless steel electrode inside, which is held parallel to the

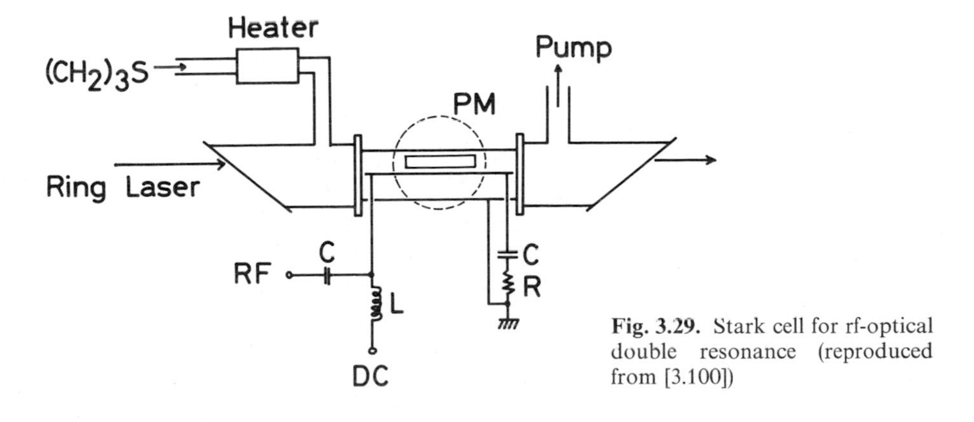

Fig. 3.29. Stark cell for rf-optical double resonance (reproduced from [3.100])

broad face of the wave guide but is insulated from the wave guide by Teflon supporters. The electrode is terminated by a 50 Ω resistance and a capacitor at one end, and the rf power and DC voltage are supplied from the other end. This configuration allows only $\Delta M_J = 0$ transitions to take place, because the laser, rf, and DC fields are all perpendicular to the surface of the electrode. An example of the Stark-resolved spectrum, the 4_{22}–4_{23} transition of H_2CS in the $\tilde{A}^1 A_2$, $v = 0$ state, is reproduced in Fig. 3.30.

Generally speaking, the metal wave guide and cavity cells may be applied to only a limited number of transient molecules, because the metal surface easily deteriorates most of them. A novel design is required of the MODR cell suitable for each transient molecule, to maintain its high concentration in the cell.

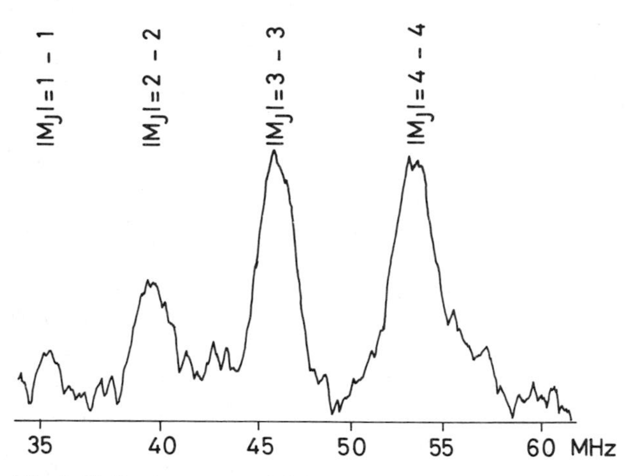

Fig. 3.30. Stark pattern of the 4_{22}–4_{23} transition of H_2CS in the $\tilde{A}^1 A_2 v = 0$ state observed by rf-optical double resonance (reproduced from [3.100])

3.4.3 Infrared-Optical Double Resonance

Infrared-optical double resonance (IODR) allows us to observe excited vibronic states which cannot be reached because of symmetry selection rules or are difficult to access due to unfavorable Franck-Condon factors from the ground vibronic state by a single-photon process. An example is a vibrational state of a $D_{\infty h}$ molecule in an excited state of Π_u symmetry which has an odd quantum number in the normal mode of Σ_u symmetry. This state is not reached by a single-photon step from the ground vibronic state of Π_g symmetry, because its vibronic symmetry is also Π_g. One may observe this excited vibronic state by pumping the $\Pi_u - \Pi_g$, 0–0 band, while simultaneously applying an IR radiation corresponding to the vibrational transition within the excited Π_u state. No such studies have, however, been reported.

Two types of IODR have been reported; one is a combination of a pulsed UV laser and a pulsed IR laser, and the other achieves a sub-Doppler resolution by employing a cw CO_2/N_2O laser and a cw dye laser. In the pulsed experiments, the IR radiation was generated by a parametric oscillation in two $LiNbO_3$ crystals [3.101], a pulsed F-center laser [3.102], or a pulsed CO_2 laser [3.103, 104]. The double resonance experiment with pulsed lasers provides us with information on molecular energy transfer. It has also been successfully applied to observe weak transitions and/or transitions among heavily overlapped spectra [3.105]. *Amano* et al. [3.106] have reported a high-resolution IODR experiment on NH_2, in which they employed the LMR technique (Sect. 3.2.2) to observe vibrational transitions in an excited electronic state. *Muenchausen* and *Hills* [3.107] have also observed IODR signals of ND_2, but without using an external magnetic field; they exploited exact coincidence of the vibrational transitions in the ground electronic state with CO_2 laser lines and observed the intensity change upon CO_2 laser irradiation of the fluorescence induced by a dye laser. An IODR experiment using a tunable diode laser as an IR source has been described by *Kawaguchi* [3.108], who observed the change in absorption due to the ν_3 vibration rotation transition of BO_2 when a cw dye laser excited the $\tilde{A}^2\Pi_u\,(000)$–$\tilde{X}^2\Pi_u\,(000)$ band.

This section describes in detail the IODR experiment performed by *Amano* et al. [3.106]. The CO_2/N_2O laser delivers an output power sufficient to saturate most IR transitions, but its oscillation frequency is fixed. So magnetic field tuning is employed, as in LMR. A cw dye laser with a spectral width as narrow as a few MHz selectively pumps a limited range of Doppler components of an electronic transition, which are simultaneously subjected to the IR radiation derived from a CO_2/N_2O laser. The DR signal is observed as a change in fluorescence intensity.

The underlying principle of the experiment is illustrated in Fig. 3.31. The fluorescence/absorption cell was placed inside the CO_2/N_2O laser cavity and between the pole caps of a Varian 15″ electromagnet which delivers a magnetic field up to 22 kG. The dye laser beam was introduced into the cell through a ZnSe mirror at the end of the cell. In some experiments the cell was displaced outside the laser cavity by simply moving the reflector to position M', as shown in Fig. 3.32. In the extracavity arrangement, the IR laser beam was focused on the center of the pole caps by an antireflection coated ZnSe lens with 38 cm focal length, and the dye laser beam was focused at the same point. This arrangement was suitable for detecting weak IR transitions, because the power density is an order of magnitude larger in the extracavity arrangement than in the intracavity arrangement.

The output power of the dye laser was about 80 mW with a spectral width of about 10 MHz, whereas the IR laser delivers an output of about 500 mW for a typical laser line. The IR beam waist was less than 1 mm in diameter when focused in the extracavity cell, and thus the power density was larger than 50 W/cm^2. In the intracavity arrangement, the power density was estimated to be about 5.8 W/cm^2, because the reflectivity of the ZnSe end mirror was 92 %.

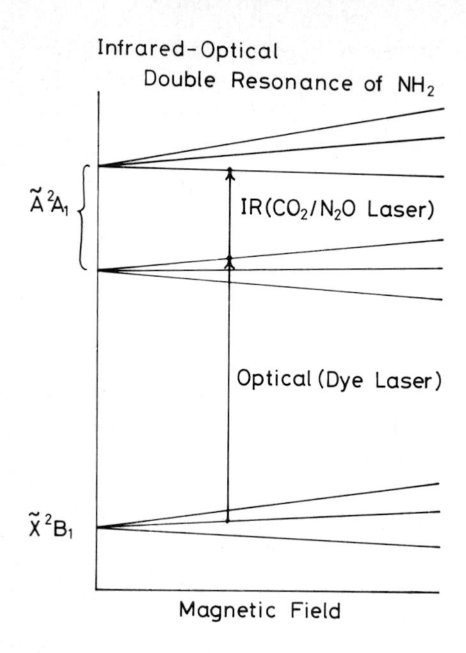

Fig. 3.31. Illustration of the principles underlying an infrared-optical double resonance experiment

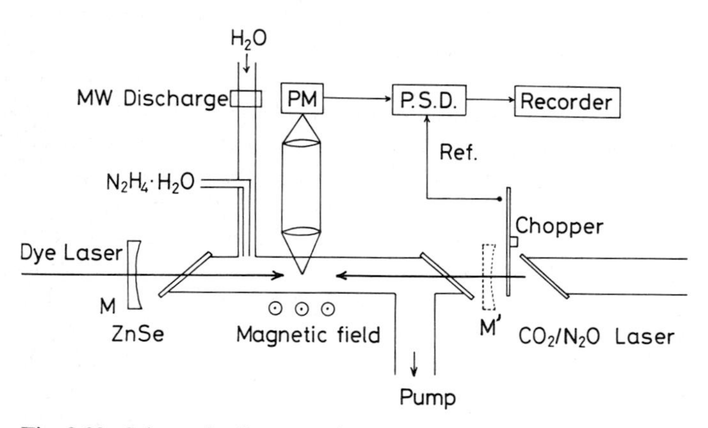

Fig. 3.32. Schematic diagram of a spectroscopic arrangement for infrared-optical double resonance (reproduced from [3.106])

For a strong infrared transition like the $v_2 = 10 \leftarrow 9$ transition of NH_2 in the $\tilde{A}^2A_1$ state, the extracavity arrangement enabled the vibrational transition to be saturated, and, in fact, the observed IODR signal showed saturation broadening. However, the S/N ratio for such a signal was found to be better in the intracavity than in the extracavity arrangements, mainly because the number of molecules interacting with the two laser beams simultaneously was much larger in the former than in the latter.

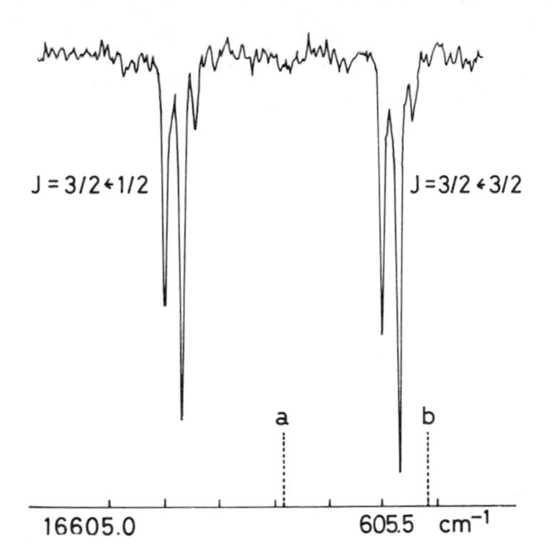

Fig. 3.33. The $v_2 = 10,\ 2_{11} - v = 9, 2_{20}$ transition of NH_2 in the $\tilde{A}$ state, recorded by infrared-optical double resonance with dye laser scanning (reproduced from [3.106])

Laser-excited fluorescence was collected through a telescope on a photomultiplier placed well outside the magnet so as to be unaffected by the magnetic field. The IR beam was mechanically chopped at 690 Hz, and the fluorescence signal was demodulated by a lock-in amplifier operated at the chopping frequency and was displayed on a strip chart recorder.

Figure 3.33 shows an example of the observed DR spectra, obtained by sweeping the dye laser frequency. The IR transition $(0, 10, 0)\, 2_{11} \leftarrow (0, 9\ 0)\, 2_{20}$, $J, M_J = 3/2, 1/2 \leftarrow 3/2, 3/2$ of NH_2 in the $\tilde{A}$ state is in resonance with the CO_2 P(6) laser line when the magnetic field of 4380 G is applied. The higher- and lower-frequency spectral lines shown in Fig. 3.33 correspond, respectively, to the $J, M_J = 3/2, 3/2 \leftarrow 3/2, 1/2$ and $J, M_J = 3/2, 3/2 \leftarrow 1/2, 1/2$ components of the $\tilde{A}(0, 9, 0)\, 2_{20} \leftarrow \tilde{X}(0, 0, 0))\, 1_{10}$ electronic transition, each exhibiting triplet hyperfine splittings. The dotted lines *a* and *b* denote the positions of the $J = 3/2 \leftarrow 1/2$ and $J = 3/2 \leftarrow 3/2$ transitions at the zero field, respectively.

Figure 3.34 illustrates how the DR signals are observed, using a triplet hyperfine structure as an example. It clearly shows that when the IR frequency (or the magnetic field) is swept, the observed hyperfine splitting $\Delta\nu^{\mathrm{IR}}(\mathrm{obs})$ is given by

$$\Delta\nu^{\mathrm{IR}}(\mathrm{obs}) = \Delta\nu^{\mathrm{IR}}(1 \pm v/c) \pm (\nu^{\mathrm{IR}}/\nu^{\mathrm{OP}})\,\Delta\nu^{\mathrm{OP}}, \tag{3.4.1}$$

where $\Delta\nu^{\mathrm{IR}}$ and $\Delta\nu^{\mathrm{OP}}$ denote the actual hyperfine splittings in the IR and optical (OP) transitions which appear at the frequencies ν^{IR} and ν^{OP}, respectively. The $+$ and $-$ signs apply to the two laser beams propagating antiparallel and parallel, respectively. The ratio $\nu^{\mathrm{IR}}/\nu^{\mathrm{OP}}$ is 1/16 to 1/20, so that the second term of (3.4.1) is less than 10% of the first term in most cases, but certainly exceeds the measurement uncertainty. The intracavity arrangement thus gives two sets of spectra, the average of which leads to the real hyperfine splittings.

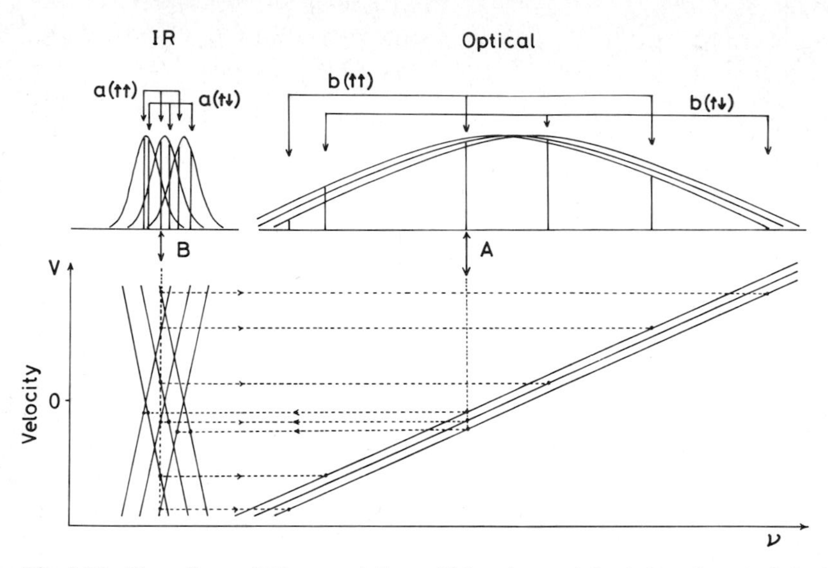

Fig. 3.34. Hyperfine splittings and line widths observed by infrared-optical double resonance; an optical and an infrared transition are used as examples, both of which consist of three hyperfine components. The upper part reproduces Doppler-broadened line shapes and the lower part shows the Doppler shifts (reproduced from [3.106])

When the dye laser is swept, the observed hyperfine splitting $\Delta\nu^{\mathrm{OP}}(\mathrm{obs})$ corresponds to

$$\Delta\nu^{\mathrm{OP}}(\mathrm{obs}) = \Delta\nu^{\mathrm{OP}} \pm (\nu^{\mathrm{OP}}/\nu^{\mathrm{IR}})\,\Delta\nu^{\mathrm{IR}}, \tag{3.4.2}$$

in which the second term is predominant. Figure 3.33 shows such an example. This sort of spectrum is useful in making M_J assignments in an intermediate state.

Equations (3.4.1, 2) may be applied as they are to the discussion of the linewidths observed by DR. Namely, the linewidth in the case of IR laser scan is close to the homogenous width (the pressure broadening width), whereas the dye laser scan increases it by the factor of $\nu^{\mathrm{OP}}/\nu^{\mathrm{IR}}$.

3.5 Generation of Transient Molecules

The chemistry of transient molecules still involves many unknown factors, which often make high-resolution spectroscopic studies of these molecules really difficult and sometimes even impossible. The spectroscopic methods discussed in Sect. 3.1, 2, for example, require molecules to be abundant by as much as 10^6-10^8 molecules/cm^3 [3.47], a number which is quite small from the spectroscopic point of view and yet formidable to attain for a number of most reactive species like molecular ions. The diversity of chemical properties of molecules further increases the difficulty of preparing highly concentrated reactive species. In view of these circumstances, any informating obtainable from other physical and chemical methods is of some use in finding chemical systems for efficient generation of the transient molecule under study. One should search for appropriate references in the fields of chemical kinetics, especially photochemistry using flash photolysis techniques, mass spectrometry, and matrix studies using conventional IR, uv, and ESR spectroscopy.

Table 3.3 lists the transient molecules which the author's group has so far investigated at the Institute for Molecular Science using various kinds of high-resolution spectroscopic techniques described in Chap. 3. It also includes the methods we employed in producing the molecules. The molecules are classified into three groups, according to the schemes of their production. Group I comprises molecules of relatively long lifetimes, i.e., 100 ms or longer; they may be generated outside the cell by either electrical discharge or pyrolysis and continuously pumped through the cell, e.g., CS, NS, HNO, and H_2CS. When the lifetime is as short as a few ten ms or even close to 10 ms, as for molecules in Group II, they must be prepared directly in the cell by reacting activated discharge products of one reactant species with the second species inside the cell. An increase of the pumping speed decreases the lowest limit of the lifetime of the molecule to be detected. As shown in Table 3.3, many diatomic and polyatomic molecules belong to Group II.

For molecular species of even shorter lifetimes, e.g., less than 1 ms, they must be generated uniformly over the entire cell, in order to make the effective path length long enough for absorption spectroscopy. A dc or ac discharge inside the cell has been employed for this group (Group III) of molecules; CCl, SiN, FeO, HSO, and CH_3 are examples. The shortest lifetime observed for Group III molecules is perhaps less than 1 ms. The direct discharge in the cell is, however, not suitable for molecules that are easily decomposed in discharge plasma, e.g., FSO, NCO, and CH_3O. The method is also difficult to apply when some of the precursors are too susceptible to electrical discharges; before they react with other molecules to produce the desired species, they may all be decomposed to useless fragments. In these cases we have to increase the pumping speed, by either shortening the cell length or introducing a high-speed pump. At the end of Table 3.3 three examples are listed which are generated by methods other than electrical discharge. The CH_3 radical was also obtained by pyrolyzing di-*tert*-butylperoxide, the singlet ($\tilde{a}^1 A_1$) CH_2 molecule by photolyzing ketene with a uv line from a high-power Ar^+ laser, and the NO_3 radical by the chemical reaction $NO_2 + O_3$.

The optimum sample pressure is different for different spectroscopic means. For microwave specroscopy, for example, the linewidth is mainly determined by the pressure broadening, so that the sample pressure is normally maintained at a few ten mTorr. This is in contrast with IR spectroscopy, for which the Doppler width predominates over the pressure width, as long as the sample

Table 3.3. Production of transient molecules investigated at the Institute for Molecular Science

Molecule	Reaction employed [a]	Spectroscopic method [b]
I) *Production outside the cell*		
CS	$(CS_2)_{\text{ac disch.}}$	IR [3.109]
NS	$(N_2 + S_2Cl_2)_{\text{MW disch.}}$	IR [3.110]
HNO	$C_3H_6 + NO + (O_2)_{\text{MW disch.}}$	MODR [3.97]
H_2CS	$[(CH_2)_3S]_{\text{Pyrolysis}}$	MODR [3.98]
II) *Discharge outside the cell*		
Diatomic molecule		
SO ($^3\Sigma^-$, $^1\Delta$)	$OCS + (O_2)_{\text{MW disch.}}$	IR LMR [3.111, 112]
SF	$OCS + (CF_4)_{\text{MW disch.}}$	MW [3.113], IR [3.38]
PO	P (red) + $(O_2 + H_2)_{\text{MW disch.}}$	FIR LMR, MW [3.114]
PD	P (red) + $(D_2$ or $D_2O)_{\text{MW disch.}}$	FIR LMR [3.61]
AsH	As + $(H_2$ or $H_2O)_{\text{MW disch.}}$	FIR LMR [3.115]
Triatomic molecule		
HO_2	Allyl alcohol + $(O_2)_{\text{MW disch.}}$	MW [3.116], IR
SF_2	$OCS + (CF_4)_{\text{MW disch.}}$	MW [3.117]
FSO	$OCS + (O_2 + CF_4)_{\text{MW disch.}}$	MW [3.118], IR
NCO	$HNCO + (CF_4)_{\text{MW disch.}}$	MW [3.119]
PO_2	P (red) + $(O_2 + H_2)_{\text{MW disch.}}$	FIR LMR, MW [3.120]
HPO	$PH_3 + (O_2 + H_2)_{\text{MW disch.}}$	MW [3.121]
NH_2	$N_2H_4 + (H_2O)_{\text{MW disch.}}$	IR LMR [3.48]
CD_2	$CD_2CO + (CF_4)_{\text{MW disch.}}$	IR [3.44]
PH_2	P (red) + $(H_2)_{\text{MW disch.}}$	LES [3.122]
HSO	$H_2S + (O_2)_{\text{MW disch.}}$	LES [3.70]
CHF	$CH_3F + (CF_4)_{\text{MW disch.}}$	LES [3.123]
CHCl	$CH_3Cl + (CF_4)_{\text{MW disch.}}$	LES [3.124]
SiH_2	$SiH_4 + (CF_4)_{\text{MW disch.}}$	LES [3.125]
CCN	$CH_3CN + (CF_4)_{\text{MW disch.}}$	LES [3.126]
Other polyatomic molecules		
HCCN	$CH_3CN + (CF_4)_{\text{MW disch.}}$	MW [3.127]
CH_2F	$CH_3F + (CF_4)_{\text{MW disch.}}$	IR, MW [3.128]
CH_2Cl	$CH_3Cl + (CF_4)_{\text{MW disch.}}$	MW [3.129]
CH_3O	$CH_3OH + (CF_4)_{\text{MW disch.}}$	MW [3.130]
CH_2CHO	$CH_3CHO + (O_2)_{\text{MW disch.}}$	MW [3.131]
CH_3S	$CH_3SH + (CF_4)_{\text{MW disch.}}$	MW [3.132]
III) *Discharge inside the cell*		
Diatomic molecule		
CF	$(C_2F_4)_{\text{ac disch.}}$	IR [3.36]
	$(CF_4 + CH_3F)_{\text{dc disch.}}$	MW [3.29]
CCl	$(CCl_4)_{\text{dc disch.}}$	IR [3.32], MW
SiN	$(SiH_4 + N_2)_{\text{dc disch.}}$	MW [3.133]
SiF	$(SiF_2)_{\text{dc disch.}}$	MW [3.134]
SiCl	$(SiCl_4)_{\text{dc disch.}}$	MW [3.135]
GeF	$(GeF_4)_{\text{dc disch.}}$	MW [3.136]
FeO	$[(C_5H_5)_2Fe + O_2]_{\text{dc disch.}}$	MW [3.28]

Table 3.3. (continued)

Molecule	Reaction employed[a]	Spectroscopic method[b]
BCl	$(BCl_3)_{dc\ disch.}$	MW [3.137]
NCl	$(N_2 + Cl_2)_{ac\ disch.}$	IR, MW [3.138]
PF	$(PH_3 + CF_4)_{dc\ disch.}$	MW [3.139]
PCl	$(PCl_3)_{dc\ disch.}$	MW [3.140], IR
FO	$(O_2 + CF_4)_{ac\ disch.}$	IR [3.141]
BrO	$(Br_2 + O_2)_{ac\ disch.}$	IR [3.142]
SCl	$(S_2Cl_2)_{ac\ disch.}$	IR [3.45]
Polyatomic molecules		
DO_2	$(CH_3OD + O_2)_{dc\ disch.}$	MW [3.143]
HSO	$(H_2S + O_2)_{dc\ disch.}$	MW [3.27]
ClSO	$(S_2Cl_2 + O_2)_{dc\ disch.}$	MW [3.144]
PH_2	$(PH_3 + O_2)_{dc\ disch.}$	MW [3.145]
PF_2	$(PH_3 + CF_4)_{dc\ disch.}$	MW [3.146]
FCO	$(C_2F_4 + CO)_{ac\ disch.}$	IR [3.147]
BO_2	$(BCl_3 + O_2)_{ac\ disch.}$	IR [3.148]
ClBO	$(BCl_3 + O_2)_{ac\ disch.}$	IR [3.149]
FO_2	$O_2 + (CF_4)_{MW\ disch.}$	IR [3.150]
CF_3	$[(CF_3CO)_2O, CF_3COOCH_3, CF_3CH_2OH, CF_3C_6H_5, CF_3I]_{ac\ disch.}$	IR [3.151]
	$[(CF_3CO)_2O]_{dc\ disch.}$	MW [3.152]
CH_3	$\{[(CH_3)_3CO]_2, CH_3I, (CH_3)_2CO, CH_3SH, CH_3OH, CH_3I, CH_4\}_{ac\ disch.}$	IR [3.37]
IV *Others*		
CH_3	$\{[(CH_3)_3CO]_2\}_{pyrolysis}$	IR [3.37]
CH_2	$(CH_2CO)_{UV}$	LES [3.153]
NO_3	$NO_2 + O_3$	IR [3.154]

[a] $(M)_{MC\ disch.}$ and $(M)_{dc/ac\ disch.}$ denote products of a microwave discharge in M outside the cell and of a dc or ac discharge in M inside the cell, respectively

[b] MW = Microwave Spectroscopy, FIR = Far-Infrared Spectroscopy, IR = Infrared Spectroscopy, LMR = Laser Magnetic Resonance, LES = Laser Excitation Spectroscopy, MODR = Microwave Optical Double Resonance

pressure is lower than a few Torr. Since the IR absorption increases with the sample pressure until the pressure broadening becomes comparable with the Doppler width, a fairly high sample pressure of a few hundred mTorr to a few Torr is preferred in most cases. The difference in the optimum sample pressure often forces one to choose different chemical systems even if one plans to generate identical species.

In the first high-resolution spectroscopic investigation of HO_2 by *Radford* et al. [3.155] using FIR LMR, they generated the radical by the following methods: (1) O + small hydrocarbon compounds, (2) microwave discharge in water, (3) $H + O_2 + M$, (4) $O_2 + HCO$, and (5) $H_2O_2 + F$. However, none of these reactions are suitable for microwave specroscopy. *Evenson* suggested that the reaction of allyl alcohol with microwave-discharged oxygen might be the best source of HO_2 when the total sample pressure had to be maintained at a relatively low value, and this was, in fact, confirmed by *Saito* [3.116]. This method is, however, difficult to apply to DO_2, since fully deuterated allyl alcohol is not readily available; all hydrogen atoms in allyl alcohol have been known to participate in the reaction generating HO_2. Microwave spectroscopy of DO_2 was thus carried out using a dc glow discharge in a CH_3OD/O_2 mixture in 3.5 m long absorption cell [3.143].

As the case of HO_2 indicates, free radicals of relatively long lifetimes may be produced in a number of ways, and occasionally interfere with the detection of other new transient species. The CH_3 and CF_3 radicals are further such examples; they are prevalent in a number of chemical reaction systems.

As mentioned earlier, absorption spectroscopy like microwave and IR diode laser spectroscopy may be applied to the study of transient molecules only when these species are produced uniformly over the entire absorption cell. The situation is quite different in laser excitation and LMR spectroscopy, for which molecules need to be generated in a confined space in the cell. This difference requires different methods of producing transient species to be exploited for each spectroscopic method.

The HSO radical may be used to illustrate such a situation. The first identification of this radical was by *Schurath* et al. [3.156], who observed chemiluminescence emitted from a reaction system composed of H_2S, atomic oxygen, and O_3 and assigned it to HSO. *Kakimoto* et al. [3.70] found that the laser excitation spectrum of HSO can be observed with a good signal-to-noise ratio using the reaction of H_2S with microwave discharge products of O_2, O_3 being eliminated. The laser excitation spectrum was optimized when the partial pressure of H_2S was high enough, in comparison with that of O_2, to quench the SO_2 chemiluminescence, namely 40 mTorr and 10 mTorr for H_2S and O_2, respectively. These reaction conditions optimized for laser excitation spectroscopy could not be readily transferred to microwave spectroscopy as they were. They did not generate sufficient HSO molecules either in the 40 cm long parallel-plate cell or in the large free space cell for the microwave observation of the rotational spectrum. A dc discharge in a H_2S (15 mTorr) and O_2 (10 mTorr) mixture in the free space cell permitted *Endo* et al. [3.27] to observe the microwave spectrum of HSO.

The presence of reactants and products always interferes with the observation of spectra of transient molecules which are much less abundant. This problem again bears different features for different spectroscopic methods; it may be solved in favorable cases by simply increasing the pumping speed, but, in many cases, it is necessary to introduce other chemical systems.

The CF_3 radical is such an example. Its IR spectrum was successfully observed by using a simple ac discharge in CF_3I [3.151]. However, this method of generation was found to be inexpedient for the microwave study of CF_3, because the precursor CF_3I gives strong absorptions spread over several hundred MHz in every 3 GHz, making the observation of much weaker CF_3

spectra almost impossible. Several molecules containing CF_3 groups were then tested, using the IR spectrum of CF_3 as a monitor; $(CF_3CO)_2O$, CF_3COOCH_3, CF_3CH_2OH, and $CF_3C_6H_5$ were subjected to an ac discharge in an IR absorption cell. Of these molecules, $(CF_3CO)_2O$ and $CF_3C_6H_5$ were found to give a slightly stronger CF_3 IR spectrum than CF_3I, and finally $(CF_3CO)_2O$ was chosen because it produced much fewer interference lines than $CF_3C_6H_5$. A dc discharge in $(CF_3CO)_2O$ gave the microwave spectrum of CF_3 as reported in [3.152].

It is interesting and perhaps of practical importance to note that the ac (60 Hz) discharge in the cell has been successfully employed in IR diode laser spectroscopy, whereas it is hardly ever used anymore for microwave spectroscopy because it induces serious electrical disturbance in the detection system of the microwave spectrometer. A dc glow discharge is much preferred for the latter method.

Once a spectral line is found which is most likely to be due to the molecular species searched for the reaction conditions are optimized or some other new production methods are tried, by using the observed line as a monitor. These processes are explained using the SiF radical as a typical example [3.134]. A search for the microwave spectrum of SiF was first made by using a dc glow discharge in SiF_4 contained in the 3.5 m long free space absorption cell, as described in Sect. 3.1. Two paramagnetic lines were found and assigned to the $J = 7/2 - 5/2$ transition of SiF in the $^2\Pi_{3/2}$ state. Then, a small amount (2 mTorr) of SiH_4 was added to 15 mTorr of SiF_4, and when the mixture was subjected to a dc glow discharge, the signal intensity of SiF was increased by a factor of 2. Finally another factor of 2 was gained by the discharge in SiF_2 in the 1.1 m long free space cell which was generated by flowing SiF_4 over silicon powder at 1050 °C. This result shows that SiF seems to be more efficiently generated from SiF_2 than from SiF_4. However, this rule does not hold for SiCl [3.135]; the method of generating SiCl so far found to be most efficient is a simple dc glow discharge in $SiCl_4$ in the 3.5 m long cell.

A careful examination of chemical reactions is often rewarded with the unexpected discovery of a new and more efficient means to obtain the molecule. It is well known that OCS reacts with O to produce SO and SO_2 [3.4] and with F to generate SF [3.157] and SF_2 [3.158]. In the course of studying the FSO radical by microwave spectroscopy [3.118], we found a new method that produces SO and SF more efficiently than those known before. *Radford* et al. [3.159] suggested that the FSO radical could be synthesized by the reaction of CS_2 with microwave discharge products of a CF_4/O_2 mixture. We have found that OCS gives a stronger FSO spectrum than CS_2. We have further pursued the reaction system by using the microwave spectral lines of SO, SO_2, SF, SF_2, FSO, and F_2SO as monitors, and noticed that a small addition of CF_4 to O_2 increases the SO $(X^3\Sigma^-)$ line intensity by a factor of 10 and a small amount of O_2 added to CF_4 makes the SF $(X^2\Pi_i)$ lines several times stronger. A similar improvement in efficiency was observed in generating CF [3.29] and SiF [3.134]. A dc glow discharge in CF_4 contaminated by a small amount of CH_3F

resulted in a 10-fold increase in the CF spectral intensity from a dc discharge in pure CF_4.

As shown in Table 3.3, the same chemical reaction often produces two or more transient molecules: HCCN and CCN from $CH_3CN + (CF_4)_{MW\,disch.}$, CF, CHF, and CH_2F from $CH_3F + (CF_4)_{MW\,disch.}$, and PF and PF_2 from $(PH_3 + CF_4)_{MW\,disch.}$. The optimum partial pressures of CH_3CN and CF_4 are quite different for HCCN and CCN, whereas PF and PF_2 are produced under nearly the same conditions; the optimum partial pressures are 10 mTorr CH_3CN and 30 mTorr CF_4 for HCCN [3.127], 3 mTorr CH_3CN and 150 mTorr CF_4 for CCN [3.126], and 10 mTorr PH_3 and 10 mTorr CF_4 for both PF and PF_2 [3.139, 146]. The conditions for HCCN and CCN may be understood if one considers that the two molecules are generated by a series of stepwise hydrogen abstraction reactions of CH_3CN (or CH_2CN) with CF_4 microwave discharge products. The hydrogen abstraction reactions by fluorine atoms have been successfully employed in generating various transient molecules, as listed in Table 3.3.

Another interesting example of chemical systems which is worth mentioning concerns the reactions generating a few phosphorus-containing transient molecules including PO, PO_2, and HPO. Both PO and PO_2 are produced by the P(red) + $(O_2 + H_2)_{MW\,disch.}$ reaction, the abundance of PO_2 being increased with the increase of the O_2 partial pressure. An important fact is that H_2 is indispensable in generating both PO and PO_2; the oxygen atom does not seem to react with red phosphorus directly. This fact accounts for the failure of many attempts for over more than 20 years by microwave and gas-phase EPR spectroscopists to observe the rotational spectrum of PO; most of them used the P + O reaction. Replacing P(red) with PH_3 generates another unstable molecule HPO [3.121]. For further details of these reactions, see Sect. 5.1.

4. Individual Molecules

A number of free radicals and transient molecules have been investigated at the Institute for Molecular Science and a few other places using such high-resolution spectroscopic methods as described in Chap. 3, and their spectra thus observed have been analyzed in terms of theoretical expressions for energy levels, as expounded in Chap. 2, to derive molecular parameters which characterize these short-lived molecules well. This chapter is devoted to the results for individual molecules, most of which have been studied at the Institute for Molecular Science.

The transient molecules to be discussed are grouped into a few classes according to their structures: diatomic, linear, bent XYZ or XY_2, and symmetric and other polyatomic, and they are treated in the first four sections. The fifth section discusses the fine and hyperfine structures observed for these free radicals and the significance of the interaction constants in terms of electronic structures of the molecules. In most cases the molecules thus far examined are in ground electronic states. The last section, however, is devoted to excited electronic states, in particular metastable states.

4.1 Diatomic Free Radicals

4.1.1 $^2\Sigma$ and $^3\Sigma$ Diatomic Molecules

Diatomic molecules with an even number of electrons have ground electronic states which are either $^1\Sigma$ or $^3\Sigma$. When the sum of the group numbers of the two atoms is equal to 6 or 12, the electronic state is $^3\Sigma$, whereas it is $^2\Sigma$ for the sum of 3 or 9. Exceptions to these rules are molecules containing transition elements. The present subsection is primarily devoted to the diatomic molecules investigated by the author's group.

CS ($X\,^1\Sigma^+$). Because of its relatively long lifetime, the CS molecule served as a test molecule in applying IR tunable diode laser spectroscopy to transient molecules. *Yamada* and *Hirota* [4.1] have observed the vibration-rotation bands of $^{12}C^{32}S$, $^{12}C^{34}S$, $^{12}C^{33}S$, and $^{13}C^{32}S$ and the $v = 2 - 1$ hot band of the normal species. Figure 4.1 shows a part of the observed spectrum. *Todd* and *Olson* [4.2] made a similar observation at nearly the same time, by generating

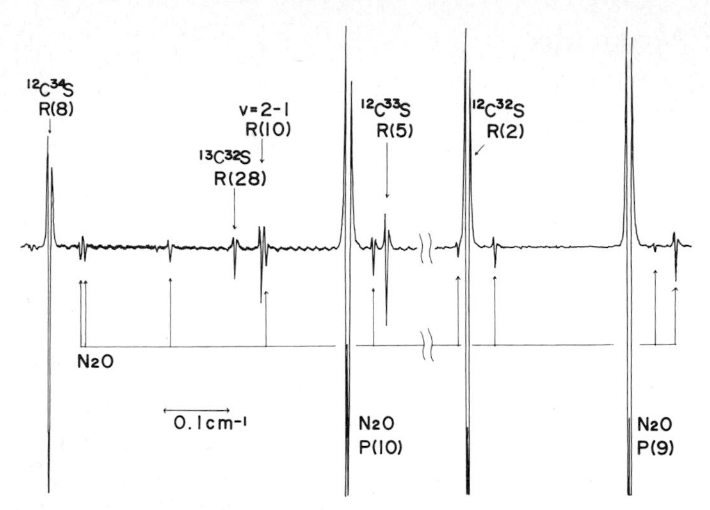

Fig. 4.1. A part of the IR diode spectrum of CS

CS by the discharge in CS_2. *Yamada* and *Hirota* have analyzed all the observed band origins by assuming isotopic relations to yield the harmonic frequency and the anharmonicity constant of $^{12}C^{32}S$ to be $\omega_e = 1285.1392\,(40)$ cm^{-1} and $\omega_e x_e = 6.4864\,(20)$ cm^{-1}, respectively.

SiN ($X\,^2\Sigma^+$). It is an analog of the well-known $^2\Sigma$ radical CN, and is of some interest not only for spectroscopy but also for astronomy [4.3]. In 1913 *Jevons* [4.4] reported the first observation of the SiN spectrum in emission, which was induced by introducing $SiCl_4$ in active nitrogen afterglow. *Mulliken* [4.5] has substantiated that the emitter of this spectrum is SiN on the basis of the silicon isotope shifts of the observed spectra. Subsequently, many spectroscopic studies have been carried out, and the band *Mulliken* observed is now known as the $B\,^2\Sigma^+ - X\,^2\Sigma^+$ system. *Bredohl* et al. [4.6] investigated the *B-X* system and have determined ground-state molecular constants. Besides $B \rightarrow X$ they also observed and analyzed much weaker bands, $D\,^2\Pi \rightarrow A\,^2\Pi$, $L^2\Pi \rightarrow A\,^2\Pi$, and $K\,^2\Sigma \rightarrow A\,^2\Pi$, but the *B-A* and *A-X* systems, which are well known for CN, have not been detected, and the locations of the $A\,^2\Pi$, $D\,^2\Pi$, $K\,^2\Sigma$, and $L^2\Pi$ states relative to the ground state have remained rather uncertain.

Saito et al. [4.7] have recently succeeded in observing the microwave spectrum of SiN in a dc glow discharge in a $SiCl_4/N_2$ or SiH_4/N_2 mixture; three rotational transitions of $N = 2-1$, $3-2$, and $4-3$ were recorded in 87–175 GHz region. Figure 4.2 reproduces the $N = 4 \leftarrow 3$, $J = 7/2 \leftarrow 5/2$ transition of SiN, exhibiting well-resolved hyperfine structure. Table 4.1 lists B_0, D_0, and γ_0 derived from the observed spectrum, along with the optical values [4.6] for comparison. The hyperfine constants are given below in Table 4.20.

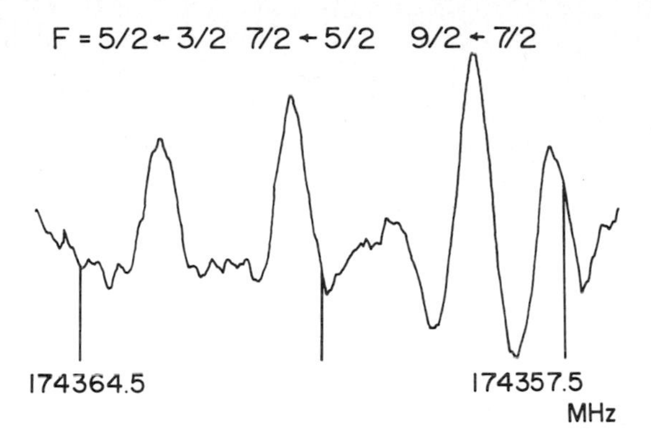

Fig. 4.2 The $N = 4 \leftarrow 3$, $J = 7/2 \leftarrow 5/2$ transition of SiN [4.7]

Yamada and *Hirota* [4.8] have recently detected three different chemical species in a SiH_4/N_2 electrical discharge plasma, by means of IR diode laser spectroscopy. One of them has been assigned to SiH on the basis that the observed spectral lines coincided with those previously reported [4.9]. The second species has recently been identified to be SiN; its 5 μm spectrum is the *A-X* system searched for by many people, but not found heretofore. Figure 4.3 shows a part of this system. All the observed lines are extremely Zeeman sensitive, as expected from the different Zeeman behavior of the two states. Because the upper-state constants of the 5 μm band agree with those *Bredohl* et al. derived for the A $v = 1$ state, this band is assigned to $v = 1 \leftarrow 0$ of the *A-X* system. Some lines which are attributed to the $v = 0 \leftarrow 0$ band are observed in the 10 μm region. When the observed $v = 1 - 0$ band origin 1972.44370(24) cm^{-1} is combined with the ω_e, $\omega_e x_e$, and $\omega_e y_e$ values of *Bredohl* et al., the electronic term value of the A state is yielded to be $T_{00} = 940.40 \pm 0.12$ cm^{-1}.

Table 4.1. Molecular constants of the SiN radical in the $X\,^2\Sigma^+$ state (MHz)[a]

	MW		optical[b]	
B_0	21 827.7987	(113)	21 827.9	(75)
D_0	0.03577	(42)	0.036	
γ_0	505.109	(43)	516	(105)

[a] Values in parentheses denote 2.5 times standard deviation and apply to the last digits of the constant

[b] [4.6]

The term value of the A state has previously been estimated by *Linton* [4.10] and *Bredohl* et al. [4.6] to be 2500–4000 cm^{-1} and 8000 cm^{-1} from perturbations observed in the *D-X* and *B-X* systems, respectively. *Foster* [4.11] has recently reinterpreted the $K\,^2\Sigma \rightarrow A\,^2\Pi$ band to be the $B\,^2\Sigma^+ \rightarrow A\,^2\Pi$, in order

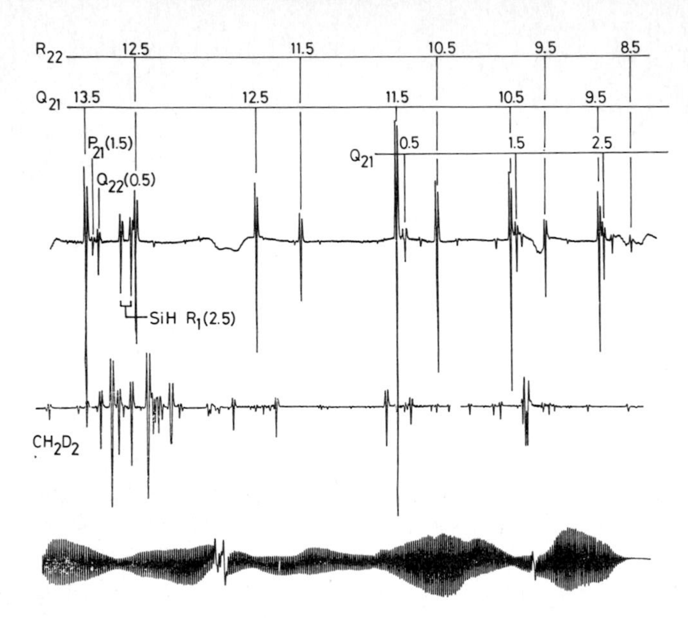

Fig. 4.3. A part of the SiN $A^2\Pi_{1/2} \leftarrow X^2\Sigma^+$ system in the 2020 cm^{-1} region [4.8]

to explain the perturbations mentioned above, and has obtained a term value for the *A* state that coincides with that of *Yamada* and *Hirota.*

N_2^+ ($X\,^2\Sigma_g^+$). It is rather surprising that only a limited number of studies have been reported on the $A\,^2\Pi_u - X\,^2\Sigma_g^+$ transition of N_2^+. *Meinel* [4.12] made the first observation of the *A-X* band in aurora glow (and thus this band is called the Meinel band), and *Douglas* [4.13] has rotationally analyzed the band. The resolution available to *Douglas* was 0.25 Å, and his data have remained to be the only source of information on the $A\,^2\Pi_u$ state of N_2^+. *Benesch* et al. [4.14] published extensive tables of line positions in the Meinel bands, but without rotational analysis. *Cook* et al. [4.15] reported the first LIF (laser-induced fluorescence) spectrum of the *A-X* band, but the pulsed dye laser employed had a spectral resolution of 1 cm^{-1} which is insufficient to resolve the rotational structure. In a recent paper, *Grieman* et al. [4.16] reported an application of a different type of laser excitation experiment on N_2^+, i.e., they detected the absorption lines not by LIF but by a difference in charge-exchange cross section between the two electronic states. The cw dye laser employed was also of about 1 cm^{-1} bandwidth, and hence no rotational analysis was performed. *Hansen* et al. [4.17] subsequently improved the resolution of the experiment by reducing the bandwidth of the dye laser to 0.1 cm^{-1}. The molecular constants derived from the least-squares analysis of the observed spectrum show a modest improvement over the earlier result.

Miller et al. [4.18] investigated the Meinel band by LIF with a cw dye laser of less than 0.002 cm^{-1} bandwidth as a source. The spectroscopic system employed is identical to that described in Sect. 3.3. They generated N_2^+ by

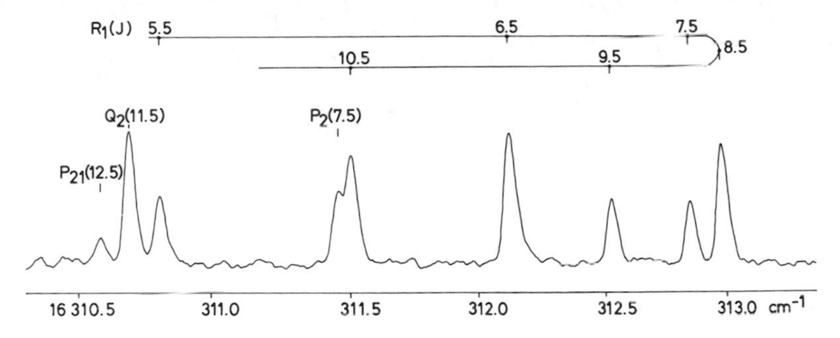

Fig. 4.4. Portion of LIF spectrum of the $A\,^2\Pi_u - X\,^2\Sigma_g^+$ Meinel system of N_2^+ [4.18]

Penning ionization of N_2 by He metastables. By passing the dye laser beam through the tip of the reaction zone, 173 lines were observed for the *A-X* $v = 4 - 0$ band in the region of 16210–16374 cm^{-1}. The observed spectrum is of Doppler-limited resolution, but this high resolution, combined with the fact that they measured over twice as many lines as *Hansen* et al., allowed Miller et al. to determine molecular constants for the $A\,^2\Pi_u$ state 5–10 times more precisely. Figure 4.4 illustrates a part of the spectrum they observed. Note that the ground-state parameters have hitherto been derived from the *B-X* system in the violet, not from *A-X* in the red. Miller et al. have thus achieved a slightly better resolution, although it is limited by the Doppler effect. They have much improved the precision of the spin-rotation coupling constant in the ground

Table 4.2. Equilibrium constants of N_2^+ for the $X\,^2\Sigma_g^+$ and $A\,^2\Pi_u$ states (in cm^{-1})

	$X\,^2\Sigma_g^+$	$A\,^2\Pi_u$
ω_e	2207.27_5	1903.45_8
$\omega_e x_e$	16.26_0	15.01_7
T_e		9167.62_0
B_e	$1.931\ 8_5$	$1.744\ 4_0$
α_e	$0.019\ 0_6$	0.0188_9

state; this constant is difficult to determine from the *B-X* system, because it is a $\Sigma - \Sigma$ transition. Miller et al. also derived molecular parameters for the $A\,^2\Pi_u$ state that are orders of magnitude more precise than those reported previously. In particular, two Λ-doubling constants p and q have been determined to three to four significant figures. To determine the equilibrium molecular parameters, Miller et al. analyzed the data of Benesch et al., Table 4.2.

SO ($X\,^3\Sigma^-$). As mentioned in Sect. 3.1.2, SO was investigated by microwave spectroscopy as early as 1964 [4.19]. Since then many studies have been carried

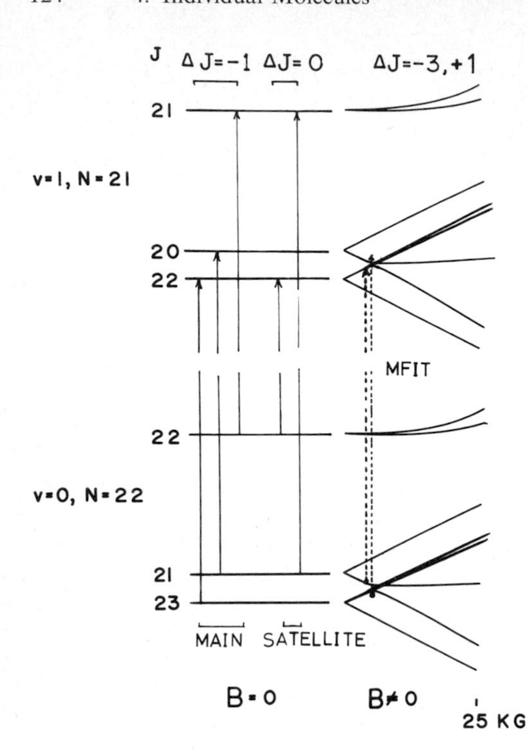

Fig. 4.5. Energie levels with $v = 1$, $N = 21$, and $v = 0$, $N = 22$ of SO in $X^3\Sigma^-$ and their Zeeman effects. Various types of transitions are indicated [4.22]

out on this molecule using microwave and electron paramagnetic resonance spectroscopy, resulting in precise molecular constants for various isotopic species not only in the ground state, but also in excited states up to $v = 8$ [4.20]. It was found in interstellar space [4.21].

Kawaguchi et al. [4.22] have applied mid-IR LMR to observe the vibration-rotation bands. As shown in Fig. 4.5, two types of transitions are allowed at zero magnetic field; those with $\Delta N = \Delta J = -1$ are referred to as main, while those with $\Delta N = -1$ and $\Delta J = 0$ are satellite. The former have large transition moments, but are weakly tuned by the external magnetic field, so that they are rarely observed by LMR. In contrast, the latter appear much more frequently in the LMR spectrum. In addition to these, the external magnetic field induces other types of transitions because of avoided crossing, Fig. 4.5; they are either of $\Delta N = -1$, $\Delta J = -3$ or of $\Delta N = -1$, $\Delta J = 1$ type. Some $\Delta N = -3$ transitions were also identified, which were allowed by the spin-spin interaction of $\Delta N = \pm 2$ type.

Wong et al. [4.23] recently observed the $v = 3 \leftarrow 0$ transition of SO by difference frequency laser spectroscopy and obtained molecular parameters in the $v = 3$ state, which, when combined with the mid-IR LMR data, lead to the harmonic frequency, the vibrational anharmonicity constants, and the equilibrium rotational, spin-spin and spin-rotation interaction constants. A recent excimer SO_2 photolysis experiment by *Kanamori* et al. [4.24] provided additional data on the vibration-rotation bands of v up to 6-5, Sect. 5.1.

NCl ($X\,^3\Sigma^-$). The first spectroscopic detection of the NCl radical was made by *Milligan* [4.25], who photolyzed chlorine azide trapped in an argon matrix at 4.2 K and observed two IR bands at 824 and 818 cm^{-1} with an intensity ratio of 2.5–2.7, which were assigned to $N^{35}Cl$ and $N^{37}Cl$, respectively. Later *Milligan* and *Jacox* [4.26] extend the measurement and observed two more bands at 816.5 and 810 cm^{-1}, which were also ascribed to the two isotopic species, but trapped in a different site of the matrix. *Briggs* and *Norrish* [4.27] reported transient absorption spectra at 2400 Å, by flash photolyzing a gaseous mixture of NCl_3, Cl_2, and N_2, and ascribed them to the NCl $^3\Pi \leftarrow X\,^3\Sigma^-$ transition based on photochemical arguments. By assigning the 2400 Å spectra to the (0, 0) band and the nearby continuous spectra to the (2, 0), (1, 0), and (0, 1) bands, they obtained the vibrational frequency as 870 and 550 cm^{-1}, respectively, for the upper and lower states. *Colin* and *Jones* [4.28] have photographed an orange afterglow downstream from a microwave discharge in a mixture of N_2 and Cl_2. The spectrum they observed consists of double-headed bands, with the strongest band at 6684.6 Å. The rotational analysis of this band assigned to (0, 0) showed that the observed spectrum was due to the $b\,^1\Sigma^+ - X\,^3\Sigma^-$ transition of NCl. From this analysis, they derived the vibrational and rotational constants of the upper and lower states. The lower state $\Delta G''(1/2)$ value 816.8 cm^{-1} is in good agreement with the result of Milligan and Milligan and Jacox, but not with that of Briggs and Norrish.

Yamada et al. [4.29] recently observed many paramagnetic absorption lines near 820 cm^{-1} in a glow discharge in a N_2/Cl_2 mixture, induced directly in a cell by IR diode laser spectroscopy. Before analyzing these spectral lines, *Yamada* et al. [4.30] observed microwave spectra to determine ground-state parameters precisely. The method of generating NCl was similar: a dc glow discharge in a 1:1 mixture of N_2 and Cl_2 with a total pressure of 15–20 mTorr. The efficiency of the production was rather insensitive to the dc discharge current in the range 30–50 mA. Because both N and Cl have nonzero nuclear spins, the microwave spectrum exhibits very complicated hyperfine structures. Figure 4.6 shows the $N = 4 \leftarrow 3$ $J = 3 \leftarrow 2$ transition split into hyperfine com-

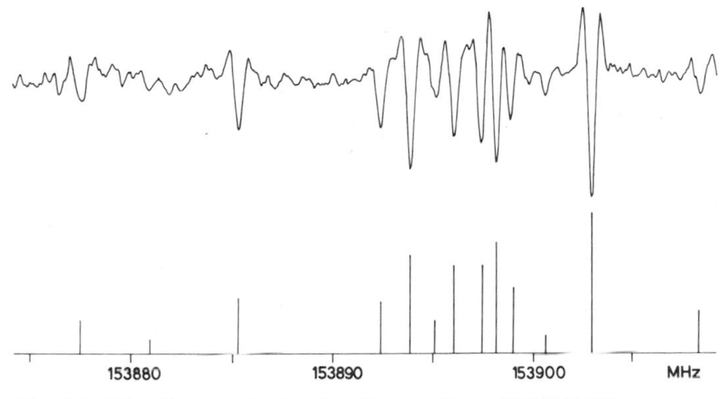

Fig. 4.6. The $N = 4 \leftarrow 3$, $J = 3 \leftarrow 2$ transition of NCl [4.30]

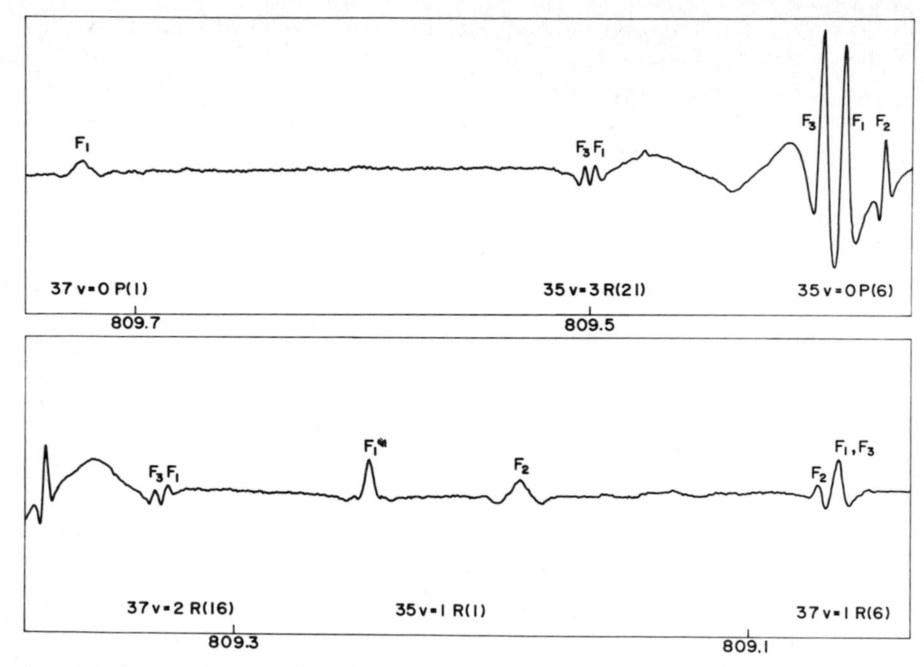

Fig. 4.7. A part of the IR diode laser spectrum of NCl in the 809 cm^{-1} region [4.29]

ponents. They observed 130 lines assigned to 11 electric-dipole rotation/fine-structure transitions, which were least-squares analyzed using the Hamiltonian described in Sect. 2.3 to yield molecular constants in the ground vibrational state. The hyperfine constants obtained are discussed in Sect. 4.5.

Using the ground-state parameters thus determined, *Yamada* et al. [4.29] analyzed the diode laser spectra. It was straightforward to assign the fundamental band of $^{14}N^{35}Cl$, but many lines presumably due to $^{14}N^{37}Cl$ and to hot bands were left unassigned. A part of the observed spectrum is reproduced in Fig. 4.7. Finally they assigned the $v = 1 \leftarrow 0$, $2 \leftarrow 1$, $3 \leftarrow 2$, and $4 \leftarrow 3$ bands of $^{14}N^{35}Cl$ and the $v = 1 \leftarrow 0$, $2 \leftarrow 1$, and $3 \leftarrow 2$ bands of $^{14}N^{37}Cl$. In making these assignments, isotope relations proved to be of great help, but Yamada et al. noted a slight breakdown of the relations, which has prompted them to examine critically the microwave spectra of isotopic species in excited vibrational states; $v = 1, 2$ and 3 for $^{14}N^{35}Cl$ and $v = 0$ and 1 for $^{14}N^{37}Cl$. All these observed spectra corrected for hyperfine structures were fitted to expressions for the vibration-rotation and fine-structure energies in terms of Dunham-type coefficients. The breakdown of Born-Oppenheimer separation has been established for $Y_{10}(\sim \omega_e)$, $Y_{01}(\sim B_e)$, λ_e, and γ_e.

Watson [4.31, 32] has given the following expression for the Dunham coefficient, which takes into account the breakdown of the isotope relation:

$$Y_{ij} = \mu^{-(i+2j)/2} U_{ij} [1 + m_e \Delta_{ij}^{A}/M_A + m_e \Delta_{ij}^{B}/M_B], \tag{4.1.1}$$

where μ stands for the reduced mass given by $M_A M_B/(M_A + M_B)$ with M_A and M_B denoting the masses of the atoms A and B, respectively, U_{ij} for an isotope invariant constant, m_e for the electron mass, and Δ_{ij}^A and Δ_{ij}^B represent the correction factors for the breakdown. A similar relation may hold for the spin-spin and spin-rotation interaction constants. The Δ factors thus determined are 0.095(23) and 0.0993(88) for ω_e and B_e, respectively, where the values in parentheses denote 2.5σ.

PF($X\,^3\Sigma^-$). *Douglas* and *Frackowiak* [4.33] observed an emission from a microwave discharge in PF_3 in the region 2200–5000 Å, and assigned a band at 3300 Å to the $B\,^3\Pi \rightarrow X\,^3\Sigma^-$ transition of the PF radical. From an analysis of the observed spectrum, they determined the rotational and the spin-spin and spin-rotation interaction constants and the vibrational frequency in the $X\,^3\Sigma^-$ state, together with molecular parameters in the upper electronic state. *Colin* et al. [4.34] identified another emission band $b\,^1\Sigma^+ \rightarrow X\,^3\Sigma^-$ around 7450 Å.

Saito et al. [4.35] have investigated this radical by microwave spectroscopy. As described in Sect. 3.5, the dc glow discharge in a PH_3/CF_4 mixture is one of the most efficient ways of generating the radical; only slow pumping is required for the reaction mixture to flow though a cell, presumably because the reaction is rather slow and the radical is long lived. The PF_2 radical has also been identified in the same reaction system. The PF spectrum has been analyzed using the Hamiltonian matrix elements from *Ryzlewicz* et al. [4.36], including those of the hyperfine interactions of the two nuclear spins, which were derived using Hund's case $(b)_{\beta J}$ functions as bases, to yield precise molecular constants. The hypefine parameters thus derived are discussed in Sect. 4.5.

PCl($X\,^3\Sigma^-$). *Huber* and *Herzberg* [4.37] have compiled two bands observed for PCl; one is the $B \leftarrow X$ diffuse band at 41 234 cm^{-1} and the other the $b\,^1\Sigma^+ \rightarrow X\,^3\Sigma^-$ emission band at 12 087 cm^{-1}. The harmonic frequency in the ground state has been reported to be $\omega_e = 577$ cm^{-1} with the anharmonicity $\omega_e x_e = 3.5$ cm^{-1}.

Minowa et al. [4.38] have recently detected the PCl radical by microwave spectroscopy and derived molecular parameters including the b and c hyperfine coupling constants for both P and ^{35}Cl nuclei in the ground vibronic state. A dc discharge in PCl_3 produced PCl. *Kanamori* et al. [4.39] also investigated this radical by using IR tunable diode laser spectroscopy. They found that adding H_2 to PCl_3 increases the generation efficiency. Source frequency modulation (Sect. 3.2.1) was employed to observe absorption lines, 111 in total, which were assigned to $v = 1 \leftarrow 0$ (60 lines) and $2 \leftarrow 1$ (30 lines) of $P^{35}Cl$ and to $v = 1 \leftarrow 0$ (21 lines) of $P^{37}Cl$. A combined analysis of the microwave and IR data yields molecular constants of high precision. The hyperfine constants are $b(\mathrm{Cl}) = 19.22(6)$, $c(\mathrm{Cl}) = -37.7(2)$, $eQq(\mathrm{Cl}) = -37.0(8)$, $b(\mathrm{P}) = 274.47(7)$, and $c(\mathrm{P}) = -480.3(2)$, in MHz with standard errors in parentheses.

PH, PD($X\,^3\Sigma^-$). The $A\,^3\Pi_i - X\,^3\Sigma^-(0,0)$ band of PH and PD was observed in emission by *Ishaque* and *Pearse* [4.40], who generated the radicals by a continuous discharge in a H_2 or D_2 and phosphorus mixture. *Legay* [4.41] analyzed the (1, 0) band of PH photographed in absorption in a flash photolysis of phosphine. *Rostas* et al. [4.42] rotationally analyzed six bands of $A - X$ including the (0, 1) band of PH and the (1, 0) and (1, 1) bands of PD. *Davies* et al. [4.43, 44] observed and analyzed the $N = 5 \leftarrow 4$ rotational transition of PH in the ground vibronic state, in addition to the spectrum in the $a^1\Delta$ state, by FIR LMR. *Uehara* and *Hakuta* [4.45] reported observing the vibration-rotation spectrum of PD in $X\,^3\Sigma^-$ by CO laser LMR.

Table 4.3. Equilibrium structure and potential constants of PH and PD in $X\,^3\Sigma^-$

Fitted parameter		Derived parameter	
B_0(PH)	8.412 524 (21) cm^{-1}	ω_e(PH)	2366.79 (16) cm^{-1}
B_0(PD)	4.362 8675 (77)	ω_e(PD)	1700.35 (12)
B_1(PD)	4.269 248 (65)		
		B_e^{ad}(PH)	8.532 7 (14)
ν_0(PH)	2276.2 (1)	B_e^{ad}(PD)	4.408 02 (58)
ν_0(PD)	1653.284 91 (51)		
		r_e^{BO}	1.421 40 (22) Å
		Δ_r^H	1.70 (39)
		a_1	−2.3797 (18)[a]
		a_2	3.461 (14)[a]

[a] Anharmonic potential constants

Ohashi et al. [4.46] extended the FIR LMR measurements to PD and also to PH. They observed more than 60 resonances of PD for 12 laser lines, assigned to both $v = 0$ as well as $v = 1$. Each resonance was found to be split into two components by the ^{31}P hyperfine interaction, and two resonances showed further splittings due to the D nucleus. For PH five laser lines were employed to observe the $N = 2 \leftarrow 1$, $5 \leftarrow 4$, and $7 \leftarrow 6$ transitions. Molecular constants of the two isotopic species thus obtained were analyzed to obtain equilibrium constants. Three types of corrections were applied to the observed B_0 constants to derive the Born-Oppenheimer equilibrium internuclear distance r_c^{BO}: (1) the Dunham correction, (2) the electron slipping correction, and (3) the adiabatic correction, summarized in Table 4.3. *Anacona* et al. [4.47] recently measured the $v = 1 \leftarrow 0$ band of PH by IR tunable diode laser spectroscopy and improved the band origin by a factor of 30, namely 2276.2106 (29) cm^{-1}. Since this value agrees well with the old one listed in Table 4.3, the derived parameters need not be revised. It must, however, be noted that if the new band origin of PH is used to its full accuracy, a new Born-Oppenheimer correction factor must be included.

AsH ($X\,^3\Sigma^-$). Only one high-resolution study of AsH has been reported; *Dixon* and *Lamberton* [4.48] have investigated the $A\,^3\Pi_i - X\,^3\Sigma^-$ system observed in

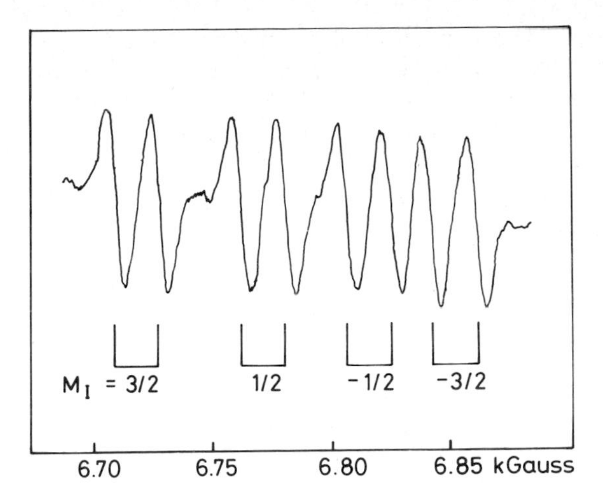

Fig. 4.8. FIR LMR spectrum of the AsH $N,J\ M_J = 3,4,2 \leftarrow 2,3,1$ transition [4.49]

absorption of AsH and AsD generated by flash photolysis of AsH_3 and AsD_3, respectively. They have photographed and rotationally analyzed the (0, 0) and (1, 0) bands. Although they noted that the $^3\Pi$ state was perturbed, they succeeded in deriving molecular parameters of the ground vibronic state and some effective values for the upper state.

Kawaguchi and *Hirota* [4.49] observed five rotational/fine-structure transitions of AsH in the region 639–1799 GHz by FIR LMR, and obtained precise molecular constants including hyperfine coupling constants for the ground vibronic state. As described in Sect. 3.5, the radical was generated by the reaction of microwave discharge products of a H_2/O_2 mixture with metallic arsenic placed near the center of the magnet pole caps. It is interesting to note that oxygen is necessary to initiate the reaction, but later the reaction is sustained without oxygen. Each resonance is split into eight hyperfine components spread over 200 Gauss, which are ascribed to hyperfine interactions of both As and H nuclei, the interaction being larger for As than for H. The nuclear electric quadrupole interaction makes the separations between successive hyperfine components unequal, as shown in Fig. 4.8. The hyperfine parameters derived are $\alpha(\mathrm{As}) = -11.5(14)$, $\beta(\mathrm{As}) = -159.4(13)$, $eQq(\mathrm{As}) = -97.6(72)$, $\alpha(\mathrm{H}) = -49.80(66)$, and $\beta(\mathrm{H}) = 4.15(60)$, in MHz with three standard errors in parentheses. The α and β constants are related to Frosch-Foley parameters as $\alpha = b + c/3$ and $\beta = c/3$.

4.1.2 $^2\Pi$ Diatomic Molecules

Hund's coupling case (a) is a good approximation of a $^2\Pi$ state, which is split into two spin states $^2\Pi_{1/2}$ and $^2\Pi_{3/2}$ by the spin-orbit interaction. The case of a positive spin-orbit coupling constant A is referred to as regular and a negative A as inverted; the $^2\Pi_{3/2}$ state is higher and lower, respectively, for the regular and inverted cases.

Diatomic molecules with the $^2\Pi$ electronic ground states may be formed from the following combinations of atoms: I–IV (regular), I–VI (inverted), IV–VII (regular), V–VI (regular), and VI–VII (inverted), where I, IV, V, ... denote the groups in the periodic table. Positive ions may be obtained by similar combinations: I–V$^+$, I–VIII$^+$, and so on.

Most spin-orbit coupling constants hitherto reported have been determined by observing electronic spectra like the $\Sigma - \Pi$ transition. The direct magnetic transition between the $^2\Pi_{3/2}$ and $^2\Pi_{1/2}$ states has recently been observed in the IR or FIR regions for NO, BrO, AsO, SeH, and SeD. When the A constant is not much larger than the B rotational constant, as in many hydrides, the spin uncoupling mixes the $^2\Pi_{1/2}$ and $^2\Pi_{3/2}$ states to a considerable extent, endowing electric dipole character to the transition between the two resulting states. Otherwise, the A constants determined by optical spectroscopy are assumed in analyzing IR and microwave spectra; the uncertainty in A reflects on the accuracy of the centrifugal correction term for the spin-rotation interaction A_J (and γ) among others.

Dixon and *Kroto* [4.50] have shown that the spin-orbit coupling constant of a diatomic molecule XY is approximately given in terms of those of the composite atoms as follows:

$$A_{XY} = C_X^2\,|A_X| + C_Y^2\,|A_Y|, \tag{4.1.2}$$

where C_X and C_Y denote the coefficients of the LCAO molecular orbital:

$$(\pi)_{XY} = C_X(p\pi_X) + C_Y(p\pi_Y). \tag{4.1.3}$$

Conversely, the observed A_{XY} constant may be used to estimate C_X or C_Y, i.e., the spin densities on X and Y.

As discussed in Chap. 2, each rotational level in a $^2\Pi$ state is split into two by Λ-type doubling; the splitting is larger in $^2\Pi_{1/2}$ than in $^2\Pi_{3/2}$ for low rotational levels. In fact, when the pure precession model is assumed, the p and q Λ-doubling constants, which roughly measure the splittings of the two spin states, are given by

$$p = (-1)^s 4AB/\Delta E, \qquad q = (B/A)\,p, \tag{4.1.4}$$

where only one Σ state is assumed to contribute to the Λ doubling, and ΔE denotes its excitation energy and s is 0 or 1 according to whether the Σ state is Σ^+ or Σ^-. As mentioned in Chap. 2, the sign of p is correlated with that of the hyperfine d constant. Since the d constant is always positive for a molecule with one unpaired electron, the sign of p is uniquely determined when the hyperfine structure is resolved for the Λ-type doublets and may be employed to guess the parity of the perturbing states.

The magnetic hyperfine coupling constants are extremely valuable in obtaining information on the unpaired electron distribution in the molecule. This sort of data is really characteristic of free radicals, as pointed out in Chap. 1.

There are two types of unpaired electron distributions which must be clearly distinguished. One is the orbital density provided by the hyperfine a constant, and the other the spin density which may be calculated from $d + c/3$. Furthermore, the s character of the unpaired electron orbital may be obtained from $b + c/3$. The hyperfine interaction constants are discussed in Sect. 4.5.

Hydrides. Pure rotational transitions have been mainly observed by FIR laser magnetic resonance (FIR LMR), the direct transitions between Λ doublets by microwave spectroscopy, and the vibration-rotation transitions by IR laser spectroscopy. The molecular species so far investigated by these methods include OH, OD, SH, SD, SeH, SeD, CH, and SiH, whereas HBr^+, DBr^+, HCl^+, DCl^+, and HF^+ are the only ionic species so far inverstigated by FIR LMR. Here the CH radical is employed to illustrate the high-resolution spectra of hydride radicals.

The high-resolution spectrum of CH was first observed in space by means of optical spectroscopy. After many trials failed to detect its rotational and/or Λ-doublet spectra, *Evenson* et al. [4.51] finally succeeded in observing several rotational transitions of CH in the FIR region by using FIR LMR with a H_2O laser as a source. The subsequent development of the optically pumped FIR laser has enabled *Hougen* et al. [4.52] to extend the LMR measurements, which were followed by a recent study of *Brown* and *Evenson* [4.53].

The spin-orbit coupling constant A of CH happens to be very close to $2B$, and thus its rotational level structure is well approximated by Hund's coupling case (b), i.e., well expressed by a constant times $N(N+1)$, where N denotes the rotational quantum number of an integer. The FIR LMR study in [4.53] measured the rotational transitions up to $N = 5 \leftarrow 4$. The radical was generated by the reaction of CH_4 with microwave discharge products of a F_2 and He mixture. The spectrum was quite strong, and the hyperfine structure was well resolved and often observed as Lamb dips.

The observed FIR LMR spectrum was combined with the microwave data previously observed by radioastronomy [4.54] and was subjected to least-squares analysis to yield 13 zero-field parameters and 6 g values. The accuracy of the LMR measurement was mainly determined by the uncertainty in the laser frequency which was less than 2 MHz. The standard error of the fit to be compared with was 2.2 MHz, except for the spectrum obtained using the 70.5 μm CH_3OH laser line; the origin of this poor fitting is not clear.

The Λ-doubling spectrum of CH was observed for the first time by a radiotelescope in 1974 [4.54]. It was recently observed in the laboratory by *Brazier* and *Brown* [4.55] using a microwave-optical double resonance technique and also by *Bogey* et al. [4.56], who employed an *rf* discharge in various gas mixtures in detecting relatively high J transitions of CH.

IV–VII Molecules. The CF radical has been extensively investigated by optical spectroscopy, as compiled by *Huber* and *Herzberg* [4.37]. Recently its vibrational and rotational transitions were observed by IR diode laser [4.57],

FIR LMR [4.58], FIR laser side band [4.59], and microwave spectroscopy [4.60].

Kawaguchi et al. [4.57] generated the CF radical by a high-voltage 60 Hz discharge in CF_4 or C_2F_4 contained in an absorption cell and observed the fundamental vibration-rotation band in the region 1250–1332 cm^{-1}. To avoid interference by the strong ν_3 band of CF_4 at 1284 cm^{-1} (CF_4 was found to be easily produced from C_2F_4 when discharged), Zeeman modulation was employed not only for $^2\Pi_{3/2}$, but also $^2\Pi_{1/2}$. An example of the observed spectrum is displayed in Fig. 4.9. It may be worth noting that when the spin-orbit coupling constant A is not too large in comparison with the B rotational constant, mixing takes place between $^2\Pi_{1/2}$ and $^2\Pi_{3/2}$, making the former slightly paramagnetic. In fact, (2.5.31) leads to the g_J factor for an intermediate between Hund's case (a) and case (b) as follows:

$$g_J = \{4g_L + g_S \pm [2(g_L + g_S)(Y-2) - 4g_S(J+3/2)(J-1/2)]/ [4(J+1/2)^2 + Y(Y-4)]^{1/2}\}/[4J(J+1)], \tag{4.1.5}$$

where $Y = A/B$. The + and − signs apply to the $^2\Pi_{3/2}$ and $^2\Pi_{1/2}$ states, respectively. The Zeeman energy is given by $E_Z = g_J \beta M_J H$, which leads to the Zeeman coefficient of R (3.5), $M_J = 4.5 \leftarrow 3.5$, for example, to be 7.47×10^{-3} and 1.8×10^{-3} cm^{-1}/kG, respectively, for $^2\Pi_{3/2}$ and $^2\Pi_{1/2}$. The Zeeman shift thus becomes comparable with the Doppler linewidth of 0.0029 cm^{-1}, when the magnetic field is as high as 0.39 kG and 1.6 kG for the two spin states. The modulation field of 800 G peak-to-peak normally employed for diode laser spectroscopy (Sect. 3.2.1) is large enough to modulate the $^2\Pi_{3/2}$ transition and to modulate the $^2\Pi_{1/2}$ transition partially. The observed relative intensity reflects these situations, as exemplified in Fig. 4.9; each Λ-type doublet component of $^2\Pi_{1/2}$ is expected to be 0.73 times as intense as the $^2\Pi_{3/2}$ line which is not resolved by Λ doubling when the Boltzmann distribution is assumed.

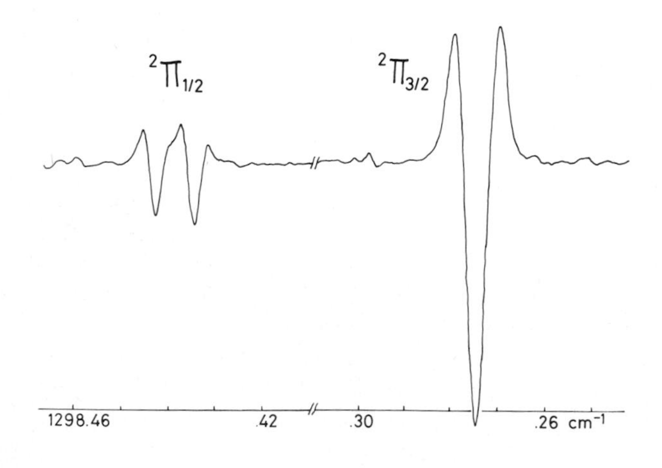

Fig. 4.9. The R(3.5) transition of CF recorded with Zeeman modulation [4.57]

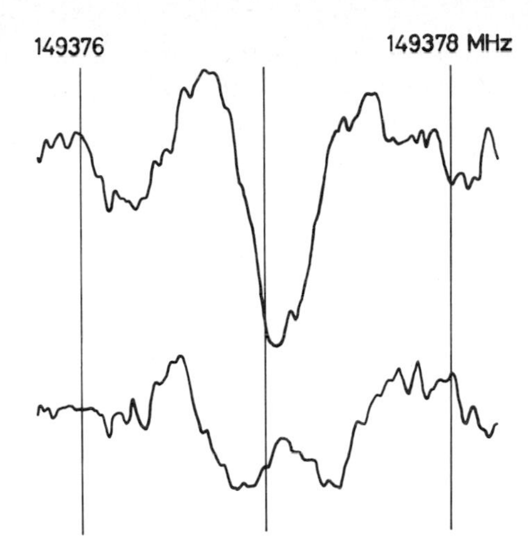

Fig. 4.10. The $J = 4.5 \leftarrow 3.5$, $F = 5 \leftarrow 4$, e transition of SF in $X\,^2\Pi_{1/2}$. The lower trace was obtained with a magnetic field of 60 G [4.62]

When $Y \gg 1$, Zeeman modulation cannot be applied to $^2\Pi_{1/2}$. For CCl, for example, the $^2\Pi_{1/2}$ lines were observed by source modulation [4.61]. High J lines of $^2\Pi_{3/2}$ are also difficult to modulate, since the Zeeman coefficient is inversely proportional to J^3. The Zeeman effect thus often provides us with qualitative information on the J value, even when the Zeeman components are not completely resolved. In microwave spectroscopy Zeeman effects have been employed in discriminating paramagnetic lines of a free radical from many diamagnetic lines of reactants and products. This technique was successfully applied even to the SF radical for which Y is as large as 700; the Zeeman coefficient of the $J = 3.5 \leftarrow 2.5$ transition in $^2\Pi_{1/2}$ is about 0.45 MHz/60 G. Because the linewidth is of the order of 0.2 MHz in the microwave region, the Zeeman effect was in fact of great use in picking up SF lines [4.62]. An example of the Zeeman effect is illustrated in Fig. 4.10.

The CF radical was also subjected to a microwave study by *Saito* et al. [4.60], who found an efficient way of producing CF, namely a dc discharge in CF_4 with a small amount of CH_3F added (Sect. 3.5). They evaluated molecular constants including hyperfine coupling constants both for the $^2\Pi_{1/2}$ and $^2\Pi_{3/2}$ spin states. Other IV–VII type molecules studied by microwave spectroscopy include CCl [4.63], SiF [4.64], SiCl [4.65], and GeF [4.66], as listed in Table 3.1. It is interesting to note that the Λ-doubling constant p is positive for CF, CCl, and SiCl, whereas it is negative for SiF and GeF. For the former the contributions of Σ^- states are dominant.

V–VI Molecules. The NO molecule represents this class and, because it is chemically stable, has been subjected to a number of spectroscopic studies. In contrast, no high-resolution IR and microwave spectroscopy has been reported on the PO radical until quite recently, in spite of many trials made over the last 20 years.

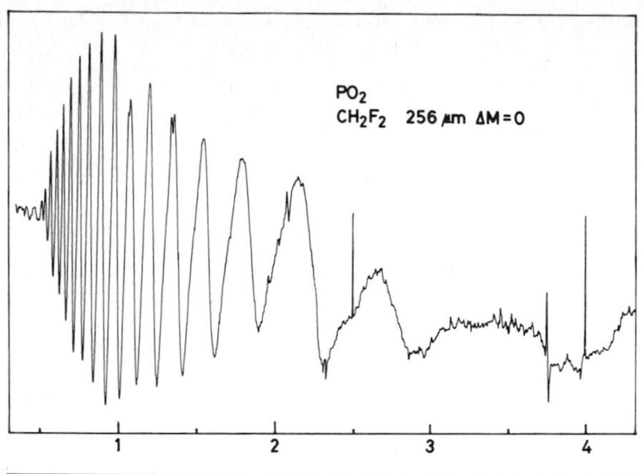

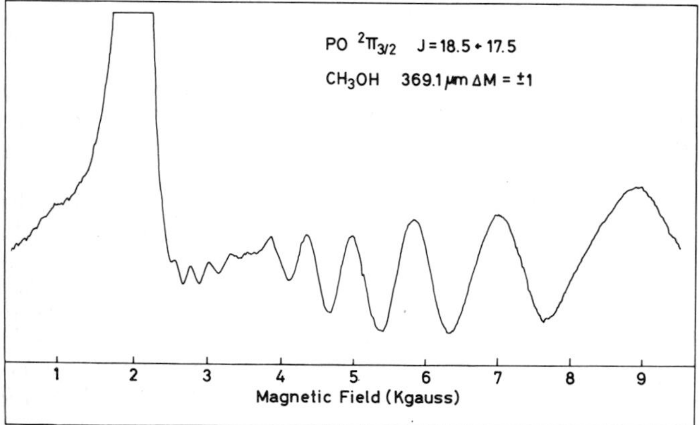

Fig. 4.11. FIR LMR spectra of PO_2 (*upper trace*) [4.235] and PO (*lower trace*) [4.67]

During the course of studying the PD radical by FIR LMR, *Ohashi* et al. [4.46] detected PO_2 and then PO by chance in the reaction of red phosphorus with microwave discharge products of D_2O. *Kawaguchi* et al. [4.67] confirmed that the spectra of both PO_2 and PO were also seen when D_2O was replaced by either H_2O, ($H_2 + O_2$), or ($D_2 + O_2$), indicating that the new species contain both oxygen and phosphorus, but not hydrogen. The spectrum assigned to PO showed a linear Zeeman pattern, as shown in the lower trace of Fig. 4.11. It is interesting to note that hydrogen was indispensable in generating PO. Two rotational transitions were thus detected by LMR and assigned to $J = 18.5 \leftarrow 17.5$ in $^2\Pi_{3/2}$ and $J = 22.5 \leftarrow 21.5$ in $^2\Pi_{1/2}$. Once a best condition for producing PO is found by the most sensitive FIR LMR method, other spectroscopic methods might be applied efficiently to detail molecular parameters; microwave [4.67] and IR diode laser [4.68] spectroscopy have in fact allowed us to observe the pure rotational and vibration-rotation spectra of the molecule. Figure 4.12 illustrates a part of the diode laser spectrum of PO.

The next member of the Group V oxides is AsO, which was investigated by *Uehara* [4.69] using CO_2 LMR spectroscopy. The LMR spectrum of AsO

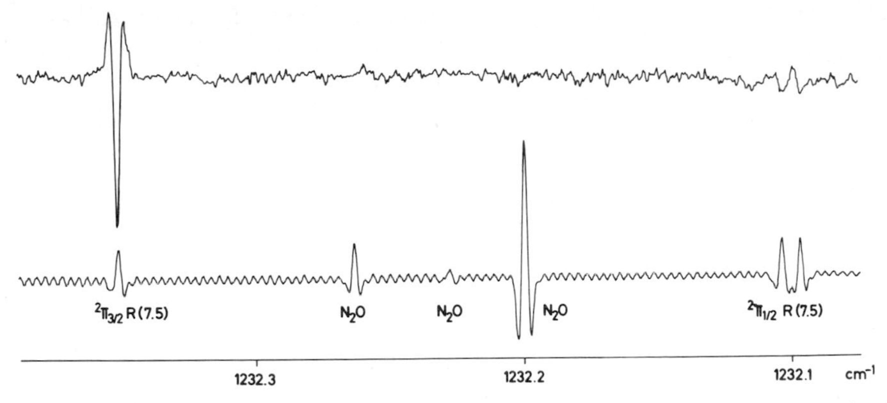

Fig. 4.12. A part of the IR diode laser spectrum of PO; the upper trace is recorded with Zeeman modulation, whereas the lower trace with source frequency modulation. The latter trace also shows a few lines of N_2O as a wavelength standard [4.68]

exhibits widely split hyperfine structures due to the As nucleus of $I = 3/2$; the nuclear quadrupole effect was easily seen by unequal spacings of the quartet structure. He also observed additional absorptions around 1025 cm^{-1} and has postulated that they are due to the $^2\Pi_{3/2} \leftarrow {}^2\Pi_{1/2}$ magnetic dipole transition, but this assignment still awaits confirmation.

The Λ-type doubling constant p has been reported to be positive for NO, NS, and PO, showing that Σ^- state contributions predominate over those of Σ^+.

VI–VII Molecules. The FO molecule was first detected accidentally by *McKellar* [4.70] using CO_2 LMR in products of a microwave discharge in CF_4 induced in a quarts tube; the spectrum thus observed was assigned to the fundamental vibration-rotation band in the $^2\Pi_{3/2}$ state. No gas phase spectra have been reported previously in any of the regions from uv down to microwave. The F atom seems to attack the inside wall of the quartz tube to form an FO molecule. When a small amount of O_2 was added, the intensity of the FO spectrum increased by about 50%, but further addition of O_2 decreased the signal. Subsequently, diode laser spectroscopy was applied to the same system, and 14 vibration-rotation lines in $^2\Pi_{3/2}$ were observed with Zeeman modulation, but $^2\Pi_{1/2}$ spectra have not been seen, because this state is located 177 cm^{-1} higher than $^2\Pi_{3/2}$ and the Zeeman effect is too small to modulate lines [4.71]. *McKellar* [4.72] later measured the dipole moment by inserting parallel electrodes into an LMR absorption cell, and found that it changed greatly with vibrational excitation: $\mu = 0.0043\,(12)\,D$ and $0.0267\,(27)\,D$, respectively, in $v = 0$ and 1 of the $^2\Pi_{3/2}$ state. The very small dipole moment of FO in the ground vibronic state accounts for the failure of many trials to detect the pure rotational spectrum [4.73, 74].

The BrO radical has been investigated by various spectroscopic techniques. *McKellar* [4.75] observed the $^2\Pi_{1/2} - {}^2\Pi_{3/2}$ magnetic dipole transition in the ground electronic state using IR LMR. *Cohen* et al. [4.76] estimated the vibrational frequency to be $\omega_e = 724(10)\ \mathrm{cm}^{-1}$ for $^{81}\mathrm{Br}^{16}\mathrm{O}$ by inserting their microwave data into

$$\omega_e = (4B_e^3/D_e)^{1/2}. \tag{4.1.6}$$

This value has prompted *Barnett* et al. [4.77] to reexamine the earlier assignment for the uv spectrum and to derive $\Delta G''_{1/2} = 722.1(11)\ \mathrm{cm}^{-1}$ for $^{81}\mathrm{Br}^{16}\mathrm{O}$. To determine the vibrational frequency much more precisely, *Butler* et al. [4.78] have observed the fundamental vibration-rotation band of BrO using an IR diode laser spectrometer. An example of the observed spectrum is displayed in Fig. 4.13. Some low-J lines showed resolved hyperfine structures, which were well explained by the microwave parameters already reported. No vibration-rotation spectrum in the $^2\Pi_{1/2}$ state has been observed because of the unfavorable Boltzmann factor. The data obtained for $^{79}\mathrm{BrO}$ and $^{81}\mathrm{BrO}$ allowed *Butler* et al. to calculate the harmonic vibrational frequency with the anharmonicity using isotope relations. They found it necessary to include higher-order terms for the centrifugal distortion constant, which are given by

$$\pm [\alpha_e^B A_J/\omega_e + \alpha_e^A D_e/(2\,\omega_e)], \tag{4.1.7}$$

where the + and − signs apply to $^2\Pi_{1/2}$ and $^2\Pi_{3/2}$, respectively. Because these correction terms depend on the A constant, they may be important for molecules with large spin-orbit interaction constants. *Cohen* et al. [4.79] considered similar terms in their analysis of the ClO submillimeter wave spectra.

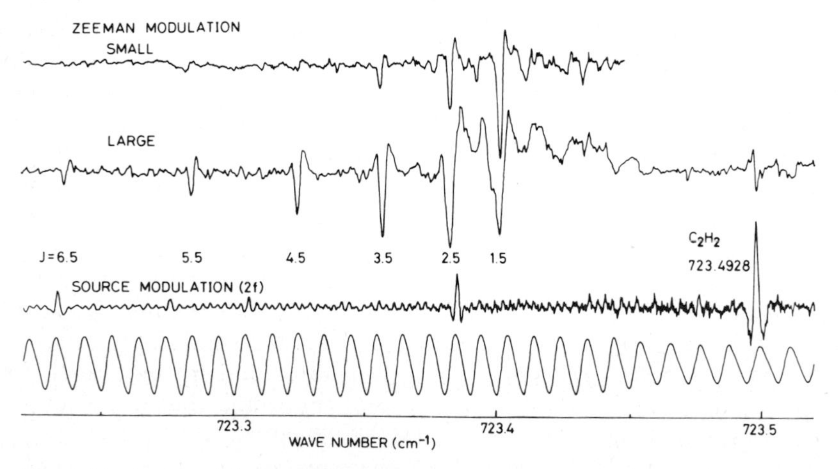

Fig. 4.13. A part of the IR diode laser spectrum of BrO in $X^2\Pi_{3/2}$; the two top traces are recorded with Zeeman modulation, the third trace shows the spectrum of C_2H_2 as a wavelength standard, and the bottom trace represents etalon fringes [4.78]

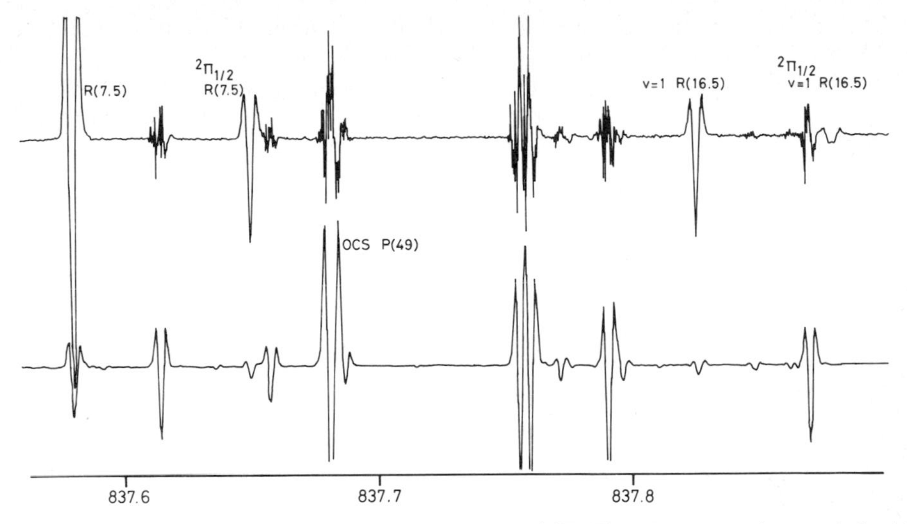

Fig. 4.14. A part of the IR diode laser spectrum of SF. The upper trace is recorded with discharge current modulation and is compared with the lower trace obtained by conventional source frequency modulation [4.80]

Endo et al. [4.80] have successfully applied discharge current modulation, Sect. 3.2.1, to the SF radical, which was produced by a discharge in a OCS and CF_4 mixture. The $^2\Pi_{3/2}$ spectrum was easily seen with Zeeman modulation, while this method was rather difficult to apply to the $^2\Pi_{1/2}$ spectrum. Furthermore, the SF spectrum was badly overlapped by the strong ν_3 band of OCS. As Fig. 4.14 shows, discharge current modulation enabled Endo et al. to distinguish the SF spectral lines. Care must, however, be taken of the fact that this modulation often also picks up hot band lines excited by the discharge.

Yamada et al. [4.81] have detected SCl by diode laser spectroscopy. They generated the radical by a discharge in either S_2Cl_2 or SCl_2 diluted with He contained in an absorption cell. Low-J transitions were detected by Zeeman modulation. Although high-J lines have Zeeman effects too small to be modulated, they were seen on the Zeeman modulated spectrum; the Zeeman field seemed to affect the discharge for producing the radical.

4.1.3 Molecules in Δ States

In contrast with Σ and Π states, only a few examples of molecules in Δ states have been reported so far. However, there is an important Δ molecule, the oxygen molecule in the metastable state $a\,^1\Delta_g$. The electronic configuration of the oxygen molecule $(1s\sigma_g)^2\ (1s\sigma_u)^2\ (2s\sigma_g)^2\ (2s\sigma_u)^2\ (2p\sigma_g)^2\ (2p\pi_u)^4\ (2p\pi_g)^2$ leads to three low-lying electronic states $^3\Sigma_g^-$, $^1\Delta_g$, and $^1\Sigma_g^+$, of which $^3\Sigma_g^-$ is the ground state and $^1\Delta_g$ the lowest excited state. Because the transition between these two states is forbidden in both orbital and spin, the $^1\Delta_g$ state is a metastable state with a long radiative lifetime. It plays a unique role in oxidation reactions of organic compounds.

The $a\,^1\Delta_g$ oxygen molecule was first identified spectroscopically by *Herzberg* and *Herzberg* [4.82] in 1947. *Falick* et al. [4.83] reported the gas phase EPR spectra of O_2 in the a state, and *Scalabrin* et al. [4.84] observed its pure rotational spectrum for the first time using FIR LMR. In the latter experiment, the metastable oxygen molecule was generated by a 2450 MHz microwave discharge in a pure oxygen sample and was introduced into a cell placed between the pole pieces of an electromagnet, and five pure rotational transitions were thus observed. *Cazzoli* et al. [4.85] have recently succeeded in observing the zero-field rotational transitions by using a millimeter-wave spectrometer with a 4 m long free space cell 12 cm in diameter and by generating the metastable oxygen by a dc discharge in oxygen contained in the cell. They have thus observed two lowest rotational transitions with an accuracy higher than that achieved with FIR LMR and have improved the molecular constants in the a state.

The rotational energy level in the $^1\Delta$ state is given by

$$E(J) = B_0[J(J+1) - 4] - D_0[J(J+1) - 4]^2 \tag{4.1.8}$$

with the Λ-doubling term

$$E_\Lambda = \pm(-1)^J\,(1/2)\,q\,J(J+1)\,[J(J+1) - 2]. \tag{4.1.9}$$

For $^{16}O_2$ one of the Λ-doubling components is missing, resulting in a staggering of the rotational spectrum. In a $^1\Delta$ state, the q constant is usually very small; it is several ten Hz in the $a\,^1\Delta_g$ O_2.

The SO molecule isovalent with O_2 also has a $^1\Delta$ metastable state which is the lowest excited state. The rotational spectrum in this state was first observed by *Carrington* et al. [4.86] using gas phase EPR. *Saito* [4.87] later observed the zero-field pure rotational spectrum using a Stark modulation spectrometer. *Yamada* et al. [4.88] reported the vibration-rotation spectrum recorded with mid-IR LMR and have discussed the Zeeman effect of SO in the $a\,^1\Delta$ state. The ground-state SO is rather easily generated by the reaction of sulfur-containing molecules like OCS with microwave discharge products of O_2. When a large excess of discharged oxygen is added, the oxygen molecule in the metastable state efficiently excites SO into the $a\,^1\Delta$ state. No Λ doubling has been resolved for SO in $a\,^1\Delta$.

Another example of a molecule in the lowest metastable $a\,^1\Delta$ state, NF, was detected for the first time by *Curran* et al. [4.89] using EPR. *Davies* et al. [4.90] recently observed its vibration-rotation spectrum using a diode laser spectrometer. They generated $a\,^1\Delta$ NF by the reaction of NF_2 with hydrogen atoms. This reaction has been known to generate primarily NF in $a\,^1\Delta$ (91%), and, in fact, $a\,^1\Delta$ was detected earlier than $X\,^3\Sigma^-$, of which the diode laser spectrum was recently reported [4.91].

When a molecule contains a transition metal, it may have low-lying electronic states of high-spin multiplicity and/or large orbital angular momentum, as demonstrated by *Field, Merer*, and their collaborators [4.92–95]. One such example, FeO, is discussed here. Although FeO has long been known to be responsible for the orange emission system, even its ground state has not been clearly identified. *Cheung* et al. [4.94, 95] have recently investigated FeO by observing laser-induced fluorescence (LIF) in the visible region and by dispersing the near IR emission with a Fourier transform spectrometer, and concluded that the ground state of FeO is $^5\Delta_i$. They have thus obtained reliable molecular constants of FeO in the ground electronic state.

These results have facilitated *Endo* et al. [4.96] to observe the pure rotational spectrum of FeO in the 154–184 GHz region. Transitions in the three lowest spin sublevels with $\Omega = 4$, 3, and 2 were seen, but the weakness of the spectrum has precluded observing the spectra in the $\Omega = 1$ and 0 levels, because these levels lie 600 and 800 cm^{-1} above the $\Omega = 4$ level, respectively. The FeO radical was generated by a dc glow discharge in a mixture of ferrocene and oxygen in a 1 m long free space cell. Although no substantial improvement was made for the ground-state molecular parameters derived from the optical spectrum, the observed microwave spectrum will be of great use in identifying FeO in interstellar space.

4.2 Linear Polyatomic Molecules

The electronic state of a linear polyatomic molecule is also classified according to the value of Λ, the molecular axis component of the electronic orbital angular momentum, as in the case of the diatomic molecule, namely the states with $|\Lambda| = 0, 1, 2, 3, \ldots$ are referred to as $\Sigma, \Pi, \Delta, \Phi, \ldots$. This section discusses spectroscopic results obtained for a few Σ and Π molecules.

4.2.1 Molecules in Σ Electronic States

Most molecules in $^1\Sigma$ states are chemically stable, and a number of high-resolution spectroscopic studies have been performed on them. There are a few chemically active $^1\Sigma$ species such as protonated molecular ions. A representative example would be HCO^+, which was first detected in interstellar space; one of the *U* lines (unidentified interstellar lines) reported by *Buhl* and *Snyder* [4.97] was suggested by *Klemperer* [4.98] to be due to HCO^+, and the assignment was later confirmed to be correct by a laboratory microwave experiment by *Woods* et al. [4.99]. The HN_2^+ and HCS^+ ions were also observed in interstellar space prior to laboratory detection. It is now well recognized that these molecular ions play specially important roles in forming molecules in interstellar space (Sect. 5.3).

There are only a few examples of linear polyatomic molecules known which are in Σ states and have finite electron spin quantum numbers: CCH, C_4H, and C_3N in $X\,^2\Sigma^+$ and HCCN in $X\,^3\Sigma^-$. They are of great interest for astronomy. In fact, CCH, C_4H, and C_3N were first detected by radiotelescopes and were identified by fine and hyperfine structures of the observed rotational spectra. Although the C_3N and C_4H spectra observed in circumstellar space of IRC + 10 216 [4.100, 101] did not show any hyperfine structure, a subsequent observation [4.102, 103] on TMC1 resolved the structure thanks to the narrower linewidth achieved. Later the three species were observed in the laboratory with the aid of FIR LMR and microwave spectroscopy. *Gottlieb* et al. [4.104] generated the C_4H radical by a liquid N_2 cooled dc discharge in a mixture of acetylene and He or Ar and found that C_4H was as abundant as CCH in the discharge plasma. *Carrick* et al. [4.105] recently observed five bands of CCH in the region 3600–4200 cm^{-1} using a color-center laser as a source; they assigned four of them to the $\tilde{A}\,^2\Pi \leftarrow \tilde{X}\,^2\Sigma^+$ system with the upper states of the three bands vibronically interacting with three high vibrational states of Π symmetry associated with the $\tilde{X}$ state, and the remaining one band was assigned to the (0 1 0) ← (0 1 0) hot band of $\tilde{A} \leftarrow \tilde{X}$.

Little is known of the gas phase spectrum of HCCN. *Merer* and *Travis* [4.106, 107] observed an extensive triplet-triplet band system of complicated structure around 3400 Å from a flash photolysis of diazoacetonitrile and ascribed it to HCCN, but were unable to analyze it because of overlap by an intense spectrum of CNC. While searching for the spectra of CCN in a reaction system of CH_3CN and microwave discharge products of CF_4, *Saito* et al. [4.108] observed the microwave spectrum of HCCN; each rotational transition was found to be split into a triplet because of the spin-spin interaction. Although no hyperfine structure was resolved, the linewidth was found to vary with spin component, Fig. 4.15. The observed spectral pattern is completely consistent with that expected for a linear molecule in a $^3\Sigma^-$ state, in agreement with an earlier matrix EPR study by *Bernheim* et al. [4.109–111]. This molecule may also be of some interest for astronomy, Sect. 5.3.

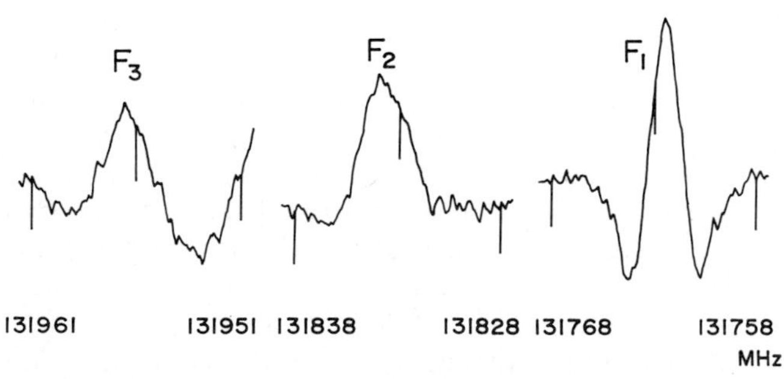

Fig. 4.15. The $N = 6 \leftarrow 5$ transition of HCCN in $\tilde{X}\,^3\Sigma^-$ [4.108]

4.2.2 Molecules in Π Electronic States

The Π electronic state is doubly degenerate unless either a bending vibration or an overall rotation is excited. The removal of the degeneracy by a bending mode is referred to as the Renner (or Renner-Teller) effect, discussed in detail in Sect. 2.4. This effect has so far been investigated mainly by optical spectroscopy, but IR spectroscopy is expected to supply more detailed information on it because it allows us to measure the Renner splitting readily. One problem with IR spectroscopy is the low sensitivity; most molecules exhibiting the Renner effect are short-lived and are thus difficult to investigate by conventional IR spectroscopy which is of low sensitivity. In fact, conventional IR spectroscopy has been successfully applied to only one case, namely the NCO $(0\,1\,0)\,\kappa\,{}^2\Sigma \leftarrow (0\,0\,0)\,{}^2\Pi$ transition in a low-temperature matrix reported by *Milligan* and *Jacox* [4.112]. Infrared lasers have widened the scope of IR spectroscopy. *Kawaguchi* and *Hirota* [4.113] observed the $(0\,1\,0)\,\kappa\,{}^2\Sigma \leftarrow (0\,0\,0)$ ${}^2\Pi_{3/2}$ band of the BO_2 radical using a tunable diode laser as a source combined with Zeeman modulation, and many more examples will follow in the future.

Gas phase electron paramagnetic resonance (EPR) and microwave spectroscopy have been applied to NCO and NCS. They provide indirect but important information on the Renner effect, namely on rotational constants, g factors, and other constants which are affected by the Renner effect.

In this section, the three species BO_2, NCO, and CCN are discussed in some detail. The Renner parameter $|\varepsilon|$ is fairly small in all three examples.

BO_2. The first high-resolution spectroscopic study of BO_2 was performed by *Johns* [4.114], who generated the radical by flash photolysis of a BCl_3 and O_2 mixture and observed two bands, $\tilde{B}\,{}^2\Sigma_u^+ - \tilde{X}\,{}^2\Pi_g$ and $\tilde{A}\,{}^2\Pi_u - \tilde{X}\,{}^2\Pi_g$. From the observed spectrum he evaluated the rotational constants in many vibronic states, the vibrational frequencies in the $\tilde{X}$, $\tilde{A}$, and $\tilde{B}$ states, and the Renner parameters in the $\tilde{X}$ and $\tilde{A}$ states.

Subsequently the $\tilde{A}\,{}^2\Pi_u - \tilde{X}\,{}^2\Pi_g$ band was reinvestigated by dye laser spectroscopy in more detail. *Russell* et al. [4.115] excited rovibronic transitions of the $\tilde{A} - \tilde{X}$ band of ${}^{11}BO_2$ and ${}^{10}BO_2$ by using the Ar^+ 514.5 nm and 448 nm laser lines and dispersed the induced fluorescence using a grating spectrometer. The vibration rotation levels in the $\tilde{X}\,{}^2\Pi_g$ state thus observed were analyzed by using the following Hamiltonian (or energy level expressions):

$$\boldsymbol{H} = \boldsymbol{H}_{\mathrm{E}} + \boldsymbol{H}_{\mathrm{V}} + \boldsymbol{H}_{\mathrm{SO}} + \boldsymbol{H}_{\mathrm{RT}} + \boldsymbol{H}_{\mathrm{ANH}} + \boldsymbol{H}_{\mathrm{anharm}} + \boldsymbol{H}_{\mathrm{ROT}}, \tag{4.2.1}$$

where

$$\boldsymbol{H}_{\mathrm{V}} = \sum_i (v_i + d_i/2)\,\omega_i, \tag{4.2.2}$$

$$\boldsymbol{H}_{\mathrm{ANH}} = \sum_{i=1}^{3} \sum_{j>i} x_{ij}(v_i + d_i/2)(v_j + d_j/2) + g_{ll} l^2, \tag{4.2.3}$$

and $\boldsymbol{H}_{\mathrm{SO}}$, $\boldsymbol{H}_{\mathrm{RT}}$, and $\boldsymbol{H}_{\mathrm{ROT}}$ are given by (2.4.16, 8, 17) respectively. The term

$\boldsymbol{H}_{\text{anharm}}$ which is essentially equivalent to (2.4.52), represents here only that part of the vibrational anharmonicity responsible for Fermi interactions; the rest of the anharmonicity effect is included in $\boldsymbol{H}_{\text{ANH}}$, (4.2.3). *Russell* et al. [4.115] numerically diagonalized the energy matrix for each Fermi polyad. They subjected the off-diagonal elements of the Renner Hamiltonian to the Van Vleck transformation to fourth order, but neglected the cross term of $\boldsymbol{H}_{\text{RT}}$ and $\boldsymbol{H}_{\text{anharm}}$, as pointed out by *Dixon* et al. [4.116]. *Russell* et al. have assigned the vibrational levels up to (4, 1, 0) and (5, 2, 0) for the Σ and Φ vibronic manifolds in the $\tilde{X}\,^2\Pi_g$ state, respectively, but for such high vibrational levels the off-diagonal matrix elements of $\boldsymbol{H}_{\text{RT}}$ and $\boldsymbol{H}_{\text{anharm}}$ can be appreciable in comparison with the vibrational energy level spacing. A direct diagonalization of the full matrix would be necessary, as also demonstrated by a few examples of $^1\Sigma$ molecules like CO_2, OCS, and N_2O [4.117, 118]. No such extensive analysis has been reported on Renner molecules.

Weyer et al. [4.119] observed the laser excitation spectrum of BO_2 in Π vibronic states using a cw dye laser as a source. *Dixon* et al. [4.116] also recorded fluorescence from single vibronic levels, again employing a cw dye laser, and have analyzed the Fermi resonance between the (1, 0, 0) and (0, 2, 0) levels more extensively than *Russell* et al. to determine two anharmonic potential constants to be $W_1 = 31.0\ (1.2)\ \text{cm}^{-1}$ and $W_2 = 0.0\ (0.8)\ \text{cm}^{-1}$.

Kim et al. [4.120] observed microwave optical double resonance signals of BO_2 under a magnetic field. They used the Ar^+ 514.5 nm laser line to excite the $\tilde{A}(1, 1, 0)\,\kappa\,^2\Sigma\ J = 8.5 \leftarrow \tilde{X}(0, 1, 0)\,\kappa\,^2\Sigma\ J = 7.5$ transition of $^{11}BO_2$ and scanned microwaves between 2 and 2.7 GHz to induce magnetic dipole transitions between Zeeman split M_J levels of the $\tilde{X}(0, 1, 0)\,\kappa\,^2\Sigma\ J = 7.5$ state. This study yielded the Δg_l value to be $-0.0089\,(55)$ in addition to the Renner parameter.

Muirhead et al. [4.121] and *Lowe* et al. [4.122] applied saturation spectroscopy with a cw dye laser to BO_2 to obtain hyperfine interaction constants for both $\tilde{A}$ and $\tilde{X}$ states. *Schulz* et al. [4.123] were able to reduce the linewidth by observing backscattered fluorescence and determined the $P - R$ separation in the $\tilde{A}\,(0, 0, 0)\,^2\Pi_{3/2} - \tilde{X}\,(0, 0, 0)\,^2\Pi_{3/2}$ band.

Kawaguchi et al. [4.124] observed the vibrational transitions of BO_2 in the $\tilde{X}\,^2\Pi_g$ state using IR tunable diode laser spectroscopy. Because of $D_{\infty h}$ symmetry, the ν_1 mode is forbidden in the infrared, so they investigated the ν_2 and ν_3 bands.

An interesting feature of the vibrational spectrum was noted in the ν_3 frequency. *Johns* [4.114] reported $2\nu_3$ in the ground state to be 2644 cm^{-1} from the observation of the $\tilde{A}\,(0, 0, 0)\,^2\Pi_u \rightarrow \tilde{X}\,(0, 0, 2)\,^2\Pi_g$ emission band. This result leads to a ν_3 frequency of about 1322 cm^{-1}, which seems to be unusually low, in comparison with $\nu_1 = 1070\ \text{cm}^{-1}$. For a linear symmetrical XY_2-type molecule, the ratio of the antisymmetric and symmetric harmonic frequencies ω_3/ω_1 is given by

$$\omega_3/\omega_1 = \{[(K - k)/(K + k)]\,[(m_X + 2m_Y)/m_X]\}^{1/2}, \tag{4.2.4}$$

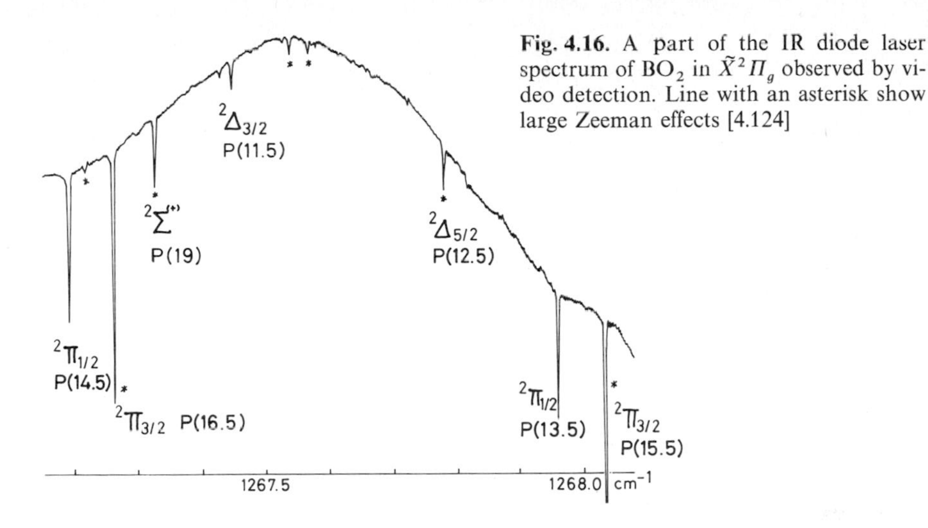

Fig. 4.16. A part of the IR diode laser spectrum of BO_2 in $\tilde{X}^2\Pi_g$ observed by video detection. Line with an asterisk show large Zeeman effects [4.124]

where K and k denote, respectively, the bond stretching and bond interaction force constants and m_X and m_Y stand for the masses of the X and Y atoms, respectively. For BO_2 the mass factor in (4.2.4) is 1.98. When K is much larger than k, as is usually the case, ω_3 must be nearly twice as large as ω_1. The above data on $\tilde{X}\,^2\Pi_g$ thus deviate considerably from this expectation, whereas the $\tilde{A}\,^2\Pi_u$ state behaves "normally" in this respect: $\nu_1 = 994\ \mathrm{cm}^{-1}$ and $(2\nu_3)/2 \cong \nu_3 = 2357\ \mathrm{cm}^{-1}$.

Kawaguchi et al. produced BO_2 either by the reaction of BCl_3 with microwave discharge products of O_2 or by a high-voltage 60 Hz discharge in an O_2 and BCl_3 mixture induced in an absorption cell, as mentioned in Sect. 3.5. The BO_2 production was optimized when the discharge emitted strong pale green fluorescence. Because the inherent intensity of the ν_3 band is large in BO_2 as in CO_2, Kawaguchi et al. could observe not only the ν_3 fundamental band, but also the ν_3 hot bands from the ν_3 and ν_2 ($^2\Delta$, $\mu\,^2\Sigma$, and $\kappa\,^2\Sigma$) states. Figures 4.16, 17 show examples of the observed diode laser spectra. The observation could be made with source frequency modulation, Zeeman modulation, or even merely chopping the laser beam; some strong lines absorb up to 40% of the incident laser power for a path length of 10 m. Zeeman modulation was useful in detecting weak lines, but could not be applied to high-J transitions of $K = v_2 + 1$, $P = K \pm 1/2$ because the Zeeman shifts were too small for the maximum modulation field of 500 G peak-to-peak which was available.

Kawaguchi et al. [4.124] determined the origins of the ν_3 fundamental and the $2\nu_3 \leftarrow \nu_3$ hot bands to be 1278.2585(4) and 1365.2265(32) cm^{-1}, respectively, from the analysis of the diode laser spectra. Therefore, the $v_3 = 2-0$ frequency is 2643.5 cm^{-1}, which agrees well with Johns' value of 2644 cm^{-1}. Note that the ν_3 mode has a large and negative anharmonicity. This fact, together with the very small ν_3/ν_1 ratio mentioned above, indicates that the

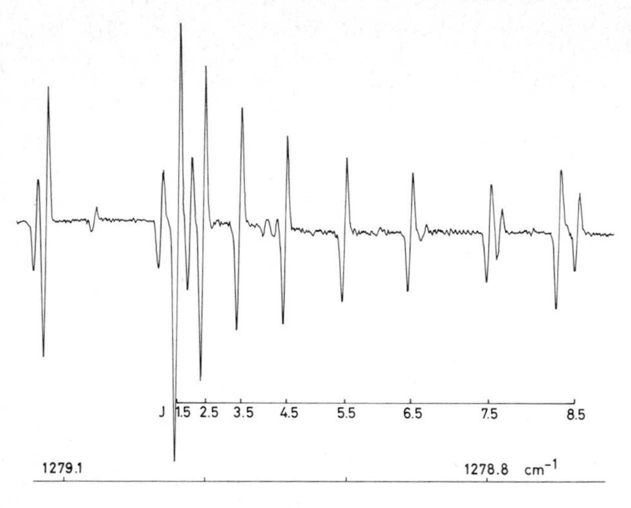

Fig. 4.17. Q-branch transitions of the ν_3 band of $^{11}BO_2$ in $\tilde{X}^2\Pi_{3/2}$, recorded with source frequency modulation [4.124]

Table 4.4. Observed frequencies of the ν_3 and ν_1 vibrations in linear YXY Type molecules (in cm^{-1})

Molecules	Electronic state	ν_3 band			ν_1 band
		ν_{1-0}	ν_{2-0}	ν_{2-0}/ν_{1-0}	
$^{11}BO_2$	$\tilde{X}\,^2\Pi_g$	1278.3 [a]	2643.5 [a]	2.085	1071 [b]
$^{11}BO_2$	$\tilde{A}\,^2\Pi_u$		4714.0 [b]		994 [b]
CO_2^+	$\tilde{X}\,^2\Pi_g$		2938.0 [c]		1280 [c]
CO_2^+	$\tilde{A}\,^2\Pi_u$		5462.0 [c]		1126 [c]
CO_2	$\tilde{X}\,^1\Sigma_g$	(2396.0) [d]		2.0	(1354) [d]

[a] Diode laser spectroscopy [4.124]
[b] Optical spectroscopy [4.114a]
[c] [4.114b]
[d] Harmonic frequencies taken from [4.117]

antisymmetric B – O stretching mode of BO_2 in $\tilde{X}\,^2\Pi_g$ is quite anomalous. As Table 4.4 shows, the CO_2^+ ion in $\tilde{X}\,^2\Pi_g$, which is isoelectronic with BO_2, behaves similarly. Kawaguchi et al. have proposed a vibronic interaction between $\tilde{X}\,^2\Pi_g$ and $\tilde{A}\,^2\Pi_u$ through the ν_3 mode to account for the anomalies.

Because the ν_3 vibration is a parallel mode, we may set $\chi_k = 0$ in (2.4.6), leaving the vibronic interaction potential of the form ax_3, where

$$a = \Sigma Z^2 e^2 \cos \chi_e / r^2 \tag{4.2.5}$$

has a matrix element between $\tilde{A}$ and $\tilde{X}$ (for simplicity, this matrix element is again designated by a in the following) and x_3 stands for the coordinate of the ν_3 mode. When the harmonic force constant of the ν_3 vibration is designated by f and the unperturbed energy difference between $\tilde{A}$ and $\tilde{X}$ by ΔE, the effective potential function is given by

$$V(\tilde{A}/\tilde{X}) = (1/2)\, f x_3^2 + (1/2)\, \Delta E \pm (1/2)\, [(\Delta E)^2 + 4a^2 x_3^2]^{1/2}. \qquad (4.2.6)$$

This potential function has only one minimum at $x_3 = 0$ in the $\tilde{A}$ state, whereas it has two minima symmetrically displaced from $x_3 = 0$ or one minimum at $x_3 = 0$ in the $\tilde{X}$ state, depending on whether $2a^2/f\,\Delta E$ is larger or smaller than 1.

Because BO_2 in $\tilde{X}$ does not show any indication that a double minimum is present, Kawaguchi et al. expanded the square root in (4.2.6) as

$$V(\tilde{X}\,^2\Pi_g) = (1/2)\,(f - 2a^2/\Delta E)\, x_3^2 + [a^4/(\Delta E)^3]\, x_3^4 \qquad (4.2.7)$$

in analyzing the observed band origins, obtaining the parameters

$$a = 0.000\,993 \text{ erg/cm},$$

$$f = 9.00 \text{ md/Å}, \quad \text{and}$$

$$2a^2/\Delta E = 5.63 \text{ md/Å}.$$

When transferred to the $\tilde{A}$ state, these parameters give the $v_2 = 2 - 0$ separation as 4702 cm^{-1}, which agrees with Johns' value of 4714 cm^{-1}, and the fundamental frequency as 2384 cm^{-1}. Figure 4.18 compares the potential functions in the $\tilde{A}$ and $\tilde{X}$ states.

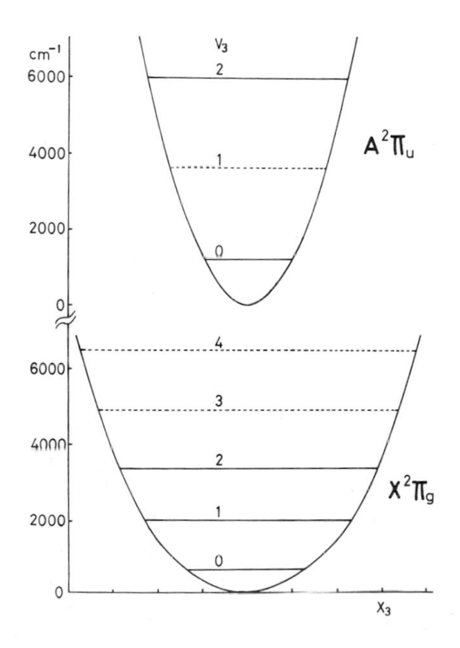

Fig. 4.18. Potential functions for the ν_3 mode of BO_2 in $\tilde{A}^2\Pi_u$ *(upper trace)* and $X^2\Pi_g$ *(lower trace)* [4.124]

Table 4.5. Observed rotational constants of linear triatomic molecules in the $v_2 = 1\ ^2\Sigma$ vibronic states (in cm^{-1})

Molecule (electronic state)	$\kappa\,^2\Sigma$	$\mu\,^2\Sigma$	$B^{\kappa}_{eff} - B^{\mu}_{eff}$	$\varepsilon\omega_2$	ε	Ref.
BO_2 ($\tilde{X}\,^2\Pi_g$)	0.3304	0.3307	−0.0003	−86.4	−0.19	[4.124]
BO_2 ($\tilde{A}\,^2\Pi_u$)	0.3125	0,3108	0.00167	−13	−0.026	[4.114]
CNC ($\tilde{X}\,^2\Pi_g$)	0.4539	0.4576	−0.0037	176	0.55	[4.107]
CCC ($\tilde{A}\,^1\Pi_u$)	0.4083	0.4159	−0.0076	165.3	0.54	[4.126]
CO_2^+ ($\tilde{X}\,^2\Pi_g$)	0.3847	0.3833	0.0014	−96.8	−0.190	[4.127]
NCO ($\tilde{X}\,^2\Pi$)	0.3901	0.3908	−0.0007	−76	−0.14	[4.128][a]

[a] The rotational levels of both the $\kappa\,^2\Sigma$ and $\mu\,^2\Sigma$ states are fitted with Hamiltonian including rovibronic interaction, expressed by (2.4.61)

It is interesting to point out that according to the present model the linear YXY molecule can be asymmetric with two unequal $X-Y$ bond lengths. This possibility may be realized when ΔE is not large, in other words, if the highest occupied π_g and π_u orbitals are not much different in energy, in comparison with the vibronic interaction. This phenomenon closely resembles the Jahn-Teller effect, as pointed out by *Öpik* and *Price* [4.125], although there is no genuine degeneracy to start with.

Table 4.5 lists the effective rotational constants determined for BO_2 in the $(0, 1, 0)\,^2\Sigma$ state together with those of other molecules subjected to the Renner effect. The effective rotational constant in the $v_2 = 1\ ^2\Sigma$ state is given by (2.4.71), which is reproduced here:

$$B^{\kappa;\mu}_{eff} = B_v[1 \pm (B_v/2r)\cos^2 2\beta] \pm s, \tag{4.2.8}$$

where s denotes the effect of the rovibronic interactions and is given by (2.4.62). If s is neglected, the rotational constant is larger in $\kappa\,^2\Sigma$ than in $\mu\,^2\Sigma$, contrary to the observations for BO_2 and a few other molecules. The observed difference between the rotational constants of the $\kappa\,^2\Sigma$ and $\mu\,^2\Sigma$ gave the s constant, which was further converted by (2.4.65) to the anharmonic constant W_2 of $3.87(37)\ cm^{-1}$ for BO_2 [4.124]. This result does not agree with that obtained from an analysis of vibronic levels [4.116], and the discrepancy is probably ascribed to the effect of higher vibronic levels, neglected in the latter analysis.

A similar analysis was applied to NCO, yielding $W_2 = 3.31(11)\ cm^{-1}$ [4.128]. For CCN the $(0, 1, 0)\,\kappa^2\Sigma$ state has not been detected, as discussed below; instead the difference between the rotational constants of the $\mu\,^2\Sigma$ and $^2\Delta$ states was employed to obtain W_2 to be $-7.58\ cm^{-1}$ [4.129]. These results indicate that the anharmonic potential constant W_2 is clearly correlated with the Renner parameter ε. The negative W_2 constant of CCN means that the anharmonic potential constant f''_{i22} ($i = 1$ and 3 for XY_2- and XYZ-type molecules, respectively) is larger than f'_{i22}, but f''_{i22} is associated with the lower energy potential curve when ε is positive.

As mentioned earlier, *Kawaguchi* and *Hirota* [4.113] have observed the $(0, 1, 0)\,\kappa\,^2\Sigma \leftarrow (0, 0, 0)\,^2\Pi_{3/2}$ transition of BO_2 by IR diode laser spectroscopy. The lines were much weaker than those of the ν_3 band primarily because of the $\cos^2\beta$ factor in (2.4.44) and of the transition moment of ν_2 smaller than that of ν_3, as in the case of CO_2. Zeeman modulation was employed, because the Zeeman effect is large enough even for high-J transitions, as discussed in Sect. 2.5. The band origin obtained for the $(0, 1, 0)\,\kappa\,^2\Sigma - (0, 0, 0)\,^2\Pi_{3/2}$ transition agrees well with the value derived from the laser-excited fluorescence study.

In the course of studying BO_2 by diode laser spectroscopy, *Kawaguchi* et al. [4.130] detected a short-lived molecule ClBO in a discharge plasma of an O_2 and BCl_3 mixture induced in an absorption cell. This is the first spectroscopic observation of this linear molecule in the gas phase.

NCO. The first high-resolution spectroscopic study of the NCO radical was carried out by *Dixon*, who rotationally analyzed the $\tilde{A}\,^2\Sigma^+ - \tilde{X}\,^2\Pi_i$ [4.131] and $\tilde{B}\,^2\Pi_i - \tilde{X}\,^2\Pi_i$ [4.132] bands observed in absorption. *Carrington* et al. [4.133] reported on the EPR spectra in the $\tilde{X}\,(0\,0\,0)\,^2\Pi_{3/2}$, $\tilde{X}\,(0\,1\,0)\,^2\Delta_{5/2}$, and $\tilde{X}\,(0\,2\,0)\,^2\Phi_{7/2}$ vibronic states and discussed a correction term Δg_l for the g factor in terms of the Herzberg-Teller interaction between the $\tilde{X}\,^2\Pi$ and $\tilde{A}\,^2\Sigma^+$ states (Δg_l arising from this origin was later referred to as $\Delta g_l^{(1)}$). The Δg_l constant is useful in discriminating between the first-order and second-order contributions to the Renner parameter ε.

Dixon [4.131] also observed the parallel band ($\Delta K = 0$) in addition to the main perpendicular band in the $\tilde{A}\,^2\Sigma^+ - \tilde{X}\,^2\Pi$ transition with the intensity ratio of 1 to 10. This subsidiary band was accounted for by *Bolman* and *Brown* [4.134] as for Δg_l, namely in terms of the Herzberg-Teller interaction. *Bolman* et al. [4.135] reinvestigated the $\tilde{A}\,^2\Sigma^+ - \tilde{X}\,^2\Pi_i$ band by placing a particular emphasis on the rotational energy level structure of $v_2 = 1$ in the ground electronic state, and revised the Renner parameter given by Dixon to -0.144 ± 0.001. They also extended the discussion of Carrington et al. on the correction term Δg_l so as to include the effect of the $\Delta v_2 = \pm 2$ matrix elements of the Renner-Teller interaction within the $\tilde{X}\,^2\Pi$ state; this part is referred to as $\Delta g_l^{(2)}$. The $\Delta g_l^{(1)}$ term has recently been reevaluated to be $-0.00689(1)$ by CO laser LMR [4.136].

Conventional microwave spectroscopy has also been applied to NCO in the $\tilde{X}\,(0\,0\,0)\,^2\Pi_i$ [4.137] and $\tilde{X}\,(0\,1\,0)\,^2\Delta_i$ [4.138] states. *Kawaguchi* et al. [4.139] recently observed the rotational spectrum of NCO in the $\tilde{X}\,(0\,1\,0)\,^2\Sigma$ state to investigate the rovibronic interaction in detail, and, more recently, they extended the measurements on the $\tilde{X}\,(0\,0\,0)\,^2\Pi_i$ and $\tilde{X}\,(0\,1\,0)\,^2\Delta_i$ states and observed the $\tilde{X}\,(0\,2\,0)\,^2\Phi_i$ state as well. The observed spectrum were analyzed using (2.4.46, 61), respectively, for the $K = v_2 + 1$ and $K = 0(\Sigma)$ states. These microwave studies have clearly revealed vibrational dependences of molecular constants; the rotational constant in the $\tilde{X}\,(0\,1\,0)\,^2\Sigma$ state differs slightly from that in the $\tilde{X}\,(0\,1\,0)\,^2\Delta$ state, the effective spin-rotation coupling constant

Table 4.6. Difference (in MHz) between the rotational constants of linear triatomic molecules in the (0, 1, 0) Δ and (0, 1, 0) Σ states

Molecule	State	Observed [a]	Calculated [b]	Ref.
NCO	$\tilde{X}\,^2\Pi$	1.99 (12)	1.64	[4.128]
BO_2	$\tilde{X}\,^2\Pi$	3.8 (14)	3.2	[4.124]
CNC	$\tilde{X}\,^2\Pi$	44	43	[4.107]
CO_2^+	$\tilde{X}\,^2\Pi$	−30 (26)	2.4	[4.127]
CCC	$\tilde{A}\,^1\Pi$	99	33	[4.126]

[a] Observed $B(\Delta) - B(\Sigma)$ values. Values in parentheses denote three standard deviations and apply to the last digits of the constants

[b] $B(\Delta) - B(\Sigma)$ calculated using the formula including rovibronic interaction [4.128]

shows a large v_2 dependence in the $K = v_2 + 1$ states, and the a hyperfine coupling constant takes quite different values in different vibronic states.

The rotational constant difference listed in Table 4.6 with those of other Renner molecules has been explained by *Kawaguchi* et al. [4.128] as an effect of higher-order rovibronic interaction including the Renner effect; the third-order correction is given by (2.4.59, 66) and the fourth-order term has also been worked out. As shown in Table 4.6, the calculated difference ($^2\Delta - {}^2\Sigma$) agrees well with the observed value for NCO, BO_2, and CNC, but does not for CCC and CO_2^+.

The effective spin-rotation coupling constants obtained for NCO [4.139] and CCN [4.129] are − 32.15 (47) and − 112.7 (42) in MHz with three standard errors in parentheses, respectively. As pointed out by *Veseth* [4.140] and also *Brown* and *Watson* [4.141], the spin-rotation coupling constant γ cannot be determined for a $^2\Pi$ molecule separately from the rotational dependence of the spin-orbit coupling A_J, unless spectroscopic data are available for more than one isotopic species. Therefore, either A_J or γ is fixed at zero in analyzing the observed spectrum. Then the derived constant is an effective constant given by

$$\gamma_{\text{eff}} = \gamma - p^*/2 - (A - 2BK)\,A_J/2B \quad \text{or} \tag{4.2.9}$$

$$(A_J)_{\text{eff}} = A_J - [2B/(A - 2BK)](\gamma - p^*/2), \tag{4.2.10}$$

depending on whether $A_J = 0$ or $\gamma = 0$ is assumed. The vibronic dependence of the spin-orbit coupling constant A given by (2.4.40 or 41) is obviously too small to account for the drastic charges of γ_{eff}. Instead, *Kawaguchi* et al. [4.128] considered higher-order effects caused by interactions with other electronic states represented by the following Hamiltonians:

$$\boldsymbol{H}_{\text{RT}} = V_{22}\,Q_2 \cos(\theta - \phi)/hc, \tag{4.2.11}$$

$$\boldsymbol{H}_{\text{LS}} = [(A + 2B)/2](L_+S_- + L_-S_+), \quad \text{and} \tag{4.2.12}$$

$$\boldsymbol{H}_{\text{JL}} = -\,B(J_+L_- + J_-L_+). \tag{4.2.13}$$

The H_{LS} and H_{JL} terms are the L-uncoupling terms in (2.4.45) which cause Λ-type doubling in the $^2\Pi$ state. A fourth-order perturbation calculation yields the following correction term for γ:

$$\gamma_{LU} = p^* \eta \Delta E(v_2 + 1) \sqrt{X}/(2hc\omega_2), \tag{4.2.14}$$

where η is given by (2.4.42), $X = (J + 1/2)^2 - K^2$, and ΔE denotes the energy difference between the $\tilde{A}$ and $\tilde{X}$ states (it is assumed that only the $\tilde{A}\,^2\Sigma^+$ state makes a substantial contribution). As remarked in Chap. 2, p^* cannot be obtained independently from other constants, but may be set equal to p when only Σ^+ excited electronic states take part in the interaction with $\tilde{X}\,^2\Pi$. This seems to be a good approximation for NCO; the observed change of γ with v_2 requires η to be $-$ 0.012, which is close to the value obtained from the vibronic dependence of the a hyperfine coupling constant. For CCN, on the other hand, $p^* = p$ is not a good approximation; the observed p constant is too small to account for the observed change of γ.

CCN. *Merer* and *Travis* [4.106] observed three systems of CCN, $\tilde{A}\,^2\Delta - \tilde{X}\,^2\Pi$, $\tilde{B}\,^2\Sigma^- - \tilde{X}\,^2\Pi$, and $\tilde{C}\,^2\Sigma^+ - \tilde{X}\,^2\Pi$, and evaluated molecular constants from the observed spectrum. The Renner parameter was found to be large, $\varepsilon = 0.44$, and vibronic interactions including the Renner effect may thus affect the properties of this radical. The bending energy levels are illustrated in Fig. 4.19.

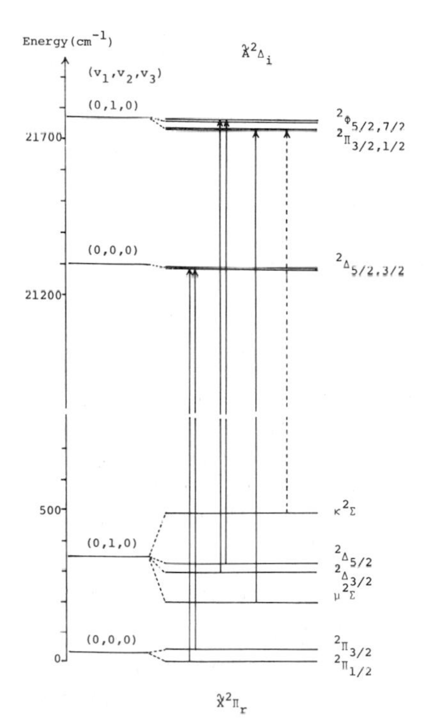

Fig. 4.19. $\tilde{A}^2\Delta_i - \tilde{X}^2\Pi_r$ transition of a linear triatomic molecule exhibiting the effect of Renner-Teller interaction

Kakimoto and *Kasuya* [4.142] and *Kawaguchi* et al. [4.129] have recently reinvestigated the (0 0 0) – (0 0 0), (0 1 0) – (0 1 0), and (0 2 0) – (0 2 0) bands of the $\tilde{A} - \tilde{X}$ transition using cw dye laser excitation spectroscopy.

An interesting feature of the CCN spectrum is that Merer and Travis failed to observe the $\tilde{A}\,(0\,1\,0)\,{}^2\Pi - \tilde{X}\,(0\,1\,0)\,\kappa\,{}^2\Sigma$ band in absorption, although the $\tilde{A}\,(0\,1\,0)\,{}^2\Pi - \tilde{X}\,(0\,1\,0)\,\mu\,{}^2\Sigma$ was clearly seen. A similar observation was reported by *Kawaguchi* et al. [4.129], who could detect even the $\tilde{A}\,(0\,2\,0)\,{}^2\Delta - \tilde{X}\,(0\,2\,0)^2\,\Phi$ band of which the initial state is higher than $\tilde{X}\,(0\,1\,0)\,\kappa\,{}^2\Sigma$. *Hakuta* et al. [4.143, 144] have, however, observed both the $\tilde{A}\,(0\,1\,0)\,{}^2\Pi \rightarrow \tilde{X}\,(0\,1\,0)\,\mu\,{}^2\Sigma$ and $\tilde{A}\,(0\,1\,0)\,{}^2\Pi \rightarrow \tilde{X}\,(0\,1\,0)\,\kappa\,{}^2\Sigma$ bands in emission with nearly equal intensities, where they excited either the $\tilde{A}\,(0\,1\,0)\,{}^2\Pi - \tilde{X}\,(0\,1\,0)\,\mu\,{}^2\Sigma$ or $\tilde{A}\,(0\,1\,0)^2\,\Phi - \tilde{X}\,(0\,1\,0)\,{}^2\Delta$ transition with a dye laser or an Ar^+ laser. *Kawaguchi* et al. [4.129] proposed a model of preferential population of CCN molecules in $\mu\,{}^2\Sigma$ when they are generated by the reaction of CH_3CN with microwave discharge products of CF_4.

4.3 Nonlinear XY_2- and XYZ-Type Triatomic Free Radicals

A number of nonlinear triatomic free radicals have been investigated by high-resolution spectroscopy, presumably because such species in doublet states serve as a prototype of polyatomic free radicals. This section is devoted to such species either of C_s symmetry (Sect. 4.3.1) or of C_{2v} symmetry (Sect. 4.3.2), while the physical significance of the spin-rotation interaction constants and the hyperfine coupling constants derived from the observed spectra are discussed in Sect. 4.5.

The molecular structure (the r_0 structure) is easily calculated from the observed ground-state rotational constants for the XY_2-type molecules, whereas more than one isotopic species needs to be studied spectroscopically for an XYZ-type molecule to determine its three structural parameters. As is well recognized, the ground-state rotational constants and the vibration-rotation constants depend not only on the structure parameters, but also on the third-order anharmonic potential constants. Thus structure can be determined only with good estimates for the anharmonicity constants. Such an analysis may be performed straightforwardly for an XY_2-type molecule, while, because of lower symmetry, additional data are required for an XYZ-type radical to be taken into account. The structures of three representative molecules of HYZ-type are briefly discussed, together with some preliminary results on HPO and FO_2 in Sect. 4.3.3.

4.3.1 Nonlinear XYZ-Type Free Radicals of C_s Symmetry

HO_2. The hydroperoxyl radical HO_2 is perhaps one of the representative triatomic free radicals belonging to C_s symmetry. It had often been postulated as an important intermediate in various chemical reactions such as the explosive recombination of hydrogen and oxygen [4.145] and the combustion of hydrocarbons [4.146].

Jacox and *Milligan* [4.147, 148] have identified the fundamental bands of the three normal modes of HO_2 and its isotopic species trapped in low-temperature Ar matrices. Its existence in the gas phase was first demonstrated by *Paukert* and *Johnston* [4.149], who observed all three vibrational bands with the aid of molecular modulation, but the resolution was not high.

Soon afterwards, *Radford* and co-workers [4.150] observed the rotational spectrum of HO_2 using FIR laser magnetic resonance (FIR LMR). They tested many reactions which had been thought to generate HO_2 and found that most of them really gave the radical. *Hougen* [4.151] has developed a method which allows us to analyze LMR spectra of free radicals of any sort in orbitally nondegenerate doublet states, and he and his co-workers [4.152] have subsequently applied this method to the FIR LMR spectrum of HO_2 to determine the rotational (A, B, and C), the symmetric-top centrifugal distortion (D_N, D_{NK}, and D_K), and the diagonal components of the spin-rotation coupling (ε_{aa}, ε_{bb}, and ε_{cc}) constants.

The chemistry for producing HO_2 and its molecular constants reported in these LMR studies facilitated the microwave observation of the pure rotational spectrum; *Beers* and *Howard* [4.153, 154] detected the $1_{01} - 0_{00}$ transitions of HO_2 and DO_2, and *Saito* [4.155] observed the a-type $N = 2 - 1$ and four b-type transitions. Figure 4.20 shows an example of the observed spectral lines, the hyperfine structure of the $2_{12} - 1_{11}$, $J = 5/2 - 3/2$ transition of HO_2. By analyzing the observed transition frequencies Saito showed that the off-diagonal component of the spin-rotation coupling tensor, $\varepsilon_{ab} + \varepsilon_{ba}$, caused a few spin-rotational levels to shift. The $\varepsilon_{ab} + \varepsilon_{ba}$ term gives the matrix elements of $\Delta N = 0, \pm 1$, $\Delta K_a = \pm 1$, and $\Delta J = 0$, which affect most of the F_2 component of the $N_{0,N}$ levels and the F_1 component of the $(N-1)_{1,N-2}$ levels, because these two rotational levels are nearly degenerate in HO_2. The splitting between them becomes smallest and changes sign between $N = 9$ and 10, as seen from Fig. 4.21. In fact, the shift due to the $\varepsilon_{ab} + \varepsilon_{ba}$ term was found to be large and opposite in sign for the $9_{09} - 8_{18}$, $J = 17/2 - 15/2$ and $10_{0,10} - 9_{19}$, $J = 19/2 - 17/2$ transitions. Saito determined the value of $|\varepsilon_{ab} + \varepsilon_{ba}|$ from the spin splittings observed for the b-type transitions by means of a second-order perturbation treatment. He also determined the proton hyperfine coupling constants and established HO_2 to be a π radical, i.e., its ground electronic state to be $^2A''$, as suggested by an earlier EPR study [4.156] of HO_2 trapped in an Ar matrix.

Subsequently, *Barnes* et al. [4.157, 158] observed the EPR spectrum of HO_2 in the gas phase and analyzed their own data by combining them with the microwave and LMR results already reported, so obtaining improved molecular constants. *Charo* and *De Lucia* [4.159] have carried out extensive measurements of the millimeter-wave spectrum of HO_2 in the region 150–500 GHz and have further refined the rotational, centrifugal distortion, and spin-rotation constants of HO_2 in the ground vibronic state.

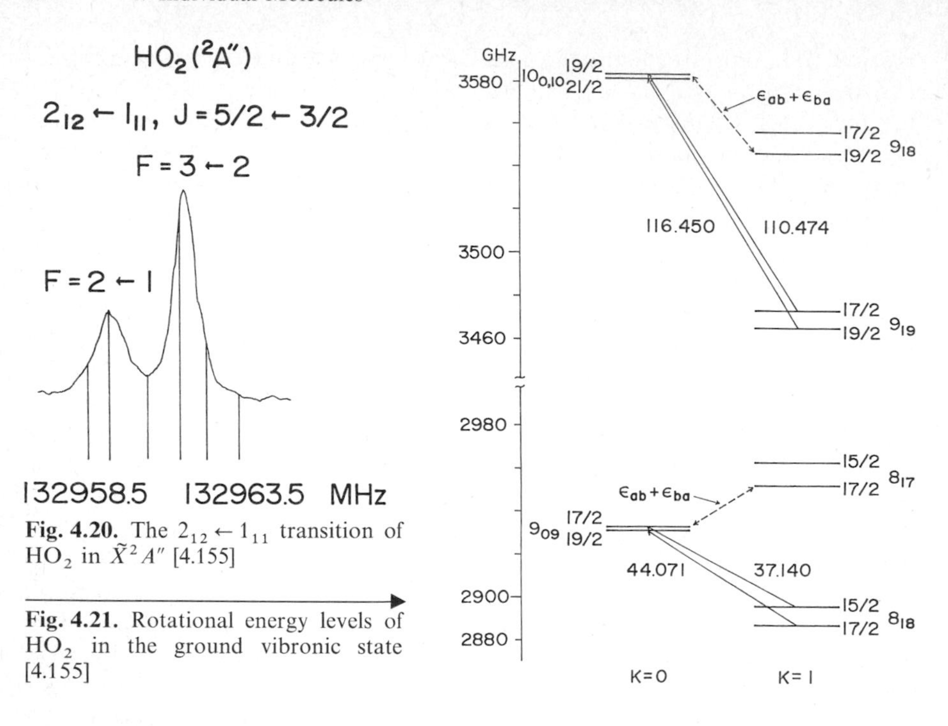

Fig. 4.20. The $2_{12} \leftarrow 1_{11}$ transition of HO_2 in $\tilde{X}^2A''$ [4.155]

Fig. 4.21. Rotational energy levels of HO_2 in the ground vibronic state [4.155]

The HO_2 radical is fairly stable even in a parallel-plate Stark cell. *Saito* and *Matsumura* [4.160] have thus measured the Stark effects of the $1_{01} - 0_{00}$ and two *b*-type transitions of HO_2 to determine the dipole moment: $\mu_a = 1.412(33)$, $\mu_b = 1.541(16)$, and $\mu_{\text{total}} = 2.090(34)$ *D*, where the unresolved Stark patterns of the *b*-type transitions were analyzed by computer simulation.

The vibration-rotation spectra of HO_2 have been studied by using mid-IR LMR (ν_3) [4.161, 162], IR diode laser (ν_2) [4.163], and difference-frequency laser (ν_1) [4.164] spectroscopy. The $\tilde{A}\,^2A' - \tilde{X}\,^2A''$ electronic transition has been observed in absorption in the near-IR region by *Jones* and collaborators [4.165, 166] using a Fourier transform spectrometer.

The deuterated species has also been subjected to quite extensive spectroscopic studies. *Barnes* et al. [4.167] have observed the gas phase EPR and FIR LMR spectra and have determined most of the ground-state molecular constants. They have combined the rotational constants of DO_2 with those of HO_2 to calculate the r_0 structure. *Saito* et al. [4.168] observed several *a*-type and *b*-type rotational transitions of DO_2 by microwave spectroscopy and have improved the ground-state parameters including the hyperfine coupling constants. They have also demonstrated that the hyperfine coupling constants determined for HO_2 and DO_2 could be explained only when the off-diagonal term T_{ab} of the dipolar interaction tensor was taken into account. Thus the tensor has been determined completely.

The vibration-rotation spectrum of DO_2 has also been investigated quite extensively. *McKellar* [4.169] has observed the ν_2 band by mid-IR LMR, which he analyzed by taking into account the Coriolis interaction with ν_3. *Uehara* et al. [4.170] have recently extended observations of the ν_2 and ν_3 bands by using IR diode laser spectroscopy. They also collaborated with *Lubic* et al. [4.171] in ν_1 band observations, where the diode laser and difference-frequency laser were employed by the two groups, respectively. Molecular constants are now available for both HO_2 and DO_2 in the ground and all the three fundamental states and allow us to calculate the equilibrium properties of the hydroperoxyl radical; the molecular structure is discussed in Sect. 4.3.3 together with third-order anharmonic potential constants.

FO_2. A few derivatives of HO_2 are known to exist, including FO_2. *Arkell* [4.172] reported the first observation of the IR spectrum of FO_2 trapped in N_2 or Ar matrices at 4 K and assigned the two stretching bands at 584 and 1494 cm^{-1}, based upon the $^{16}O/^{18}O$ isotope shifts of the frequencies. This study was followed by the observation by *Spratley* et al. [4.173] of the bending mode at 376.0 cm^{-1}. Both groups suggested that the bond lengths of FO_2 are similar to those of F_2O_2, based upon the observed stretching frequencies. *Noble* and *Pimentel* [4.174] utilized the $^{16}O/^{18}O$ isotopic shifts to establish that the molecule involves two nonequivalent oxygen atoms, namely has FOO structure. *Jacox* [4.175] has recently reexamined the FO_2 matrix spectrum.

The electron spin resonance (ESR) spectrum has been observed for FO_2 in low-temperature matrices [4.176–178], and has been interpreted in terms of the bent FOO structure, in agreement with the IR studies.

Yamada and *Hirota* [4.179] observed the ν_2 band of the FO_2 radical in the gas phase by IR diode laser spectroscopy and determined the rotational, centrifugal distortion, and spin-rotation interaction constants. The band origin they determined is 579.31839 (35) cm^{-1}, 5–7 cm^{-1} lower than those reported for FO_2 in matrices. Yamada and Hirota derived structure parameters (Sect. 4.3.3), which support the conclusions of the previous IR studies [4.172–174].

HSO. The XSO molecules constitute another group closely related with the hydroperoxyl radical. The "parent" molecule HSO plays an important role in oxidation reactions of *S*-containing inorganic as well as organic molecules. *Schurath* et al. [4.180] gave the first evidence of its existence in the gas phase by observing chemiluminescence in the visible. *Kakimoto* et al. [4.181] studied the $\tilde{A}\,^2A'(0\,0\,3) - \tilde{X}\,^2A''(0\,0\,0)$ band by using Doppler-limited dye laser excitation spectroscopy and determined molecular constants in both the ground and excited states. *Ohashi* et al. [4.182] have extended the same sort of measurements to DSO to obtain additional information on the radical, including the molecular structure. The observation of the $\tilde{A} - \tilde{X}$ transition of HSO has recently been extended by *Satoh* et al. [4.183] to $v_3' = 4$ and by *Tsukiyama* et al. [4.184] to $v_3' = 2$ and 1.

Endo et al. [4.185] have succeeded in observing the microwave spectra of both HSO and DSO using a 3.5 m glow discharge cell (Sect. 3.1.1) and have improved ground-state molecular constants. By using the hyperfine coupling constants observed for the proton and deuteron, they concluded that the ground electronic state of HSO has $^2A''$ symmetry, as for HO_2.

Webster et al. [4.186] have measured the dipole moments of HSO in the $\tilde{A}$ (0 0 3) and $\tilde{X}$ (0 0 0) states from the Stark effects of the Doppler-limited dye laser excitation spectra. *Sears* and *McKellar* [4.187] have observed the ν_3 (S – O stretching) fundamental band of HSO using mid-IR LMR and determined molecular constants in the ν_3 state and the ν_3 band origin. In the course of this study they observed additional lines around 1050 cm^{-1}, which they suspected to be due to HOS. According to ab initio calculations by *Sannigrahi* et al. [4.188, 189], the isomer HOS is even more stable than HSO. No other spectroscopic evidence has been reported for the existence of HOS, however.

FSO and ClSO. There are two halogenated derivatives of HSO known. *Radford* et al. [4.190] observed a few FIR LMR lines in the reaction of CS_2 with discharge products of an O_2 and CF_4 mixture. They thought these lines were due to FSO. *Endo* et al. [4.191] found a more efficient method to produce FSO (Sect. 3.5) and observed its microwave spectrum by using a Stark modulation spectrometer with a parallel-plate cell. The observed spectrum includes *b*-type *P*, *Q*, and *R* transitions with *J* less than 13 and *K* less than 5. Because the fine and hyperfine coupling constants are of the same order, both interactions must be taken into account simultaneously. Figure 4.22 shows the *Q*-branch transitions $N_{1,N-1} - N_{0,N}$ with $N = 2 - 7$; several extra lines with $\Delta N = \Delta F = 0$, but $\Delta J = \pm 1$ were observed for $N < 4$. Normally these transitions are weak,

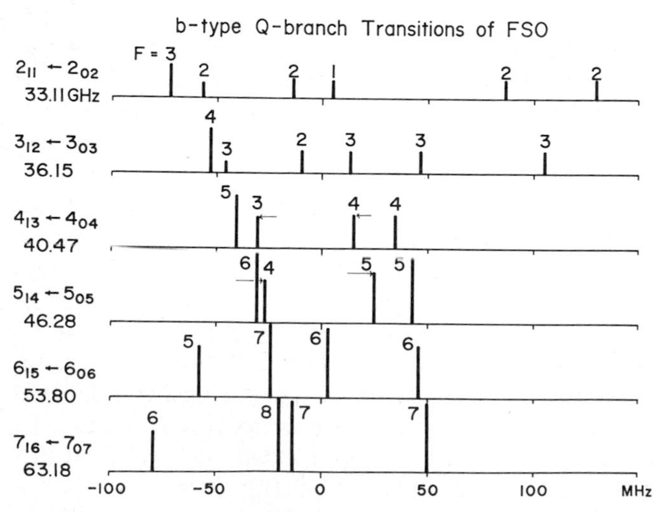

Fig. 4.22. *Q*-branch transitions of FSO in the ground vibronic state; arrows indicate frequency shifts of the lines caused by the $\varepsilon_{ab} + \varepsilon_{ba}$ term

perhaps an order of magnitude weaker than those of the $\Delta N = \Delta J = \Delta F = 0$ transitions, but, because of mixing of $N = F$ levels by the hyperfine coupling, the $\Delta J = \pm 1$ transitions gain intensity from the $\Delta J = 0$ transitions. Figure 4.22 also illustrates that the $\varepsilon_{ab} + \varepsilon_{ba}$ term causes local perturbations of the Q-branch transitions, indicated by arrows; 3_{21} and 4_{22} are the main perturbers for 4_{13} and 5_{14}, respectively. Endo et al. have determined 19 molecular parameters by a least-squares analysis of all the observed lines, not only for the normal species in the ground state, but also for the normal species in the $v_3 = 1$ state and the $F^{34}SO$ species in the ground state. The dipole moment Endo et al. evaluated from the Stark effects is $\mu_a = 0.374(12)$ D, $\mu_b = 1.624(9)$ D, and $\mu_{total} = 1.666(13)$ D with the 2.5 standard errors in parentheses.

Radford et al. [4.190] also observed resonances for six FIR laser lines from the following chemical systems: (1) $SCl_2 + (O_2)_{MW\ disch.}$, (2) $SCl_2 + O$ (O was generated by $N + NO \rightarrow N_2 + O$), (3) $Cl_2 + (O_2 + CS_2)_{MW\ disch.}$, (4) $Cl_2 + (SO_2)_{MW\ disch.}$, and (5) $SOCl_2 + (O_2)_{MW\ disch.}$, where $(M)_{MW\ disch.}$ denotes microwave discharge products of M (Sect. 3.5). They suspected these LMR signals to be due to ClSO, judged from the reactions employed. As briefly mentioned in Sects. 3.1.3, 3.5, *Saito* et al. [4.192] succeeded in observing the microwave spectrum of ClSO; the reaction used was $(S_2Cl_2 + O_2)_{dc\ disch.}$, as listed in Table 3.3.

HCO and FCO. The formyl radical HCO acts as the parent species of another family of bent triatomic free radicals. Because of its importance in many fields, numerous spectroscopic studies have already been carried out on both HCO and DCO and have provided detailed information on them. In contrast with HO_2, HCO has an unpaired electron orbital in the molecular plane, i.e., its ground electronic state is of $^2A'$ symmetry.

A few derivatives of HCO were known beforehand. Among them, the FCO radical is discussed here in some detail. *Milligan* et al. [4.193] and *Jacox* [4.194] reported the IR and uv spectra of FCO prepared in low-temperature Ar, CO, or N_2 matrices. *Wang* and *Jones* [4.195] generated FCO in the gas phase by flash-photolyzing a mixture of N_2F_4, CO, and N_2, and observed an absorption spectrum in 220–340 nm. *Adrian* et al. [4.196] and *Cochran* et al. [4.197] reported the electron spin resonance spectrum of FCO in CO matrices and concluded that the molecule is a σ radical in the ground electronic state from the observed fluorine hyperfine splittings, and that the F – C – O angle is about 110° from the ^{13}C hyperfine coupling constant. *Nagai* et al. [4.198] applied IR diode laser spectroscopy to the ν_1 and ν_2 bands of FCO and the ν_1 band of $F^{13}CO$, and derived precise values of the rotational, centrifugal distortion, and spin-rotation interaction constants for the ground, $v_1 = 1$, and $v_2 = 1$ states. The band origins have been determined to be $\nu_1 = 1861.6372(1)\ cm^{-1}$ and $\nu_2 = 1026.1283(1)\ cm^{-1}$ for FCO, which agree well with the Ar matrix values, 1857 and 1023 cm^{-1}, respectively. Nagai et al. have also calculated the molecular structure and force field by using molecular constants obtained from the observed spectra.

4.3.2 Nonlinear XY_2-Type Free Radicals of C_{2v} Symmetry

The nonlinear XY_2-type molecule with C_{2v} symmetric is simpler than the nonlinear XYZ-type molecule discussed in Sect. 4.3.1 in that fewer molecular parameters are needed for XY_2 than for XYZ in expressing the rovibronic energy levels. For example, the off-diagonal terms of the spin-rotation interaction ε_{ab} and ε_{ba} and of the dipolar hyperfine interaction T_{ab} of the central atom (X) are missing in XY_2. Furthermore, each rotational level will be weighted according to the Y nuclear spin statistics, providing us with a clue for assignments of the observed spectra. In the present section four XY_2-type molecules CH_2, NH_2, PH_2, and PO_2 are discussed in detail.

CH_2. The methylene radical was initially investigated by uv and visible spectroscopy in the gas phase [4.199–201] and by ESR spectroscopy in low-temperature matrices [4.202–205]. Remarkable progress has recently been made in spectroscopy in the IR, FIR, and microwave regions of CH_2 and CD_2 in the $\tilde{X}\,^3B_1$ state, primarily focusing on the $\tilde{a}\,^1A_1 - \tilde{X}\,^3B_1$ separation. The first FIR LMR detection of CH_2 was made by *Mucha* et al. [4.206], who ascribed the observed spectra to pure rotational transitions of CH_2 based on chemical evidence and triplet hyperfine structures of the observed spectra, but they could not assign the observed resonances. More recently, *Sears* et al. [4.207, 208] first detected the ν_2 (bending) fundamental band of CH_2 by 11 μm CO_2 LMR spectroscopy, the analysis of which then facilitated observation of a few pure rotational transitions with FIR LMR [4.209]. *McKellar* and *Sears* [4.210] extended the IR LMR measurements to the ν_2 band of $^{13}CH_2$ and *Bunker* et al. [4.211] the FIR LMR observations to the pure rotational spectrum of CD_2. These results enabled *Lovas* et al. [4.212] to observe a rotational transition at 70 GHz by microwave spectroscopy. *McKellar* et al. [4.213] applied IR diode laser spectroscopy to the ν_2 band of CD_2, which appears in the region 600–800 cm^{-1} and is thus not possible to study by IR LMR.

In the diode laser experiment CH_2 and CD_2 were prepared directly in a Zeeman-modulated multiple-reflection absorption cell 4 cm in diameter by the reaction of fluorine atoms with ketene (CH_2CO and CD_2CO, respectively) obtained through pyrolysis of anhydrous acetic acid flowing through a quartz tube held at around 500 °C. The F atoms were produced by a dc discharge in a mixture of CF_4 and He, induced directly in the cell. The ketene vapor was introduced in the cell through a concentric jacket with a series of holes allowing ketene to mix uniformly with the CF_4 discharge plasma over the entire cell. The typical partial pressure was 0.5, 0.2, and 0.1 Torr, respectively, for CF_4, He, and anhydrous acetic acid. A simple discharge in a CH_2CO and CF_4 mixture did not give CH_2 spectra. An external microwave discharge in CF_4 in place of the internal dc discharge reduced the signal by a factor of 2 to 3. Substituting CH_4 for CH_2CO gave signals 2 to 4 times weaker. A dc discharge in pure CH_4 led to signals about 5 times weaker, and adding CF_4 did not increase the signal

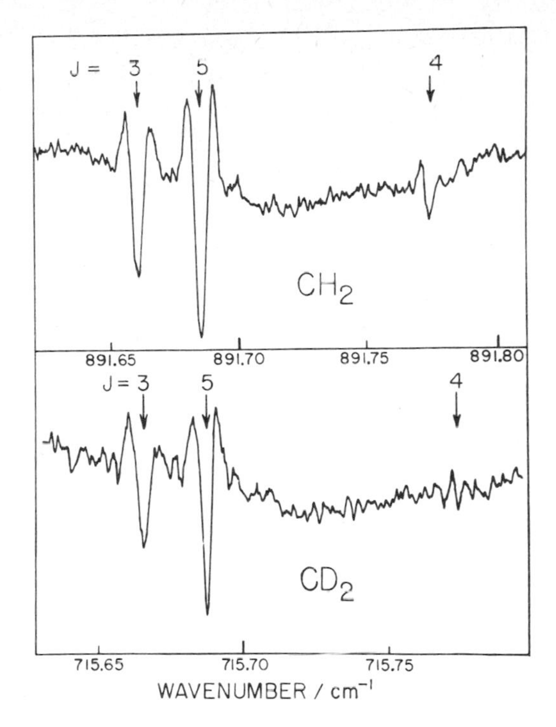

Fig. 4.23. The ν_2 $4_{04} \leftarrow 4_{13}$ transitions of CH_2 and CD_2 in $\tilde{X}\,^3B_1$ using an IR diode laser and Zeeman modulation [4.213]

strength. Neither discharges in C_2H_2 nor acetone resulted in detectable signals. A dc discharge in pyrolyzed anhydrous acetic acid gave signals 2 to 3 times weaker than that with CF_4 added.

First the $4_{04}-4_{13}$ transition of the CH_2 ν_2 band was observed at $891.7\ \mathrm{cm}^{-1}$, predicted from the LMR results [4.208]. Searching for the CD_2 spectra was guided by predictions made by *Jensen* et al. [4.214] using a nonrigid bender model, and five Q-branch transitions of $N_{0N} - N_{1N-1}$ with $N = 3-7$ were observed in the region $717.1-709.5\ \mathrm{cm}^{-1}$. Figure 4.23 reproduces the $4_{04}-4_{13}$ transition together with the same transition of CH_2. By fixing the ground-state parameters to the LMR values of *Bunker* et al. [4.211], the band origin ν_0, the rotational and centrifugal distortion constants $(1/2)(B + C)$, Δ_N, and δ_N, the spin-spin interaction constant D, and the spin-rotation interaction constant $(1/2)(\varepsilon_{bb} + \varepsilon_{cc})$ have been calculated from the observed spectra. Recently *Evenson* et al. [4.215] succeeded in observing 13 pure rotational transitions of CD_2 in the ν_2 state by FIR LMR and analyzed all available data to determine molecular parameters of CD_2 in the ν_2 state.

The ν_2 band origin $752.3795(4)\ \mathrm{cm}^{-1}$ thus obtained for CD_2 is $1.5\ \mathrm{cm}^{-1}$ higher than the predicted value of *Jensen* et al. [4.214], who employed the data on CH_2 in their nonrigid bender calculation. Later, *Bunker* et al. [4.216] repeated the analysis by taking into account all the data available including the diode laser results on CD_2 and reproduced the vibrational frequencies with the mean deviation of 0.04 cm^{-1}. Their bending potential is of double minimum type, with a Lorentzian hump at the linear configuration and with an allowance for anharmonicity. They have thus determined the potential barrier to the linearity to be 1939.27 cm^{-1}.

NH_2. The amino radical is one intermediate in chemical reactions involving ammonia and has also been identified in comet tails through its optical spectrum. It was the first example for which the Renner effect was clearly recognized. In fact, two electronic states hitherto known, $\tilde{A}\,^2A_1$ and $\tilde{X}\,^2B_1$, may be considered as resulting from a linear configuration split by a large Renner effect.

Figure 4.24 illustrates the bending potential functions in the $\tilde{A}$ and $\tilde{X}$ states calculated by *Jungen* et al. [4.217]. The $\tilde{A}\,^2A_1 - \tilde{X}\,^2B_1$ transition was observed in absorption by *Ramsay* [4.218] and later by *Dressler* and *Ramsay* [4.219] in further detail. More extensive analysis has recently been performed by *Johns* et al. [4.220] and *Birss* et al. [4.221], who have precisely evaluated molecular constants of $v_2 = 0$, 1, and 2 in the $\tilde{X}$ state. *Davies* et al. [4.222] observed the pure rotational spectrum in the ground vibronic state by FIR LMR, and *Cook* et al. [4.223] employed MODR to clarify the fine and hyperfine structures in both the ground and the excited electronic states. The NH_2 radical in the ground electronic state has also been subjected to CO and CO_2 LMR [4.224, 225] and difference frequency laser [4.226] spectroscopic studies.

The rotational, fine, and hyperfine energy levels of the ground-state NH_2 radical are well expressed by the Hamiltonians discussed in Sect. 2.3. Because the molecule is very light, higher-order centrifugal distortion terms need to be included in the rotational and the spin-rotation interaction Hamiltonians. On

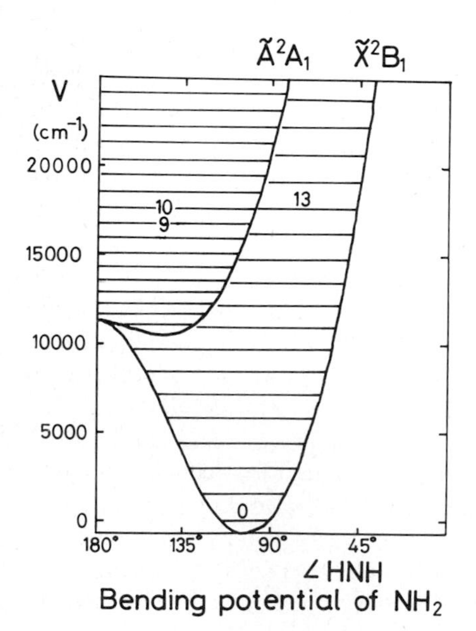

Fig. 4.24. Bending potential energy curves of NH_2 in $\tilde{A}\,^2A_1$ and $\tilde{X}\,^2B_1$

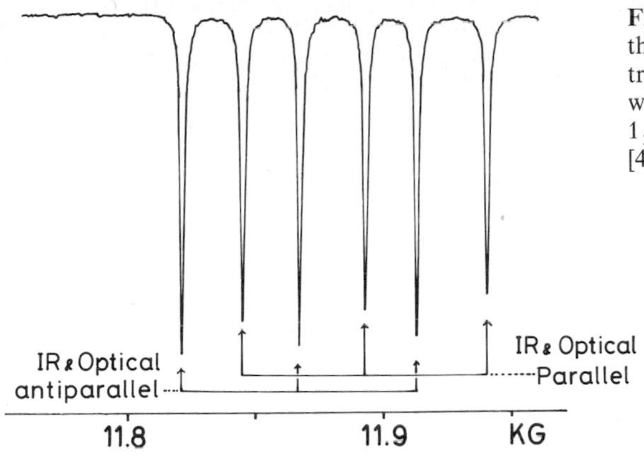

Fig. 4.25. IODR signal of NH_2; the $\tilde{A}\, v_2 = 10\ 1_{11} - \tilde{A}\, v_2 = 9\ 2_{20}$ transition is observed by LMR, while the $\tilde{A}\, v_2 = 9\ 2_{20} - \tilde{X}\, v = 0\ 1_{10}$ is pumped by a dye laser [4.227]

the other hand, the rotational energy levels in the excited electronic state deviate considerably from the ordinary expressions for an asymmetric top because of the Renner effect and other interactions primarily with excited vibrational states in the ground electronic state. The fine structure has also been observed to vary with the vibrational quantum number and the K value, the a-axis component of the rotational angular momentum.

Amano et al. [4.227] applied IR optical double resonance (IODR), Sect. 3.4.3, to the NH_2 radical to explore possible perturbations in the excited electronic state through observing the hyperfine structure and the Zeeman effect. They focused particularly on detecting highly excited vibrational states associated with the ground electronic state that interact with the excited vibronic levels.

An IODR experiment was carried out using the $\tilde{A}\,(0, 9, 0)\ 2_{20} - \tilde{X}\,(0, 0, 0)\ 1_{10}$ transition for dye laser pumping and the $(0, 10, 0)\ 2_{11} - (0, 9, 0)\ 2_{20}$ vibrational transition in the $\tilde{A}$ state as the IR transition; the latter was observed by the LMR technique using a CO_2/N_2O laser as a source. The double resonance signals were observed as an increase in fluorescence intensity, because the Franck-Condon factor is nearly the same for the $\tilde{A}\,(0, 10, 0) \leftarrow \tilde{X}$ and $\tilde{A}\,(0, 9, 0) \leftarrow \tilde{X}$ transitions. Figure 4.25 shows an example of the observed spectra recorded by sweeping the magnetic field with an absorption cell in the IR laser cavity. Because all three levels involved have their total proton spin equal to zero, only the nitrogen hyperfine structure remained, resulting in a triplet structure. Two triplets shown in Fig. 4.25 correspond to two cases where the IR and optical laser beams copropagate and counterpropagate, respectively.

Figure 4.26 illustrates the Zeeman effects of the $\Delta M_J = 1$ components of the $\tilde{A}\,(0, 10, 0)\ 2_{11} - \tilde{A}\,(0, 9, 0)\ 2_{20}$ IR transition with the CO_2 P(6), N_2O R(20), and N_2O R(21) laser lines employed as sources. The open circles denote the observed resonance signals, whereas the crosses were expected to give DR signals, but no resonances were observed in spite of careful searches.

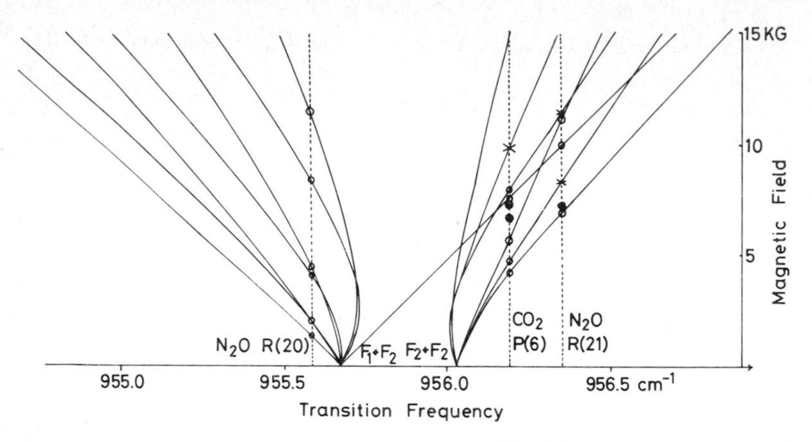

▲ **Fig. 4.26.** Zeeman shift of the $\tilde{A}\, v_2 = 10$ $2_{11} - \tilde{A}\, v_2 = 9$ 2_{20} transition of NH_2 with the $\Delta M = \pm 1$ selection rule [4.227]

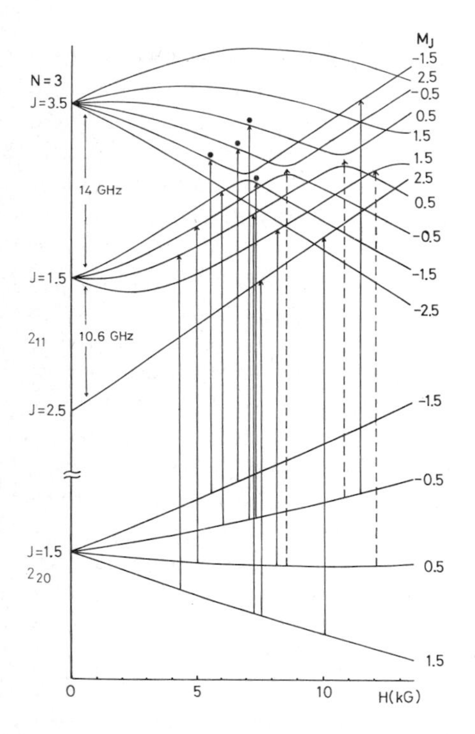

Fig. 4.27. Zeeman energy levels of $u\,N = 3\,F_1$, $\tilde{A}\, v_2 = 10$ 2_{11}, and $\tilde{A}\, v_2 = 9$ 2_{20} F_2 states of NH_2 [4.227]

Instead, a few additional lines were found, designated by closed circles. They were strong, but of opposite phases, that is, they corresponded to decreases in the fluorescence intensity. Furthermore, they showed negative Zeeman effects. These observations have been explained by assuming an unknown state (u state) which crosses with the 2_{11} level when an external magnetic field is applied, as shown in Fig. 4.27. The absence of a few resonances marked by crosses in Fig. 4.26 is ascribed to avoided crossing between the 2_{11} and u levels.

In fact, an interaction term ($\boldsymbol{H}_p$ denotes a perturbation Hamiltonian)

$$V = \langle \tilde{A}, v_2 = 10, 2_{11}, J = 2.5 | \, H_p \, | u, N = 3, J = 2.5 \rangle = 4120 \text{ MHz} \quad (4.3.1)$$

explains all the observed data. Note that avoided crossing also occurs between two levels of different J values, since the Zeeman field causes $\Delta J = \pm 1$ mixing of levels.

The observed Zeeman effects have shown the u state to have $N = 3$ with the F_1 component ($J = N + 1/2$) being lower than the F_2 component ($J = N - 1/2$). The observed hyperfine structure have led to the Fermi contact interaction constant $\sigma(N)$ for the nitrogen nucleus:

$\sigma(N) = 147.64(70)$ MHz in the $\tilde{A}$ (0, 10, 0) state,

$\sigma(N) = 10.9(48)$ MHz in the u state.

The small Fermi contact term in the u state is consistent with the assignment that it is a highly excited vibrational state in the ground electronic state manifold, because the unpaired electron orbital is extended perpendicularly to the molecular plane in the ground electronic state of $^2B_1$ symmetry. In fact, the Fermi contact term in the ground vibronic state has been determined by MODR to be 28.2 MHz [4.223]. In contrast, the $\tilde{A}$ state belongs to $^2A_1$ symmetry, and the unpaired electron orbital lies in the molecular plane, making the Fermi contact term much larger than in the $\tilde{X}$ state. The fluorescence from the u state is weak, as expected from the nature of its highly excited vibrational state.

The overall symmetry of the u state must be A_2, because it interacts with the $\tilde{A}$ (0, 10, 0) 2_{11} state. When the quantum number v_3 of the antisymmetric stretching vibration is even in the u state, the rotational level must be either 3_{30} or 3_{12}, whereas, when v_3 is odd, the rotational level must be 3_{22}.

Jungen et al. [4.217] calculated the term values for the vibrational levels with large v_2 quantum numbers in the ground electronic state manifold, but none of the calculated levels comes close to the u state. It is thus suspected that the v_1 and/or v_3 modes are excited; by referring to the vibrational anharmonicity of H_2O, the u state has been tentatively assigned to the $\tilde{X}$ (2, 8, 0) 3_{31} level. One problematic point with this assignment is that the observed spin splitting does not agree with that calculated by *Jungen* et al. [4.217] for the $\tilde{X}$ (0, 8, 0) state.

Similar u levels have been detected by MODR and laser excited fluorescence techniques. An advantage of the IODR experiment over other methods is that it allows the matrix element of the interaction between two electronic states to be determined, which is ascirbed primarily to the Renner effect.

PH_2. The $\tilde{A}\,^2A_1 - \tilde{X}\,^2B_1$ electronic transition in the visible has been the subject of a number of high-resolution spectroscopic studies. *Davies* et al. [4.228] reported FIR LMR spectra of PH_2 and determined the hyperfine coupling constants for both P and H nuclei. *Hills* and *McKellar* [4.229] reported the vibration-rotation spectrum of PH_2 observed by mid-IR LMR around 9 μm, and later *McKellar* [4.162] reanalyzed both FIR LMR and mid-IR LMR data simultaneously to improve the ground-state molecular constants. *Curl* et al. [4.230] observed dye-laser-induced fluorescence (LIF) in the visible region and estimated the P Fermi coupling constant in the $\tilde{A}$ state. *Kakimoto* and *Hirota* [4.231] resolved the hyperfine structures for 26 rotational transitions of the $\tilde{A}\,^2A_1(0\,0\,0) \leftarrow \tilde{X}\,^2B_1\,(0\,0\,0)$ vibronic band by applying intermodulated fluorescence. They also observed and analyzed the Doppler-limited LIF spectra of the same band [4.232]. *Endo* et al. [4.233] measured three Q-branch pure rotational transitions of PH_2 in the ground vibronic state by microwave spectroscopy and analyzed the observed spectrum combined with the FIR LMR, mid-IR LMR, and LIF spectra already reported to improve the rotational, spin-rotation interaction, and hyperfine coupling constants to good precision.

In the LIF and FIR LMR studies mentioned above, the PH_2 radical was produced by the reaction of red phosphorus with microwave discharge products of H_2, whereas Endo et al. found that a dc discharge in a PH_3 and O_2 mixture induced directly in an absorption cell gave 7 to 10 times stronger microwave signals of PH_2 than the red P plus H_2 discharge product reaction.

Endo et al. found it necessary to include the nuclear spin rotation interaction in analyzing the microwave spectrum they observed. They pointed out that the ratio of the nuclear spin-rotation interaction constant C_{gg} and the electron spin-rotation interaction constant ε_{gg} (g denoting one of the principal axes) is approximately equal to the ratio of the hyperfine a constant and the spin-orbit interaction constant A_{SO}, namely

$$|C_{gg}/\varepsilon_{gg}| = |a/A_{SO}|. \tag{4.3.2}$$

When a is approximated by (5/4) $T_{cc} = 780$ MHz and $A_{SO} = 300\ \mathrm{cm}^{-1}$ and $\varepsilon_{aa} = -8430$ MHz are employed, C_{aa} is calculated to be 0.73MHz, in qualitative agreement with the observed value of 0.96(5) MHz.

PO_2. This molecule is expected to be similar in many respects to the well-known NO_2 molecule; however, little has been reported on the molecular structure of PO_2 presumably because of its short lifetime. The continuum emission in the visible from the phosphorus and oxygen reaction has long been inferred to be due to PO_2, but neither vibrational nor rotational structures have been resolved, making spectrum identification rather uncertain. An exception is a very recent study by *Verma* and *McCarthy* [4.234] of the $2\,^2B_2 \leftarrow \tilde{X}\,^2A_1$ (?) band observed in absorption for a system consisting of PCl_3, O_2, and He with a 1:10:10 ratio which was flash photolyzed. They proposed a vibrational assignment, but have not resolved the rotational structure.

The pure rotational spectrum was recently observed by *Kawaguchi* et al. [4.235] whilst studying the PD radical with FIR LMR, Sect. 4.1.1. An example of the observed LMR spectrum is shown in the upper part of Fig. 4.11. Because the LMR spectrum appeared only at low magnetic field, the spin splitting was estimated to be small. Many Zeeman components were not completely resolved. Some of the observed spectra exhibited hyperfine structure due to the phosphorus nucleus. A preliminary assignment led to the A rotational constant of 3.4 cm^{-1} or 4.1 cm^{-1}.

Subsequently, microwave spectroscopy was applied to obtain more detailed information on the molecule [4.235]. First, $K_a = 1 \leftarrow 0$ Q-branch transitions were observed in the region 99–160 GHz and the two parameters $A-C$ and $B-C$ were obtained from the observed frequencies. Then the R-branch and P-branch transition frequencies were estimated by assuming a reasonable value for the inertia defect and some of them were identified. Only $K_a K_c = ee$ or oo levels were observed, indicating that the molecule has C_{2v} symmetry.

The observed spectrum was analyzed by using the Hamtilonians described in Sect. 2.3. However, the hyperfine structure was not well explained by the

Table 4.7. Vibrational frequencies and molecular structures of a few simple N/P-containing molecules[a]

N-containing molecule		Constant	Ref.	P-containing molecule		Constant	Ref.
NH	ω_e	3282.2_7	[4.37]	PH	ω_e	2365.2	[4.37]
	$\omega_e x_e$	78.3_5	[4.37]		$\omega_e x_e$	44.5	[4.37]
	ν_0	3125.5724	[4.236]		ν_0	2276.2106	[4.47]
	r_e	1.03756	[4.236]		r_e	1.4223_4	[4.37]
NO	ω_e	1904.12_2	[4.37]	PO	ω_e	1233.34	[4.37]
	$\omega_e x_e$	14.08_8	[4.37]		$\omega_e x_e$	6.56	[4.37]
					ν_0	1220.24901	[4.68]
	r_e	1.15077	[4.37]		r_e	1.476370	[4.68]
NH_2	ν_1	3219.371	[4.226]	PH_2	ν_1		
	ν_2	1497.3216	[4.225]		ν_2	1101.9086	[4.229, 162]
	ν_3	3301.110	[4.226]		ν_3		
	r_0(N – H)	1.0245_8	[4.222]		r_0(P – H)	1.418	[4.237]
	θ_0(HNH)	103.3_3	[4.222]		θ_0(HPH)	91.7	[4.237]
NO_2	ν_1	1319.795_4	[4.238]	PO_2	ν_1	(1088)	[4.235]
	ν_2	749.653_7	[4.238]		ν_2	(377)	[4.235]
	ν_3	1616.853_5	[4.238]		ν_3	(1278)	[4.235]
	r_e(N – O)	1.19389	[4.239]		r_0(P – O)	1.4665	[4.235]
	θ_e(ONO)	133.857	[4.239]		θ_0(OPO)	133.3	[4.235]
HNO	ν_1	2683.95210	[4.240]	HPO	ν_1	2095	[4.242]
	ν_2	1565.3481	[4.240]		ν_2	1188.043	[4.243]
	ν_3	1500.8192	[4.240]		ν_3	985.542	[4.243]
	r_e(H – N)	1.062_8	[4.241][b]		r_e(H – P)	1.42_0	[4.244]
	r_e(N – O)	1.205_9	[4.241][b]		r_e(P – O)	1.48_2	[4.244]
	θ_e(HNO)	109.0_9	[4.241][b]		θ_e(HPO)	102_9	[4.244]

[a] Vibrational frequencies are given in cm^{-1}, bond lengths in Å, and bond angles in degrees
[b] See Table 4.11

Hamiltonians (2.3.80, 82). Both the electron and nuclear spin-rotation interactions are small in PO_2. On the other hand, the Fermi contact term σ is large and its centrifugal distortion correction has been taken into account via

$$\sigma = \sigma_0 + \sigma_N N(N+1) + \sigma_K K^2 + \sigma_{KK} K^4. \qquad (4.3.3)$$

The correction terms σ_K was found to be most important and to be negative, in other words, as the molecule is bent by centrifugal distortion, the Fermi coupling constant decreases. This is contrary to what a simple consideration predicts, because, in the limit of a linear configuration, the unpaired electron occupies a π_u orbital and the Fermi constant would become zero. The negative K^2 dependence is rather explained by the mixing of the $^2A_1$ ground state with excited states of $^2B_1$ symmetry through the electronic Coriolis interaction of the form $-2AJ_aL_a$; the Fermi interaction will be very small in the $^2B_1$ states.

The rotational constants observed for the ground state lead to an r_0 structure as follows:

$$r_0(\mathrm{P-O}) = 1.4665(41)\,\text{Å} \quad \text{and} \quad \theta_0(\mathrm{OPO}) = 133^\circ 17'(50').$$

The molecular structures and the vibrational frequencies of a few phosphorus-containing molecules are summarized in Table 4.7, along with those of the corresponding nitrogen compounds. The two groups of molecules behave in a parallel way, with some variations.

4.3.3 Molecular Structure and Anharmonic Potential Constants of Nonlinear XYZ-Type Molecules of C_s Symmetry

It is straightforward to calculate two structure parameters [$r(X-Y)$ and $\theta(Y-X-Y)$] for a nonlinear XY_2-type molecule from its rotational constants, as done for PO_2 (Sect. 4.3.2). This does not hold for the nonlinear XYZ-type molecule. Although three rotational constants may be derived from the observed spectrum, they are insufficient to determine three structure parameters [$r(X-Y)$, $r(Y-Z)$, and $\theta(X-Y-Z)$], because the planarity of the molecule imposes a relation on the three rotational constants, leaving only two independent. It is customary to use the data on isotopic species, but, as well known, one must properly account for isotope effects creeping in the ground-state rotational constants through the zero point vibration.

Thanks to advances in rotational and vibrational spectroscopy, the rotational constants have been determined even for transient molecules, not only in the ground states, but also in excited vibrational states. If such data are available for two or more isotopic species in all the fundamental states, the equilibrium structure is readily calculated. This is really the case for the hydroperoxyl radical, as mentioned in Sect. 4.3.1. The results [4.171] are given in Table 4.8, where other values are cited for comparison. However, such a detailed study is generally very difficult to perform on short-lived molecules.

Table 4.8. Molecular structure of the hydroperoxyl radical

	r_e [4.171]	r_0 [4.166]	r_0 [4.167]	r_0 [4.245]
H – O [Å]	0.9707 (20)	0.9754 (63)	0.9774 (H) 0.9738 (D)	[0.982][a]
O – O [Å]	1.33054 (85)	1.3291 (18)	1.3339	1.3405
H – O – O [Å]	104.29 (31)	104.02 (72)	104.15	99.1

[a] Assumed

Hirota [4.241] has demonstrated that even when not all vibration-rotation constants are available, the equilibrium structure parameters may still be derived. The method is essentially based upon the fact that the vibrational potential constants are independent of isotopes when they are expressed in terms of the internal coordinates, rather than the normal coordinates. *Hoy* et al. [4.246] have developed a method of calculating the cubic (Φ_{rst}) and quartic (Φ_{rstu}) anharmonic potential constants (which are defined in terms of the normal coordinates) from the second-order or harmonic (F_{ij}), third-order (F_{ijk}) and fourth-order (F_{ijkl}) anharmonic potential constants (which are given in terms of the internal or the curvilinear coordinates). The vibration-rotation constant expression (2.2.20), which contains the cubic potential constant ϕ_{rrs}, is thus easily transformed to that in terms of third-order potential constants. The vibrational anharmonicity x_{rs} involves third- and fourth-order potential constants, but is rarely available for transient molecules. Therefore, the fundamental frequencies have been taken as harmonic frequencies in calculating the harmonic force field, which is needed for the nonlinear transformation mentioned above. We may get additional information on the harmonic force field from the centrifugal distortion constants and also from the Coriolis coupling constant, if two normal modes happen to be in Coriolis resonance.

The inertia defects, especially those in excited vibrational states, have often provided invaluable information on the harmonic force field. For a nonlinear XYZ-type molecule, the three Coriolis coupling constants satisfy the sum rule $(\zeta_{12})^2 + (\zeta_{23})^2 + (\zeta_{31})^2 = 1$. Hence the inertia defects, not only in the ground state but also in excited vibrational states, are all linear functions of $(\zeta_{12})^2 + d(\zeta_{23})^2$, where d stands for a constant common to all the vibrational states. Therefore, the inertia defects are not very useful here and have not been employed explicitly.

Instead, all three rotational constants of each vibrational state have been used to calculate the structure parameters and third-order anharmonic potential constants by the least-squares method. When the Coriolis resonance has been taken into account, the resulted rotational constants have been transformed to those corresponding to the second-order perturbation expressions mentioned in Sect. 2.2.

The method was first applied to three short-lived molecules HO_2/DO_2, HCO/DCO, and HNO/DNO. The internal coordinates are numbered as 1: $\Delta r(H - X)$. 2: $\Delta r(X - Y)$, and 3: $\Delta\theta(H - X - Y)$. There are ten third-order anharmonic potential constants: F_{111}, F_{222}, F_{333}, F_{112}, F_{122}, F_{113}, F_{133}, F_{223}, F_{233}, and F_{123}. The correlations among these parameters preclude determining all of them simultaneously, even for HO_2/DO_2 for which all 2×9 vibration-rotation constants have been determined. The three "diagonal" terms F_{iii} are exceptional; they are by far the most important and are relatively unaffected by whichever "off-diagonal" terms have been taken into account in the fitting. One to three off-diagonal terms can and must be included, but in most cases their values obtained from the fitting largely depend on the selection of the off-diagonal terms. In this respect, it is interesting to supplement the experimental data with ab initio calculated results. This procedure has been tested on HNO/DNO; four or five off-diagonal terms have been fixed to the values calculated by *Botschwina* [4.247]. Table 4.9 shows that the fitting has been considerably improved, indicated by the standard errors given (compare sets

Table 4.9. Third-order anharmonic potential constants of the HNO molecule[a]

	Set					
	A	B	C	D	E	F
F_{111}	−26.72 (15)	−27.464(95)	−26.73 (50)	−27.42 (10)	−26.71 (16)	−27.442(95)
F_{222}	−66.6 (63)	−67.6 (35)	−62 (15)	−77.5 (75)	−69 (13)	−74.1 (72)
F_{333}	−1.465(98)	−1.200(51)	−1.50 (14)	1.136(66)	−1.46 (11)	−1.173(59)
F_{112}	0.5 (24)	7.1 (14)	0.7 (26)	6.4 (14)	0.4 (26)	6.6 (15)
F_{122}	[0]	[0]	−0.7 (21)	1.6 (11)	[0]	[0]
F_{113}	[0]	[0]	[0]	[0]	[0]	[0]
F_{133}	[0]	[0]	[0]	[0]	[0]	[0]
F_{223}	[0]	[0]	[0]	[0]	−0.2 (10)	−0.59 (59)
F_{233}	−4.23 (18)	[0]	−4.28 (26)	[0]	−4.22 (19)	[0]
F_{123}	[0]	6.34 (16)	[0]	6.20 (20)	[0]	6.32 (16)
	A′	B′	C′	D′	E′	F′
F_{111}	−25.363(73)	−25.344(79)	−25.415(82)	−25.419(82)	−25.416(79)	−25.421(79)
F_{222}	−72.6 (47)	−68.5 (51)	−77.9 (62)	−76.5 (62)	−78.1 (62)	−77.0 (59)
F_{333}	−1.121(44)	−1.143(44)	−1.081(52)	−1.076(43)	−1.101(43)	−1.107(43)
F_{112}	1.57 (98)	1.0 (11)	2.4 (12)	2.2 (12)	2.5 (11)	2.3 (11)
F_{122}	[−5.58]	[−5.58]	−4.56 (82)	−3.94 (85)	[−5.58]	[−5.58]
F_{113}	[0.09]	[0.09]	[0.09]	[0.09]	[0.09]	[0.09]
F_{133}	[−0.08]	[−0.08]	[−0.08]	[−0.08]	[−0.08]	[−0.08]
F_{223}	[−1.93]	[−1.93]	[−1.93]	[−1.93]	−2.48 (43)	−2.80 (39)
F_{233}	−2.470(66)	[−3.01]	−2.427(75)	[−3.01]	−2.456(66)	[−3.01]
F_{123}	[−0.69]	−1.47 (11)	[−0.69]	−1.59 (12)	[−0.69]	−1.523(98)

[a] Units are md/Å^2 for F_{iii} ($i = 1, 2$), F_{112}, and F_{122}, md/Å for F_{113}, F_{223}, and F_{123}, md for F_{133} and F_{233}, and mdÅ for F_{333}. Values in square brackets are assumed, of which nonzero values are taken from [4.247]. Values in parentheses denote one standard deviation

Table 4.10. Third-order anharmonic potential constants of HO_2, HCO, and HNO

Molecule	i	F_{iii} [a]			F_{ii} [b]	$-F_{iii}/F_{ii}$ [c]
		obs [d]	calc(1) [e]	calc(2) [f]	obs	
HO_2	1	−46.16 (57)	−37.6	−48.5	6.69	6.91
	2	−39.7 (66)	−87.9	−33.8	6.15	6.46
	3	−0.539 (10)				0.56
HCO	1	−28.8 (60)	−26.6	−17.9	3.49	8.26
	2	−93 (30)	−136	−92	14.47	6.43
	3	−0.631 (55)				0.92
HNO	1	−25.40 (25)	−39.3	−23.6	4.03	6.31
	2	−75 (19)	−122.8	−73.3	11.07	6.77
	3	−1.11 (16)				0.84

[a] In units of md/Å^2 for $i = 1$ and 2 and mdÅ for $i = 3$
[b] In units of md/Å for $i = 1$ and 2 and mdÅ for $i = 3$
[c] In units of Å^{-1} for $i = 1$ and 2, but dimensionless for $i = 3$
[d] Values in parentheses denote estimated uncertainties
[e] Obtained assuming a diatomic model (4.3.5)
[f] The same as calc. (1), except that the harmonic frequency in (4.3.7, 8) is replaced by the fundamental frequency of the triatomic molecule

$A-F$ with $A'-F'$). It also shows that the different sets of anharmonicity constants lead to results that are more consistent with each other.

The three diagonal third-order anharmonic potential constants thus derived for the three molecules are reproduced in Table 4.10, where the diagonal harmonic potential constants are also included. It is interesting to note that the ratio F_{iii}/F_{ii} lies in the range 6.3–7.0 Å^{-1} for the stretching modes except for $\Delta r(\mathrm{H-C})$ of HCO, whereas it is less than 1.0 for the bending modes. For a diatomic molecule it is well known that the following relations exist:

$$F_{ii} = 2a_0/r_e^2, \tag{4.3.4}$$

$$F_{iii} = 6a_0 a_1/r_e^3, \tag{4.3.5}$$

where the two potential constants appear in the vibrational potential function as

$$V(\xi) = a_0 \xi^2 (1 + a_1 \xi + \ldots), \tag{4.3.6}$$

with $\xi = (r - r_e)/r_e$. One may use spectroscopic constants to calculate a_0 and a_1, and hence F_{ii} and F_{iii}, i.e.,

$$a_0 = \omega_e^2/4B_e, \tag{4.3.7}$$

$$a_1 = -(\alpha_e \omega_e/6B_e^2) - 1, \tag{4.3.8}$$

where ω_e denotes the harmonic frequency, B_e the rotational constant, and α_e

the vibration-rotation constant. The data on the respective diatomic molecules lead to the F_{iii} values for the stretching modes of the three molecules, as shown in Table 4.10. It may not be surprising that they do not agree well with the observed values, because the chemical bond would be quite different in the diatomic and triatomic molecules. It is, however, very striking that when the harmonic frequency in (4.3.7, 8) is replaced by the fundamental frequency of the triatomic molecule, the calculated F_{iii} agrees remarkably well with the observed value, except for Δr(H – C) of HCO. The present analysis indicates that the H – C bond of HCO is anomalously anharmonic. The F_{iii}/F_{ii} ratio of it is also exceptional, as noted above.

Although all the nine vibration-rotation constants have not been determined for HNO/DNO and HCO/DCO (α_3 of DNO and α_1 of HCO are missing and α_3 of DCO is not very precise), the equilibrium structure has been derived for HNO and HCO, Tables 4.11, 12 respectively, where the values already reported are given for comparison.

For a transient molecule, it is often difficult to study isotopic species to determine the molecular structure, because a large amount of expensive isotopes is required. For a nonlinear XYZ-type molecule, however, data other than the rotational constants may be exploited to estimate structural parameters; the fundamental frequencies of isotopic species (often measured in low-temperature matrices), the inertia defect, the centrifugal distortion constants, and others can give information on the geometry of the molecule. It is therefore possible to derive three structure parameters of an XYZ molecule, even when the rotational constants have been determined only for one isotopic species,

Table 4.11. Molecular structure of the HNO molecule

	r_e [4.241]	r_0 [4.248]	r_0 [4.245]
H – N [Å]	1.0628 (30)	1.062_8	1.09026 (H) 1.0795 (D)
N – O [Å]	1.2059 (33)	1.211_6	1.2090
H – N – O [A°]	109.09 (30)	108.5_8	108.047

Table 4.12. Molecular structure of the formyl radical

	r_e [4.241]	r_0 [4.249]	r_0 [4.245]
H – C [Å]	1.1182 (50)	1.125 (5)	1.1514 (H) 1.1474 (D)
C – O [Å]	1.1756 (20)	1.175 (1)	1.17708
H – C – O [A°]	124.26 (30)	124.95 (25)	123.01

provided that other data such as those mentioned above are amply available and are employed in least-squares analysis with proper weights.

Yamada and *Hirota* [4.179] applied this technique to the FO_2 radical, obtaining the following structure parameters: $r_e(F - O) = 1.649 \pm 0.013$ Å, $r_e(O - O) = 1.200 \pm 0.013$ Å, and $\theta_e(F - O - O) = 111.19 \pm 0.36°$, which support the view of *Pimentel* and his co-workers [4.173, 174] that the O – O bond in FO_2 is more or less the same as that of the oxygen molecule and the F atom is only weakly bound to the O_2 moiety. This situation contrasts with that in HO_2, in which the proton donates the electronic charge to the antibonding orbital of the oxygen molecule and thus weakens, i.e., lengthens, the O – O bond, as shown in Table 4.8.

4.4 Symmetric Top and Other Polyatomic Free Radicals

For many diatomic and triatomic free radicals, conventional optical spectroscopy has already provided high-quality data [4.37, 250] which are of great use in carrying out further detailed studies in the IR and microwave regions. Such spectroscopic data are, however, scarce for molecules consisting of more than three atoms. Many triatomic molecules may be considered to be affected by a large Renner-Teller effect, which splits the electronic state of a fictitious linear molecule into two, one reducing to the ground state and the other to an excited state located in the visible or near uv region, and the transition between the split levels may be observed with ease. In contrast, more than three atomic, nonlinear free radicals lack such an effect, and their electronic transitions appear in the uv region with the upper states often affected by predissociation, which makes it difficult to derive detailed information on the molecules. The CH_3 and CF_3 radicals are such examples; both of them are transparent through the entire visible and uv regions. Complications of the spectra obviously increase with the number of atoms in the molecule, and the resolution of optical spectroscopy, even when it reaches the Doppler limit, becomes insufficient to resolve the observed spectrum into finer details. Spectroscopy in a longer-wavelength region such as IR and microwave will be more useful in this respect, as exemplified by a few examples discussed in this section.

4.4.1 The Methyl Radical CH_3

In 1956, *Herzberg* and *Shoosmith* [4.251] reported observing the uv absorption spectrum of the methyl radical. *Herzberg* [4.200] later extended spectrum measurements to both CH_3 and CD_3. Because most of the observed bands which appear in the wavelength region shorter than 214 nm are broad, presumably due to predissociation, it was quite difficult to obtain detailed information on the radical from the observed spectra. Only a few CD_3 bands showed resolved rotational structures, leading *Herzberg* to conclude that the radical was most likely to be planar in the ground electronic state, although *Walsh* [4.252] predicted a pyramidal structure for the radical based on molecular orbital considerations. From the analysis of the CD_3 spectra, Herzberg estimated the ν_2 out-of-plane bending frequency of CH_3 to be about 580 cm^{-1}. *Tan* et al. [4.253] later observed an absorption peaking at 600 cm^{-1} by using a rapid scan IR spectrometer and assigned it to the ν_2 band.

The planarity of the methyl radical has also been subjected to spectroscopic studies in condensed phase. *Karplus* [4.254] and *Fessenden* [4.255] concluded from the proton and ^{13}C hyperfine structures of matrix electron paramagnetic resonance spectra that the methyl radical is essentially planar, although deviations of 10–15° from the planarity could not be excluded. The IR spectrum has also been observed in low-temperature matrices [4.256–259], and *Riveros* [4.256] derived a potential function for the out-of-plane bending mode from the observed spectra, which is consistent with a planar structure.

Because the methyl radical is expected to be transparent in the visible and uv regions (up to 44,000 cm^{-1}) [4.260] and to have no dipole moment because of the planarity, the IR spectrum may be the only source which provides detailed structural information on the radical. *Yamada* et al. [4.261] observed the vibration-rotation bands of $v_2 = (n + 1) \leftarrow n$ with $n = 0$, 1, and 2 by diode laser spectroscopy. The ν_3 degenerate C – H stretching band around 3200 cm^{-1} has also been observed by difference frequency laser spectroscopy [4.262].

In [4.261, 262], the methyl radical was produced either by a pyrolysis of di-*tert*-butyl peroxide at a temperature between 300–400 °C or by a glow discharge in various hydrocarbon compounds such as CH_4, CH_3I, $(CH_3)_2CO$, CH_3SH, CH_3CN, CH_3OH, and di-*tert*-butyl peroxide induced directly in a cell. Both methods are equally efficient in producing the radical, but the glow discharge has an advantage that it may be combined with a multiple-reflection cell. The glow discharge in di-*tert*-butyl peroxide was found to give the best S/N ratio for diode laser spectroscopy [4.261], while the discharge in either CH_4 or CH_3CN was favored in the difference frequency laser study [4.262], because the discharge in CH_4 and CH_3CN gave less interference lines in the ν_3 band region than that in other molecules.

Figure 4.28 illustrates some of the observed absorption spectrum in the 600 cm^{-1} region. The upper trace shows the Zeeman effects of the methyl radical lines. The strongest line absorbs as much as 50% of the incident power for a path length of 10 m. Figure 3.12 shows spectra in the same wavelength region as in Fig. 4.28 recorded with Zeeman modulation and source frequency modulation; the Zeeman-modulated spectrum is obviously preferred. However, *Amano* et al. [4.262] found that some lines of the ν_3 band were not observed by Zeeman modulation, because the spin splittings were much smaller than the Doppler linewidths for such transitions and the Paschen-Back effect prevents them from being modulated. The wavelengths were measured many cases using the source frequency modulated spectrum after an unambiguous assignment

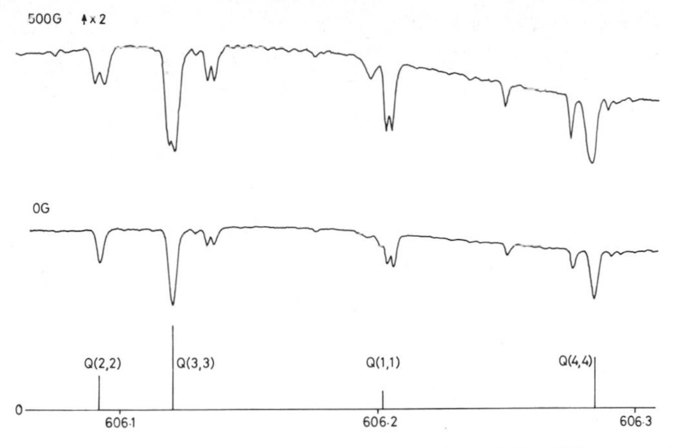

Fig. 4.28. $Q(N, K)$, $N = K = 1 - 4$ transition of the CH_3 ν_2 band. The upper trace is recorded with a magnetic field of 500 G [4.261]

was made. Accurate values of the spin splittings were also obtained from the source frequency modulation spectrum.

The observed spectrum of ν_2 and its hot bands shows that when $K = 0$, only odd and even N levels exist, respectively, in the even and odd v_2 states. This observation implies that the methyl radical is essentially planar, in other words, each v_2 state is nondegenerate and has a definite parity with respect to the inversion (or reflection on the molecular "plane"). Molecular constants were derived from the observed spectrum as follows. The ground-state C_0 constant was fixed at a value which, when combined with the B_0 constant, reproduced the inertia defect calculated from a force field, and the D_K and H_K constants were chosen so as to satisfy the planarity conditions. The ground-state B_0 and C_0 constants are 9.57789(16) and 4.74202(67) cm^{-1}, respectively, with one standard error in parentheses. The vibrational term values are given below in Table 4.14, in connection with the discussion of the out-of-plane bending vibration (Sect. 4.4.3). The observed spin splittings give spin-rotation interaction constants in the ground vibrational state $\varepsilon''_{bb} = -0.01168(61)$ cm^{-1} and $\varepsilon''_{cc} = -0.0001(21)$ cm^{-1} (the signs of these constants are reversed from those in [4.261], see [4.262]). When the ν_2 spectrum is observed by a time-resolved technique such as discharge current modulation (Sect. 3.2.1), some kinetic information may be obtained [4.263, 264], as described in Sect. 5.1.

Absorption lines observed in the region 3070–3250 cm^{-1} by difference laser spectroscopy have all been assigned to the ν_3 fundamental band. The analysis was performed by fixing the ground-state parameters to those derived from the ν_2 band, and the upper-state molecular constants were evaluated; the band origin thus obtained is 3160.8212(12) cm^{-1}. The ν_3 band spectrum may be employed to detect the methyl radical in interstellar and stellar space, because the 3 μm region is one of the atmospheric windows.

4.4.2 The Trifluoromethyl Radical CF_3

The molecular structure of the CF_3 radical has been investigated by electron paramagnetic resonance (EPR) spectroscopy in low-temperature matrices [4.265–270]. *Fessenden* and *Schuler* [4.266] deduced the FCF angle to be 111.1° by interpreting the observed hyperfine coupling constant of ^{13}C in terms of a semiempirical relation proposed by *Karplus* et al. [4.270, 271]. *Carlson* and *Pimentel* [4.272] observed three vibrational bands ν_1, ν_2, and ν_3 for CF_3 in the gas phase by using a rapid scan IR spectrometer and concluded that CF_3 has a pyramidal structure. They assigned these bands based on the matrix spectra in low-temperature matrices observed by *Milligan* et al. [4.273] and *Milligan* and *Jacox* [4.274].

The pyramidal structure of the CF_3 radical makes it a polar molecule, and, in fact, *Endo* et al. [4.275] have succeeded in observing the pure rotational spectrum in the microwave region. A high-resolution IR spectroscopic study was also performed by *Yamada* and *Hirota* [4.276] concurrently with the microwave investigation, and provided information on the molecular structure of CF_3 which is complementary to that obtained from the microwave study. Figures 4.29, 30 show examples of the diode laser spectra recorded with Zee-

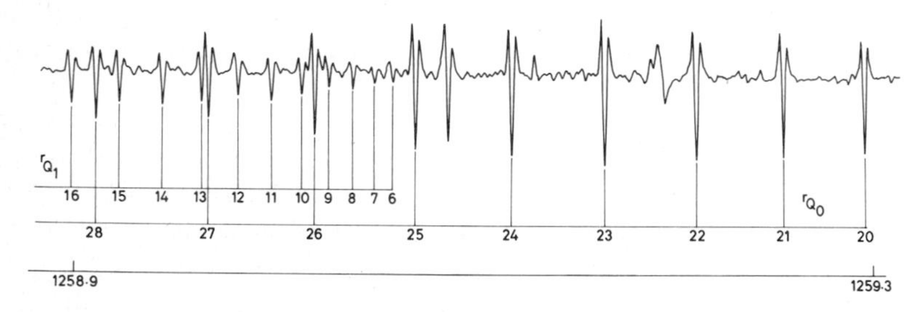

Fig. 4.29. $^rQ_0$ and $^rQ_1$ transitions of the CF_3 ν_3 band, recorded with Zeeman modulation [4.276]

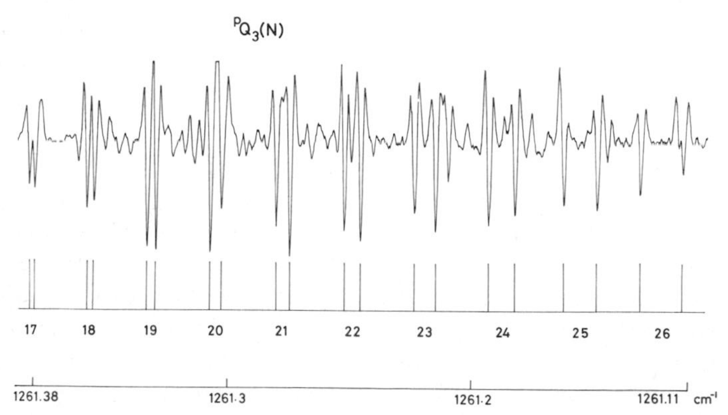

Fig. 4.30. $^pQ_3$ transitions of the CF_3 ν_3 band, recorded with Zeeman modulation; all lines exhibit $A_1 - A_2$ splitting [4.276]

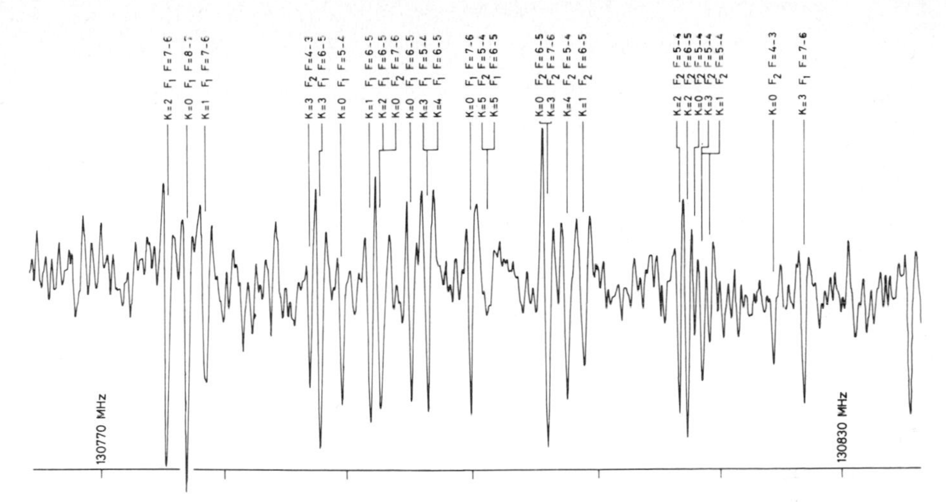

Fig. 4.31. The $N = 6 \leftarrow 5$ transition of CF_3 in the ground vibronic state [4.275]

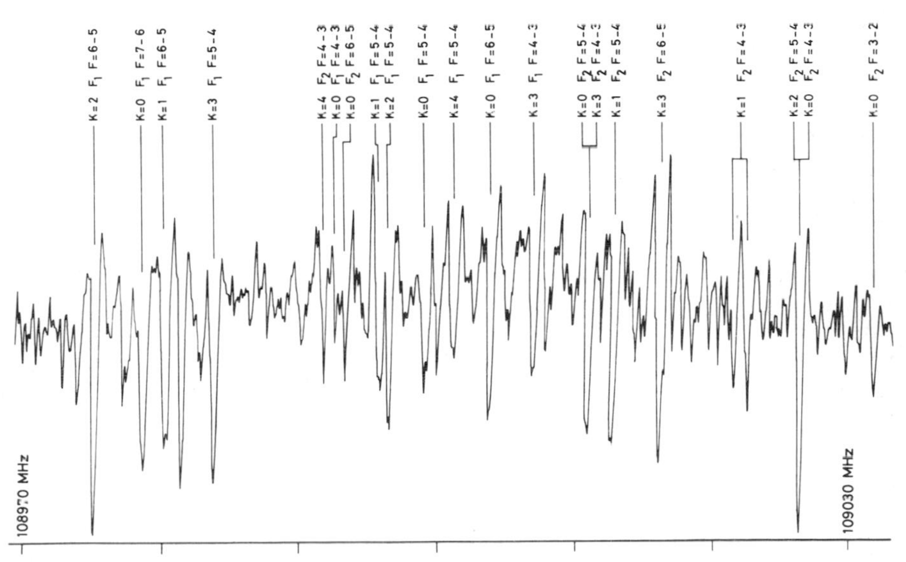

Fig. 4.32. The $N = 5 \leftarrow 4$ transition of CF_3 in the ground vibronic state [4.275]

man modulation. A few Q-branch series were first assigned, but P- or R-branch lines were found to be difficult to identify. Then pure rotational transitions were observed by microwave spectroscopy; Figs. 4.31, 32 illustrate the spectra of the two successive rotational transitions. As seen from the figures, each rotational transition was split into many components by the spin-rotation and hyperfine interactions. A computer-simulation method based on simplified combination sum and difference relations was employed to make assignments for these components [4.275]. As Fig. 4.33 shows, $K = 1$ lines were found to be

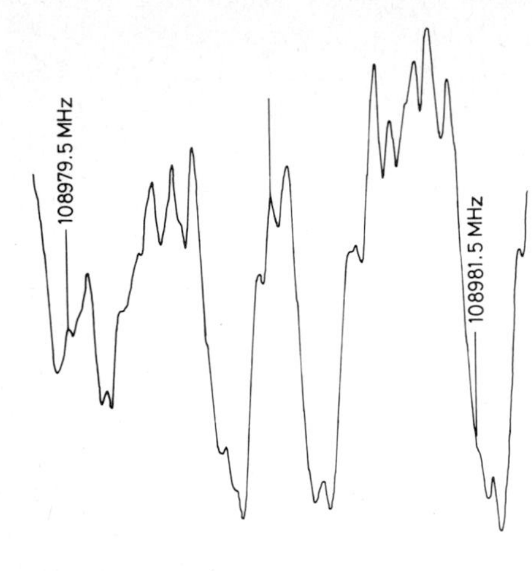

Fig. 4.33. Hyperfine splitting caused by $T_{aa} - T_{bb}$ for the $N = 5 \leftarrow 4$, $K = 1$, $J = 5.5 \leftarrow 4.5$, $F = 6 \leftarrow 5$ transition of the CF_3 in the ground vibronic state [4.275]

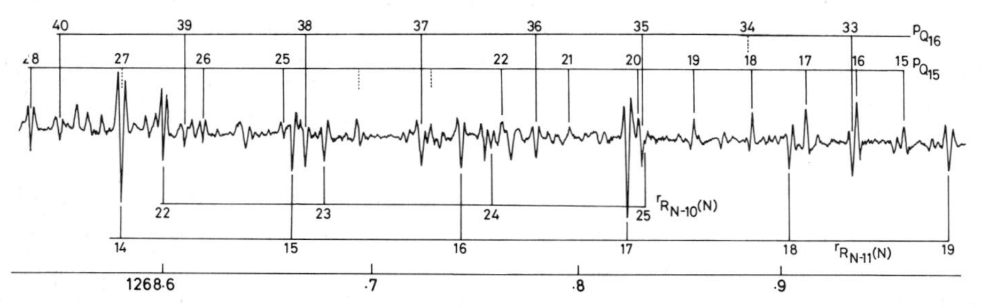

Fig. 4.34. $^rR_{N-10}$, $^rR_{N-11}$, $^pQ_{15}$, and $^pQ_{16}$ transitions of CF_3, recorded with Zeeman modulation [4.276]

split into two components of nearly equal intensities, as explained in Sect. 2.3 in terms of the $|T_{aa} - T_{bb}|$ constant.

The ground-state B_0, D_N, and D_{NK} constants thus determined facilitated assigning P and R branch lines in the diode laser spectrum. As shown in Fig. 4.34, transitions with the same $N - K$ values tend to form series with the intensity pattern strong-weak-weak. Two types of vibration-rotation interactions. Δl, $\Delta k = \pm 2, \pm 2$ and $\pm 2, \mp 1$ [referred to as the (2, 2) and (2, $-$ 1) interactions, respectively], were taken into account in analyzing the diode laser spectrum. The latter interaction has enabled the C_0 constant to be determined, which is otherwise impossible to obtain. The rotational constants thus obtained are $B_0 = 10\,900.9118\,(21)$, $C_0 = 5653.9\,(102)$, $B_3 = 10\,880.6442\,(76)$, $C_3 = 5639.1\,(102)$, and $C\zeta_3 = 4130.9\,(102)$ in MHz, with standard errors in parentheses.

The B_0 and C_0 rotational constants thus obtained have yielded the following r_0 structure: $r_0(\mathrm{C-F}) = 1.318\,(2)$ Å and $\measuredangle\,(\mathrm{F-C-F}) = 110.76\,(40)°$. The latter supports the EPR result 111.1° mentioned in [4.266], and indicates that the valence orbitals of the carbon atom are sp^3 hybridized.

The hyperfine coupling constants obtained from the microwave spectrum may be compared with the EPR results previously reported. The Fermi cou-

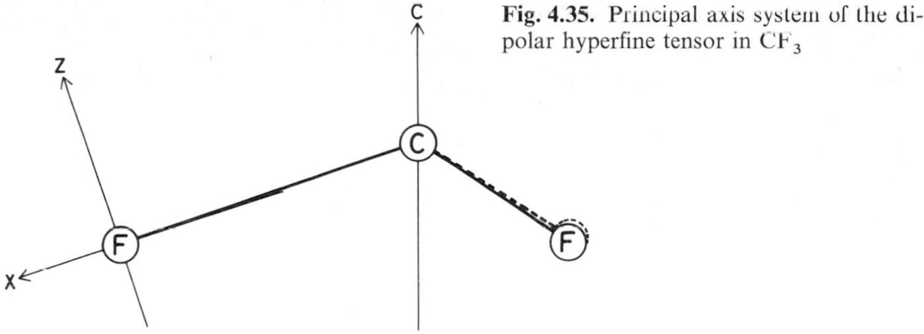

Fig. 4.35. Principal axis system of the dipolar hyperfine tensor in CF_3

pling constant of the F nucleus was determined by EPR to be 140–145 G, or 392–406 MHz [4.266–269] in various matrices, in agreement with the gas phase value of 408.5 MHz. The anisotropic hyperfine coupling constants have also been determined by EPR [4.267–269], but may be compared with the microwave values only after some appropriate transformation of the axes is performed for the latter. By assuming that one of the principal axes makes an angle of 17.8° with the molecular symmetry axis [4.268, 269], the microwave results are converted to the principal values of $T_{zz} = 372.9$ MHz, $T_{yy} = -180.1$ MHz, and $T_{xx} = -192.8$ MHz (Fig. 4.35 defines the principal axes). These values are compared with the EPR results in Table 4.13. The

Table 4.13. Dipolar hyperfine coupling constants of CF_3 (in MHz)

Method	T_{xx}	T_{yy}	T_{zz}	Ref.
EPR	304.5	−146.7	−157.9	[4.267]
EPR	336.3	−158.3	−178.0	[4.268, 269]
MW	372.9	−180.1	−192.8	[4.275]

unpaired electron spin density of the F atom is calculated to be 10.6% by comparing the above T_{zz} value with the atomic value of 3520 MHz [4.277].

The observed spin-rotation interaction constants of CF_3 scaled by the rotational constants are a few times larger than those of CH_3. This difference is at least in part ascribed to the contribution of the fluorine atom; although the spin density of the F atom is only 10.6%, the spin-orbit coupling constant 269 cm^{-1} of the F atom is almost ten times larger than that of the C atom, which is 28 cm^{-1} (Sect. 4.5.2).

4.4.3 Structures and Internal Motions of the Methyl Radical and Its Derivatives

As mentioned in Sect. 4.4.1, the out-of-plane bending vibration of the CH_3 radical is unusually low, i.e., about 606.5 cm^{-1}, which may be compared with the CH_3 deformation frequencies of about 1400 cm^{-1} in many organic molecules and with 1380 cm^{-1} of the corresponding band in CH_3^+ [4.278]. The out-of-plane bending frequency is correlated closely with the planarity of the molecule, in other words, to establish the planarity or nonplanarity of a molecule, it is imperative to determine the potential function for the out-of-plane bending mode.

Yamada et al. [4.261] observed the ν_2 and its two hot bands in the CH_3 radical by IR diode laser spectroscopy and thus evaluated the energy separations between four lowest bending states: $E_1 - E_0 = 606.4531$, $E_2 - E_1 = 681.6369$, and $E_3 - E_2 = 731.0757$ in cm^{-1}. These data clearly indicate that there is a large, negative anharmonicity. Furthermore, since $E_3 - E_2$ is larger than $E_2 - E_1$, any possibility that the potential function is double minimum is definitely excluded. In fact, by fitting these data to a potential function of the form

$$V_2(z) = (1/2)\,k_{22}\,z^2 + (1/4!)\,k_{2222}\,z^4 + (1/6!)\,k_{222222}\,z^6, \tag{4.4.1}$$

they obtained curve A shown in Fig. 4.36 where $z > 0.5$ Å must be discarded; it clearly shows that there is no hump at $z = 0$. The out-of-plane bending coordinate z is taken to be the distance between the C atom and the plane made by three hydrogen atoms, and thus the effective mass is given by $\mu = 3m_H m_C/(3m_H + m_C)$.

The anomalously low bending frequency and the large, negative anharmonicity may be accounted for by a vibronic interaction between the electronic ground state ($\tilde{X}\,^2A_2''$) and excited electronic states of $^2A_1'$ symmetry through the out-of-plane bending mode (a_2'' symmetry). This interaction is referred to by *Öpik* and *Pryce* [4.125] as the pseudo-Jahn-Teller effect. When this interaction is simply expressed as az, the effective out-of-plane potential function in the $\tilde{X}$ state is given by

$$V_2(z) = (1/2)\,k_{22}^{(0)}\,z^2 + (1/2)\,\Delta E - (1/2)\,(\Delta E^2 + 4a^2z^2)^{1/2}, \tag{4.4.2}$$

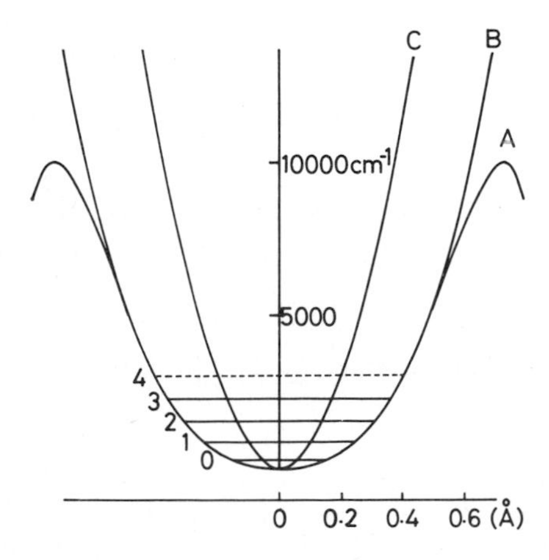

Fig. 4.36. Potential function for the CH_3 ν_2 mode [4.261]

where ΔE denotes the separation between the $\tilde{X}$ state and an excited state of A_1' symmetry (only one such state is taken into account) and $k_{22}^{(0)}$, the unperturbed quadratic force constant. When we expand the square root of (4.4.2), $V_2(z)$ reduces to

$$V_2(z) = (1/2)\,(k_{22}^{(0)} - 2a^2/\Delta E)\,z^2 + (a^4/\Delta E^3)\,z^4 - (2a^6/\Delta E^5)\,z^6 + \ldots, \quad (4.4.3)$$

and explains the low fundamental frequency (the small quadratic force constant), the large, negative anharmonicity (the large positive quartic force constant), and the negative sextic term. When the perturbing state is identified as the $\tilde{B}\,^2A_1'$ state at 46 205 cm^{-1}, the observed vibrational term values in the $\tilde{X}$ state are reproduced within 0.1 cm^{-1} by adjusting $k_{22}^{(0)}$ and a, as shown in Table 4.14. The potential function thus obtained (curve B) is compared with

Table 4.14. Out-of-plane bending vibration term values of the CH_3 radical

v	E_v (obs.)	E_v (calc.)[a]
1	606.4530 cm^{-1}	606.3601 cm^{-1}
2	1288.0900	1288.2088
3	2019.1657	2019.1172
	$k_{22}^{(0)}$ (md/Å)	2.7888 (29)[b]
	a (md)	1.0599 (77)[b]

[a] Calculated using (4.4.2) together with $k_{22}^{(0)}$ and a listed

[b] Values in parentheses denote standard deviation and apply to the last digits of the constants

that (curve A) from (4.4.1) in Fig. 4.36; the two curves agree with each other for $z < 0.5$ Å. The separation $E_4 - E_3$ is calculated to be 766.962 and 768.289 cm^{-1}, respectively, from (4.4.1, 2), which may be compared with the low-resolution value 772 ± 4 cm^{-1} reported by *Hermann* and *Leone* [4.279]. The unperturbed $k_{22}^{(0)}$ constant leads to a hypothetical harmonic frequency of 1404.35 cm^{-1}, which may be compared with the ν_2 frequency 1380 cm^{-1} of CH_3^+ [4.278]. It is certainly within the range of the CH_3 deformation frequencies normally encountered in many organic molecules. The "unperturbed" harmonic potential $(1/2)\,k_{22}^{(0)} z^2$ is curve C in Fig. 4.36.

It is not possible to test the model for the $\tilde{B}$ excited state, because, due to predissociation, only the $v_2 = 0 - 0$ band of the $\tilde{B} - \tilde{X}$ system has been observed for CH_3 and the $v_2 = 0 - 0, 0 - 2$, and $1 - 1$ bands of $\tilde{B} - \tilde{X}$ for CD_3 [4.200]. The latter set of data gives the ν_2 frequency in the $\tilde{B}$ state as $627 + \nu_2''$ for CD_3, which, when combined with the ground-state ν_2'' frequency 460 cm^{-1} estimated by *Hirota* and *Yamada* [4.280], leads to 1087 cm^{-1} (cf. [4.281]). This value is of the order of magnitude expected from the model, but a critical comparison must be postponed until the ν_2 band of CD_3 is observed and analyzed.

In view of the delicate balance between planarity and nonplanarity in the CH_3 radical, it is interesting to examine the effect of substituents on the planarity. When the three hydrogen atoms are all replaced by F, the CF_3 radical arises, which has been shown to have a pyramidal structure, as mentioned in Sect. 4.4.2. We may then expect that either or both of the two "intermediate" species CH_2F and CHF_2 show the effect of inversion doubling in its high-resolution spectra. *Fessenden* and *Schuler* [4.266] observed the hyperfine structures in the electron spin resonance (ESR) spectrum of CH_2F in low-temperature matrices, and *Karplus* et al. [4.270, 271] interpreted the ^{13}C hyperfine constant in terms of an "out-of-plane" angle of less than 5°.

The IR spectra of CH_2F trapped in Ar matrices were reported by *Jacox* and *Milligan* [4.282, 283] and *Raymond* and *Andrews* [4.284]. *Mucha* et al. [4.285] observed the gas phase spectrum of CH_2F for the first time using FIR LMR. However the spectrum observed was too complicated to be assigned; they estimated $B + C$ to be 1.953 cm^{-1} by plotting the frequencies of laser lines giving resonances versus $N(N + 1)$. Some of the observed LMR spectra exhibited doublet and sextet hyperfine structures, which *Mucha* et al. took as evidence that the molecule was planar.

Endo et al. [4.286] and *Yamada* and *Hirota* [4.287] investigated the CH_2F radical by microwave and IR diode laser spectroscopy, respectively. The ground-state rotational spectrum observed by microwave spectroscopy clearly shows the effect of the nuclear spin statistical weight which well conforms to C_{2v} point-group symmetry. This spectrum is accompanied by a set of vibrational satellites which also shows the effect of the spin statistical weight, but opposite to that of the ground state. This set has thus been assigned to the first excited state of the out-of-plane bending, or ν_4 mode. The relative intensity measurement leads to the excitation energy of about 300 cm^{-1}. The inertia defect in this state is $-$ 0.1862 uÅ^2, in accordance with the above assignment that this mode is the lowest "out-of-plane" mode. This vibration makes a dominant (negative) contribution of $-$ 0.0886 uÅ^2 to the ground-state inertia defect, which is, in fact, slightly negative, $-$ 0.0090 uÅ^2. The low ν_4 frequency, together with the negative ground-state inertia defect, suggests that there may be a small hump at the planar configuration which is perhaps lower than the ground state.

Endo et al. [4.288] have recently observed the pure rotational spectrum of CH_2Cl by microwave spectroscopy, and found the observed spectrum again to exhibit the effect of the spin statistics. The ground-state inertia defect was found to be positive (0.0333 uÅ^2), in contrast with the negative value of CH_2F. Although no vibrational satellites have been assigned, *Milligan* and *Jacox* [4.289] and *Andrews* and *Smith* [4.290] have reported the out-of-plane bending mode at 400 cm^{-1} in low-temperature matrices, which is about 100 cm^{-1} higher than that of CH_2F. These observations indicate that the CH_2Cl radical is closer to the planarity than CH_2F.

The above results for CH_3 and its derivatives suggest that more electronegative substituents for hydrogen(s) tend to make the molecule more nonplanar than less electronegative substituents, in qualitative agreement with ab initio results [4.291, 292].

4.4.4 The NO_3 Radical

The NO_3 radical has long been known to exist. It absorbs in the visible, but high-resolution spectroscopic investigations, either conventional absorption [4.293, 294] or laser-induced fluorescence [4.295–297], have failed to resolve rotational structures of the visible bands. It was suspected in a few cases that the molecule predissociates in the upper electronic state, but recent measurements of the radiative lifetime [4.295, 296] have precluded this possibility, at least for a few lowest vibrational levels of the upper electronic state. This molecule has recently attracted much attention in the field of atmospheric chemistry in addition to that of molecular science, because it is expected to play an important role in chemical reactions in the upper atmosphere (Sect. 5.2).

Based on an MO consideration, *Walsh* [4.298] predicted that the NO_3 radical has D_{3h} symmetry in the ground electronic state with the electron configuration

$$(e')^4\,(e'')^4\,(a_2'):\ \tilde{X}\,^2A_2',$$

and there are two low-lying electronic states

$$(e')^4\,(e'')^3\,(a_2')^2:\ \tilde{A}\,^2E'',$$

$$(e')^3\,(e'')^4\,(a_2')^2:\ \tilde{B}\,^2E'.$$

The visible system was thus interpreted to correspond to the $\tilde{B}\,^2E' - \tilde{X}\,^2A_2'$ transition. *Walker* and *Horseley* [4.299] performed an ab initio calculation on NO_3, whereby the $\tilde{A} - \tilde{X}$ and $\tilde{B} - \tilde{X}$ separations were 0.36 eV (2866 cm^{-1}) and 1.77 eV (14165 cm^{-1}), respectively, the latter being in fair agreement with the observed value (the 0 – 0 band at 15110 cm^{-1}).

Ishiwata et al. [4.300] succeeded in observing the ν_3 NO degenerate stretching band by IR tunable diode laser spectroscopy. They generated the NO_3 radical by the reaction of NO_2 with excess O_3 (the partial pressures being 30 mTorr and 500 mTorr, respectively). They confirmed that NO_2 ν_1 spectral lines completely disappear upon addition of O_3. The reaction mixture was pumped through a White-type multiple reflection cell. They have thus observed more than 700 lines in the region 1466–1504 cm^{-1} (about 60% of this region was scanned), of which 217 lines were assigned to the ν_3 band. The initial assignment was made for the $^rQ_0(N)$ branch, which was found to consist of only N = odd members. The next Q branch was then assigned to $^pQ_3(N)$ and the third one to $^pQ_6(N)$. It was thus confirmed that only $K = 3n$ levels are present in one state of the transition, which is presumably the ground vibronic state. These observations can be explained only if the molecule belongs to D_{3h} symmetry, i.e., it is planar with a C_3 axis.

There are several anomalous features in the observed spectrum, however. A good fitting of the spectrum was obtained only when the centrifugal distortion terms were included up to sixth order. The quartic centrifugal distortion constants obtained are an order of magnitude or more larger than the values calculated from an appropriate force field. A similar discrepancy exists between the observed and calculated first-order Coriolis coupling constant. A higher-order vibration-rotation interaction term of $\Delta l, \Delta k = \pm 2, \mp 4$ was required to explain "staggering" observed for the ${}^{P}Q_3(N)$ branch; the interaction constant obtained is much larger than those of PF_3 [4.301] and CHF_3 [4.302]. Finally, the ε_{cc} spin-rotation coupling constant derived is difficult to understand; only excited electronic states of A_1' symmetry can contribute to ε_{cc}, but no such states are expected to exist below 40 000 cm^{-1}. If the observed value of ε_{cc} is ascribed to one excited state of A_1', it must be located as low as 1000 cm^{-1}. These anomalies are ascribed to interactions with overtone and/or combination states (note that ν_3 is the highest normal mode) and to those with the lowest excited electronic state of E'' symmetry, namely the $\tilde{A}$ state. It is interesting to note that the ν_3 spectrum of ${}^{15}NO_3$ behaves more normally, suggesting that vibration-rotation interactions, either Coriolis or Fermi, may be responsible for some part of the perturbations.

4.4.5 The Methoxy Radical CH_3O

In contrast with three previous examples of symmetric top free radicals in nondegenerate electronic ground states, the methoxy radical has a doubly degenerate ground state of ${}^2E$ symmetry. As *Jahn* and *Teller* [4.303] pointed out, the degenerate electronic state of a symmetric top is unstable, the potential minimum being displaced from the symmetric configuration. Although the existence of such an effect (the Jahn-Teller effect) was first pointed out as early as 1937, it is only quite recently that detailed spectroscopic data clearly demonstrated the consequence of this effect. The methoxy radical is a good example of molecules exhibiting the Jahn-Teller effect [4.304].

In the 1950s the methoxy radical was suspected to arise as an intermediate in many reactions [4.305], but quite recently it was confirmed to exist really in the gas phase. *Ohbayashi* et al. [4.306] observed an electronic spectrum of methoxy in the near-uv region, and *Radford* and *Russell* [4.307] detected its pure rotational spectrum by FIR laser magnetic resonance. The latter study provided molecular parameters in the ground vibronic state, some of which reflect the Jahn-Teller effect [4.308]. The LMR spectrum these authors observed also showed partially resolved hyperfine structures, but they did not attempt to analyze them.

The rotational constant derived from the LMR study has enabled *Endo* et al. [4.309] to observe the zero-field pure rotational spectrum of CH_3O in the mm-wave region. Since the observed transitions are low J and K, the hyperfine structure has been well resolved. The zero-field measurement has disclosed the Jahn-Teller effect in the observed spectrum more clearly than the LMR study.

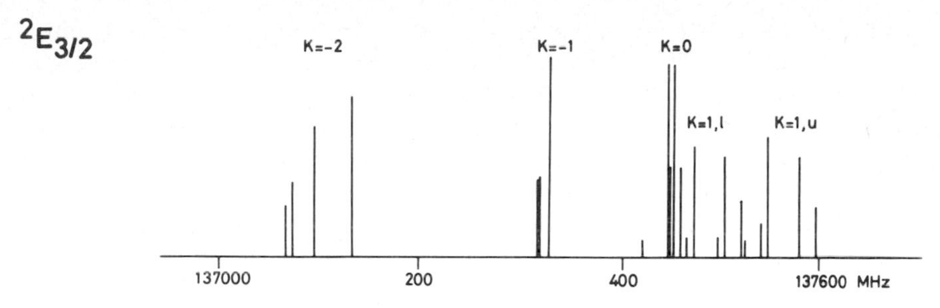

Fig. 4.37. The $J = 2.5 \leftarrow 1.5$ transition of CH_3O in the ground $^2E_{3/2}$ vibronic state [4.309]

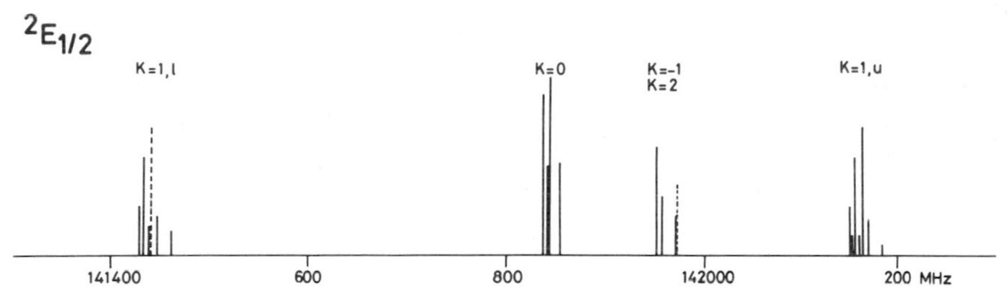

Fig. 4.38. The $J = 2.5 \leftarrow 1.5$ transition of CH_3O in the ground $^2E_{1/2}$ vibronic state [4.309]

Endo et al. [4.309] generated the methoxy radical by the reaction of CH_3OH with microwave discharge products of CF_4. The methoxy lines were discriminated against many other interference lines including those of the precursor CH_3OH by examining the Zeeman shifts when a magnetic field of a few tens of Gauss was applied. Figures 4.37, 38 show the observed spectral lines as long bars.

Because the methoxy radical has an unquenched orbital angular momentum, the rotational levels are split into two groups widely separated by the spin-orbit interaction. Figures 4.37, 38 correspond to these two spin components $^2E_{3/2}$ and $^2E_{1/2}$, respectively, where the components of the spin and orbital angular momenta along the molecular axis are parallel and antiparallel, respectively. In each spin substate, the rotational levels are further grouped by K, just as in an excited state of a degenerate vibration associated with a nondegenerate electronic state. The $K = 1$ transitions in the $^2E_{1/2}$ state are widely split into two components as a consequence of the Jahn-Teller interaction, as expected from the theoretical formulation given in Sect. 2.3.6. The $K = 1$ doubling is referred to as Λ doubling, which is a first-order effect in $^2E_{1/2}$ and a second-order effect in $^2E_{3/2}$, as for a $^2\Pi$ molecule. For CH_3O the so-called $(2, -1)$ interaction is large compared with the ordinary $(2, -1)$ vibration-rotation interaction. This effect is seen for the $K = 2$ transition of $^2E_{1/2}$, as shown in Fig. 4.38; the line is shifted to higher frequency, because the $K = -3$ levels of $^2E_{3/2}$ lie close to the $K = 2$ levels of $^2E_{1/2}$.

Each of the rotational transitions is further split into hyperfine components. The rotational levels with $K = 3n + 1$ are associated with the total spin angular momentum $I = 3/2$, whereas those with $K \neq 3n + 1$ with $I = 1/2$, and in fact, the number of the observed hyperfine components conforms well to these spin quantum numbers. There are, however, exceptions; as Figs. 4.37, 38 show, $K = 0$ transitions in both the spin states and $K = -1$ transitions in $^2E_{3/2}$ exhibited further small splittings. Because the rovibronic levels involved in these transitions are of E symmetry, these additional splittings indicate breakdown of C_{3v} symmetry, in agreement with the theoretical discussion developed in Sect. 2.3.11 which predicts that levels with $P = \pm 0.5$ ($P = K + \Sigma$) are split into two. This phenomenon is similar to the case of the CF_3 $K = 1$ levels, except that the hyperfine coupling constants responsible for the doublings are different for the two cases. The large splittings for the $K = 0$ levels are ascribed to the large Fermi contact term $\sigma_{\pm}$.

The transition frequencies measured by mm-wave spectroscopy were combined with those of LMR and subjected to least-squares analysis, where the Hamiltonian given in Sect. 2.3.6 was employed. (The hyperfine structures of the mm-wave spectra were analyzed separately.) Table 4.15 summarizes main molecular constants thus determined. Besides these constants, a few centrifugal distortion constants were required to fit the mm-wave transitions of low J and the LMR data of high J simultaneously. The (2, 2)- and (2, − 1)-type interactions observed in the mm-wave spectra necessitated that the ε_1, ε_{2a}, and ε_{2b} terms [4.310] be included in addition to the h_1 and h_2 terms taken into account in the previous LMR study [4.308] (h_1 and h_2 were denoted as X_{22} and X_{21} there).

The spin-rotation interaction constants derived are anomalously large. This is evident when these constants are interpreted using the well-known second-order perturbation expression; if the observed ε_{aa} and $(\varepsilon_{bb} + \varepsilon_{cc})/2$ constants are ascribed to the interaction with only one excited state, this state must be located as low as 2000–3000 cm^{-1} above the ground electronic state. This fact suggests that the Jahn-Teller effect substantially contributes to the spin-rotation interaction, but no detailed analysis has been carried out.

Table 4.15. Molecular constants of CH_3O (in MHz)[a]

Constant	Value	Constant	Value
$a\zeta_e d$	−1 865 980 (5030)	ε_{aa}	−40 930 (1700)
$a_D\zeta_e d$	134 (75)	$(\varepsilon_{bb} + \varepsilon_{cc})/2$	−1 428 (66)
$A\zeta_t$	54 330 (270)		
		ε_1	−171.67 (38)
A	154 960 (570)	ε_{2a}	2 420 (1080)
B	27 931.14 (46)	ε_{2b}	fixed to $(B/A)\,\varepsilon_{2a}$
h_1	75.446 (178)		
h_2	1 398 (40)		

[a] Values in parentheses denote 2.5 times standard deviations

The Jahn-Teller effect is not large for CH_3O in the ground electronic state. The observed vibronic Coriolis coupling constant $A\zeta_t$ estimates ζ_t to be 0.3506. Assuming only one vibrational mode to be Jahn-Teller active, (2.3.45 b) leads to a $(1-d)\zeta_s$ value of -0.153. Because $0 < \zeta_e < 1$ for a ^2E state, d and ζ_s must satisfy

$$0.427 < d < 0.847 \quad \text{and}$$

$$-0.267 < \zeta_s < -1 ,$$

which correspond to the Jahn-Teller coupling parameter $D(=f^2/2h\omega^3)$ of 0.05 to 0.3, namely to a relatively small Jahn-Teller effect. Both Jahn-Teller interaction and L-uncoupling affect the (2, 2) and (2, $-$1) interaction constants [4.310, 311], but no detailed analyses have been carried out.

When the C $-$ H bond length in CH_3O is the same as that of CH_3OH, the observed A and B rotational constants give the C $-$ O bond length of CH_3O as 1.376 Å. This value is smaller than the C $-$ O length of CH_3OH [4.312] and also an earlier ab initio result for CH_3O [4.313]. More recent ab initio MO studies [4.314–316] have given values much closer to the observed, however.

Table 4.16. Hyperfine coupling constants of CH_3O (in MHz)[a]

Constant	Value	Constant	Value
$\langle \Lambda \Vert a_L \Vert \Lambda \rangle \zeta_e d$	2.346 (37)	$g_S g_N \beta \beta_N \langle \Lambda \Vert T_0^2(C_0) \Vert \Lambda \rangle$	4.343 (194)
$\langle \Lambda \Vert \sigma_0 \Vert \Lambda \rangle$	119.00 (38)	$g_S g_N \beta \beta_N \langle \Lambda' \Vert T_{\pm 2}^2(C_0) \Vert \Lambda \rangle$	0.279 (25)
$\langle \Lambda' \Vert \sigma_\pm \Vert \Lambda \rangle$	153.60 (109)	$g_S g_N \beta \beta_N \langle \Lambda' \Vert T_0^2(C_\pm) \Vert \Lambda \rangle$	55.55 (109)
		$g_S g_N \beta \beta_N \langle \Lambda' \Vert T_{\mp 2}^2(C_\pm) \Vert \Lambda \rangle$	1.466 (88)

[a] Values in parentheses denote 2.5 times standard deviations. Other hyperfine constants are fixed to zero

The hyperfine coupling constants have been determined using only the mm-wave spectrum, Table 4.16. The $\sigma_\pm$, $T_{\pm 2}^2(C_0)$, $T_0^2(C_\pm)$, and $T_{\mp 2}^2(C_\pm)$ terms have $\Delta\Lambda = \pm 2$ matrix elements and are thus responsible for the hyperfine doublings. It should be noted that $g_S g_N \beta \beta_N \langle \Lambda' | \, | T_0^2(C_\pm) | \, | \Lambda \rangle$ is an order of magnitude larger than other constants, but the reason is not clear. The Fermi interaction term is the most important among the hyperfine coupling constants, because the unpaired electron is localized primarily on the O atom, making the dipolar and nuclear spin-orbit interactions small, whereas an appreciable amount of the spin density is transferred to the s orbital of the hydrogen atoms through hyperconjugation.

Although the mm-wave and LMR data have provided some interesting information on the Jahn-Teller effect, they are limited to the ground vibrational state, where the effect is small. It would be much more interesting to observe high-resolution spectra involving Jahn-Teller active states.

4.5 Fine and Hyperfine Interactions in Free Radicals

Fine and hyperfine structures caused by the electron spin and/or orbital angular momenta provide us with useful information on the electronic structures of free radicals. This section discusses the significance of fine and hyperfine interaction constants derived from high-resolution spectra.

4.5.1 Hyperfine Interaction Constants of Diatomic Molecules

For molecules without electron spin and orbital angular momenta, only a prominent hyperfine structure is due to the nuclear electric quadrupole interaction. This has been extensively investigated to provide information on the electronic structure of molecules [4.317, 318]. When a molecule possesses unquenched electron spin and/or orbital angular momenta, an additional hyperfine interaction arises between such angular momenta and nuclear spins in the molecule, which is referred to as the magnetic hyperfine interaction. It is often more important than the nuclear electric quadrupole interaction; the hyperfine structures are well resolved in most cases when the spectral resolution is better than a few MHz. The nuclear spin needs to be finite for the magnetic hyperfine interaction to occur, whereas it must be equal to or larger than 1 for the quadrupole interaction. Another important difference between the two interactions is that the magnetic hyperfine interaction provides information specific to electrons with unquenched spin and/or orbital angular momenta, whereas all electrons (and even other nuclei) in the molecule contribute to the nuclear quadrupole interaction. Usually the electrons involved in the magnetic hyperfine interaction occupy the highest orbital(s) available in the molecule, and thus govern its chemical properties to a large extent.

The so-called Frosch and Foley a, b, c, and d hyperfine coupling constants defined by (2.3.75 a–e) have been reported for a number of diatomic free radicals. Tables 4.17–20 list the results for four kinds of nuclei H, F, Cl, and N, where $b_\eta = b + c/3$, instead of b, is given because it has a clearer physical meaning of the Fermi contact term than b directly obtainable from spectroscopic experiments. All four constants exist for $^2\Pi$ molecules, but not all of them are easy to determine. For instance, in the limit of Hund's case (a), the b constant is difficult to determine accurately. Molecules containing heavy elements approximate Hund's case (a) well, making both b_η and c rather uncertain, as seen in Tables 4.17–20. It is imperative to observe both $^2\Pi_{1/2}$ and $^2\Pi_{3/2}$ to determine the four constants completely, but the large spin-orbit coupling constants of heavy molecules often preclude observing the two spin components. *McKellar* [4.72] found the dipole moment of the OF radical to be 0.0043 D in the ground vibronic state, which is too small to observe the microwave spectrum. In addition, the small magnetic moment in the $^2\Pi_{1/2}$ state left only one linear combination $a + (b + c)/2$ to *McKellar* [4.70] to determine from the mid-IR LMR.

Table 4.17. Hyperfine interaction constants of the hydrogen nucleus (in MHz)

	a	b_η	c	d	Ref.
CH ($^2\Pi_r$)	54.188	−57.71	56.98	43.481	[4.56]
NH ($^3\Sigma^-$)	–	−66.23	91.70	–	[4.319]
OH ($^2\Pi_i$)	86.112	−73.154	130.644	56.655	[4.320]
PH ($^3\Sigma^-$)	–	−48.1	16.3	–	[4.46]
SH ($^2\Pi_i$)	32.58	−52.63	32.44	27.36	[4.321]

Table 4.18. Hyperfine interaction constants of the fluorine nucleus (in MHz)

	a	b_η	c	d	Ref.
CF ($^2\Pi_r$)	705.82	151.6	−352.7	792.17	[4.60]
NF ($^1\Delta$)	753	–	–	–	[4.89]
OF ($^2\Pi_i$)	$a + (b + c)/2 = 696$				[4.70]
SiF ($^2\Pi_r$)	312.35	69	−175	359.0	[4.64]
SF ($^2\Pi_i$)	482.09	104	−317	589.76	[4.62]
PF ($^3\Sigma^-$)	–	89.433	−240.29	–	[4.35]

Table 4.19. Hyperfine interaction constants of the chlorine nucleus (in MHz)

	a	b_η	c	d	Ref.
CCl ($^2\Pi_r$)	80.199	−1.2	−39.5	82.212	[4.63]
NCl ($^3\Sigma^-$)	–	3.519	−57.764	–	[4.30]
OCl ($^2\Pi_i$)	136.235	16.3	−95.6	173.030	[4.79]
SiCl ($^2\Pi_r$)	43.67	2	−21	46.40	[4.65]

Table 4.20. Hyperfine coupling constants of the nitrogen nucleus (in MHz)

	a	b_η	c	d	Ref.
NH ($^3\Sigma^-$)	–	19.22	−67.94	–	[4.319]
CN ($^2\Sigma^+$)	–	−13.857	60.390	–	[4.322]
NO ($^2\Pi_r$)	84.205	22.517	−59.086	112.598	[4.323]
NF ($^1\Delta$)	109	–	–	–	[4.89]
SiN ($^2\Sigma^+$)	–	50.95	94.47	–	[4.7]
NS ($^2\Pi_r$)	62.38	20.7	−46.8	86.972	[4.324]
NCl ($^3\Sigma^-$)	–	22.958	−63.159	–	[4.30]

For electronic states other than $^2\Pi$, not all four constants exist; b and c in $^n\Sigma$ states with $n > 2$ and only a in $^1\Delta$. For the NF radical, the hyperfine structure has been analyzed in the metastable $a\,^1\Delta$ state, but not in the ground $X\,^3\Sigma^-$ state.

The Dipolar Interaction Constant and the Spin Density. Of the four constants, c and d are directly related to the dipolar interaction, namely they denote the axial and anisotropy components of the dipolar coupling tensor, respectively.

The angular part of the integral $\langle n'|\,(3\cos^2\theta - 1)/r^3\,|n\rangle$ in c is 4/5 and $-2/5$ for a pure p orbital around the nucleus considered when the p orbital is extended parallel (σ) and perpendicular (π) to the molecular axis, respectively. The dipolar interaction is well interpreted by considering only a p orbital of the nucleus under consideration except for the proton; the other nucleus in the molecule does not appear to contribute much to the dipolar interaction in most cases. The hyperfine coupling of the hydrogen nucleus is discussed below.

In $^2\Pi$ and $^3\Sigma^-$ states, the electron(s) causing the dipolar interaction occupies (y) p_π orbital(s), and thus the angular part of the c constant may be replaced by $-2/5$ as follows:

$$c = -(3/5)\, g_S g_N \beta \beta_N \langle r^{-3}\rangle,$$

which may be used to estimate $\langle r^{-3}\rangle$. For a $^2\Pi$ molecule, it is better to employ

$$d + c/3 = g_S g_N \beta \beta_N \langle r^{-3}\rangle.$$

Morton and *Preston* [4.277] have listed the atomic values of $g_S g_N \beta \beta_N \langle r^{-3}\rangle$, which may serve as the standards in calculating the spin density from the observed values of $g_S g_N \beta \beta_N \langle r^{-3}\rangle$, Table 4.21 for the three kinds of nuclei F, Cl, and N in various free radicals. The CN and SiN radicals require special treatment, because the unpaired electron occupies a p_σ orbital in these radicals, and $(5/6)\,c$ must be used to calculate $g_S g_N \beta \beta_N \langle r^{-3}\rangle$.

Table 4.21. Spin densities and the hyperfine interaction constants (in MHz)

		$g_S g_N \beta\beta_N \langle r^{-3}\rangle_S^{a}$	$a \equiv 2 g_N \beta \beta_N \langle r^{-3}\rangle_0$	Spin density	b_η
F in	CF	674.6	705.8	15.3%	151.6
	NF	–	753	17.1 [d]	–
	SiF	300.7	312.4	6.8	69
	SF	484.1	482.1	11.0	104
Cl in	CCl	69.1	80.2	15.7	−1.2
	NCl [b]	96.3	–	21.9	3.5
	OCl	141.2	136.2	32.2	16.3
	SiCl	39.4	43.7	9.0	2
N in	NH [b]	113.2	–	81.6	19.2
	NO	92.9	84.2	66.9	22.5
	NF	–	109	78.5 [d]	–
	NS	71.4	62.4	51.4	20.7
	NCl [b]	105.3	–	75.8	23.0
	CN [c]	50.3	–	36.3	−13.9
	SiN [c]	78.7	–	56.7	60.0

[a] For $^2\Pi$ molecules $d + c/3$ is used
[b] $-(5/3)\,c$ is used
[c] $(5/6)\,c$ is used
[d] The constant a is used to estimate the spin density

The a hyperfine constant defined by

$$a = 2g_N \beta \beta_N \langle r^{-3} \rangle$$

is compared in Table 4.21 with $g_S g_N \beta \beta_N \langle r^{-3} \rangle$. The two constants agree fairly well, indicating that the unpaired electron molecular orbital is similar in the orbital and spin distributions. (The spin average $\langle r^{-3} \rangle$ has a suffix S to differentiate it from the orbital average with suffix O.) For NF in $^1\Delta$, only the a constant is available and is used in Table 4.21 to estimate the spin density. When the hyperfine constants are determined for both nuclei in a molecule, the spin densities calculated from them may be tested against the normalization condition for their sum. The sum is 0.956 and 0.977, respectively, for NF and NCl.

As mentioned above, the hyperfine interaction constants of the second and third row atoms have been well explained by considering that the unpaired electron occupies a p orbital of the atoms involved in the interaction. However, the proton hyperfine constants need different treatment, because the p orbital of the proton is too high in energy. Note that, as Table 4.22 shows, the c constants of the proton are all positive. It is also worth noting that the a and c constants listed in Table 4.22 are approximately proportional to r_{HX}^{-3}, where r_{HX} denotes the internuclear distance of the HX radical. Therefore, the dipolar interaction of the proton is mainly ascribed to the interaction with the unpaired electron located on a p orbital of the other atom in the molecule. It is difficult to separate the angular part from the radial part of the integrals in the a, c, and d constants, because they involve two centers.

Table 4.22. Hyperfine interaction constants of the proton and the bond length for HX-type molecules (in MHz)

	a	$a^{\text{calc a}}$	c	$c^{\text{calc a}}$	r_{XH} [Å]
CH	54.188	55.92	56.98	84.83	1.1199
NH	–	–	91.70	107.09	1.0362
OH[a]	86.112	(86.112)	130.644	(130.644)	0.96966
PH	–	–	16.3	41.47	1.4214
SH	32.58	32.62	32.44	49.49	1.3409

[a] a^{calc} and c^{calc} are obtained by assuming that they are proportional to r_{HX}^{-3} and reproducing the observed values for OH

The Fermi Interaction Constant. The b_η hyperfine constant is proportional to the expectation value of the unpaired electron orbital at the nucleus considered, namely $|\psi(0)|^2$, which represents the s character of the orbital. However, its interpretation is by no means simple. For example, as already mentioned, the unpaired electron orbital has a node on a line connecting the two nuclei in the molecule, when it is in one of the $^2\Pi$, $^3\Sigma^-$, and $^1\Delta$ states, but the observed

b_η values do not vanish. This fact has been explained by spin polarization [4.325] of a p_σ orbital, caused by the unpaired electron in a p_π orbital; the part of p_σ near the nucleus studied will be polarized parallel to the spin of the unpaired electron, while the part of p_σ far from the nucleus is antiparallel to the unpaired electron spin, resulting in small positive Fermi coupling constants for heavy nuclei and small negative constants for protons, Tables 4.17–20.

As Table 4.18 indicates, the Fermi term is proportional to the spin density for F [4.62, 64], suggesting that the Fermi term is caused by spin polarization. This relation does not hold for Cl. The N Fermi term does not vary much for the species listed in Table 4.20. The Fermi term in $^2\Sigma$ molecules is more difficult to explain; it is negative in CN, whereas it is positive in SiN, although both constants are much smaller in magnitude than the atomic value 1811 MHz [4.277], indicating that the $s-p$ hybridization is relatively unimportant in these two molecules.

The Nuclear Electric Quadrupole Interaction. The discussion is limited to the Cl nuclei in the following series of molecules: BCl, CCl, NCl, OCl, and FCl, which have the electron configuration $KKL(5\sigma)^2(6\sigma)^2(2\pi)^4(7\sigma)^2(3\pi)^n$ with n of 0 to 4 and the ground electronic states of $^1\Sigma^+$, $^2\Pi_r$, $^3\Sigma^-$, $^2\Pi_i$, and $^1\Sigma^+$, respectively. Thus of the five molecules the three central ones exhibit magnetic hyperfine structures in their spectra. Table 4.23 gives the nuclear electric quadrupole coupling constants of these five molecules; for CCl and OCl the non-axially symmetric components eQq_2 are also listed.

The magnetic hyperfine interaction constants allow us to estimate the amount of π electron delocalization. When the spin density on the Cl atom is ϱ, the numbers of electrons in the 3π and 2π orbitals around the Cl atom are given approximately by $n\varrho$ and $4(1-\varrho)$, respectively, because the two orbitals are orthogonal. The total amount of p_π electrons transferred from Cl to the other atom is given by [4.30]

$$\Delta_p = 4 - [n\varrho + 4(1-\varrho)] = \varrho(4-n).$$

Table 4.23. Electric quadrupole interaction constants of the chlorine nucleus (in MHz)

	eQq_1	$eQq^{\text{corr a}}$	eQq_2	Ref.
BCl	−16.74	$(-23)^{\text{b}}$	–	[4.326]
CCl	−34.26	−60.10	84.6	[4.63]
NCl	−63.13	−87.16	–	[4.30]
OCl	−87.95	−105.73	−116.0	[4.79]
FCl	−145.87	–	–	[4.327]

[a] The correction procedure is explained in the text
[b] Calculated by assuming 0.12 π electron to be transferred to the B atom

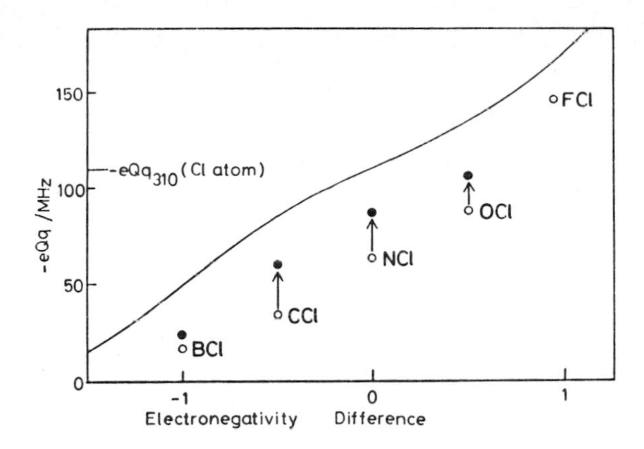

Fig. 4.39. Nuclear electric quadrupole coupling constants vs electronegativity difference for five diatomic chloride molecules [4.321]

When the spin densities listed in Table 4.21 are used, Δ_p is calculated to be 0.47, 0.44, and 0.32 for CCl, NCl, and OCl, respectively. Correcting the observed eQq_1 constant for the π-electron transfer just estimated yields eQq^{corr}, listed in Table 4.23. These values are plotted in Fig. 4.39 against the difference in electronegativity of the two atoms in the molecule; the corrections applied to the observed eQq are shown by arrows. The corrected eQq constants fall on a curve parallel to that originally proposed by *Gordy* and *Cook* [4.318]; although a discrepancy between the two curves still remains, the corrections for π-electron delocalization have reduced it considerably.

As mentioned above, the nonaxially symmetric component of the eQq tensor has been determined for CCl and OCl. Since the 1π and 2π orbitals are fully occupied, the nonaxial symmetry is primarily ascribed to the electron distribution in the 3π orbital, which will manifest itself in the d hyperfine coupling constant. In fact, we may calculate the expectation value $\langle \sin^2\theta/r^3 \rangle$ from either d or eQq_2, using (2.3.75e, 76b). The results are $\langle \sin^2\theta/r^3 \rangle_S = 7.07 \times 10^{24}\,\mathrm{cm}^{-3}$ and $\langle \sin^2\theta/r^3 \rangle_T = 10.25 \times 10^{24}\,\mathrm{cm}^{-3}$ for CCl and $\langle \sin^2\theta/r^3 \rangle_S = 14.88 \times 10^{24}\,\mathrm{cm}^{-3}$ and $\langle \sin^2\theta/r^3 \rangle_T = -14.06 \times 10^{24}\,\mathrm{cm}^{-3}$ for OCl, where $\langle\ \rangle_S$ and $\langle\ \rangle_T$ denote the averages over the unpaired electron spin distribution and the total electron distribution in the molecule, respectively [4.63]. The two averages agree with each other for the two molecules, except the negative sign for $\langle\ \rangle_T$ of OCl, which is explained by the fact that there is a hole in the 3π orbital.

4.5.2 Spin-Rotation Interaction in Nonlinear Polyatomic Molecules

For nonlinear polyatomic molecules in doublet states, the spin-rotation interaction is by far the most important, because the electronic orbital angular momentum is quenched in most of these molecules (Sect. 2.3.5). The spin-rotation interaction consists of first-order and second-order terms, the latter arising from mixing with excited electronic states through the electronic Coriolis interaction, while the spin-orbit interaction usually dominates the former [4.328]. When the spin-orbit interaction is given by

$$\boldsymbol{H}_{\mathrm{SO}} = A_{\mathrm{SO}} \boldsymbol{L} \cdot \boldsymbol{S}, \tag{4.5.1}$$

the second-order expression for $\varepsilon_{\alpha\beta}$ is [4.328–330]

$$\varepsilon_{\alpha\beta} = -2\hbar^2 \sum_n{}' \mathrm{Re}\, \langle 0|\, \mu_{\alpha\alpha} L_\alpha\, |n\rangle\, \langle n|\, A_{\mathrm{SO}} L_\beta\, |0\rangle/(E_0 - E_n), \tag{4.5.2}$$

where $\mu_{\alpha\alpha}$ denotes the $\alpha\alpha$ component of the inverse inertia tensor defined in Sect. 2.2. The spin-orbit coupling constant A_{SO} of the molecule may be approximated by that of an appropriate atom in the molecule or a weighted mean of those when the unpaired electron spin distribution in the molecule is known (Sects. 4.5.1, 3). For most bent triatomic molecules, the ground electronic state and the first excited electronic state constitute a Renner-Teller pair, i.e., they are degenerate in a fictitious linear configuration. The first excited state then makes a dominant contribution to the spin-rotation interaction in the ground state, and the ε_{aa} component is the largest in magnitude.

Table 4.24 lists the spin-rotation interaction constants so far reported. As expected, the ε_{aa} component is the largest in magnitude. When a molecule has an unpaired electron in an orbital extending perpendicularly to the molecular plane, ε_{cc} is very small, because no one electron excitation can produce the cc component. For a molecule of C_s symmetry, there are four determinable components of the ε tensor, ε_{aa}, ε_{bb}, ε_{cc}, and $|\varepsilon_{ab} + \varepsilon_{ba}|/2$, and they are all available for HO_2, HSO, and FSO.

Table 4.24. Spin-rotation interaction constants (in MHz)

		ε_{aa}	ε_{bb}	ε_{cc}	$\lvert\varepsilon_{ab} + \varepsilon_{ba}\rvert/2$	Ref.
HO_2	$^2A''$	−49571.41	−422.76	8.61	194.0	[4.159]
HSO		−10365.99	−426.66	0.23	378.0	[4.185]
FO_2		−822	$\varepsilon_{bb} + \varepsilon_{cc} = 5$			[4.179]
FSO		−339.54	34.90	1.86	207.95	[4.191]
NH_2	$^2B_1$	−9267.41	−1353.89	11.83		[4.331]
PH_2		−8427.48	−2458.62	−7.12		[4.228]
NF_2		−951.79	−92.86	4.49		[4.332]
ClO_2		−1388.19	−216.90	4.60		[4.333]
CH_2F		−1075.96	−185.77	−1.41		[4.286]
CH_2Cl		−3149.45	−237.62	11.81		[4.288]
HCO	$^2A'$	11624.5	18.96	−205.69		[4.334]
FCO		1831	$\varepsilon_{bb} + \varepsilon_{cc} = 12$			[4.198]
NO_2	$^2A_1$	5406.54	7.71	−95.27		[4.335]
CH_3	$^2A_2'$		−350	3		[4.261]
CF_3	$^2A_1$		−36.50	3.35		[4.275]

Table 4.25. Normalized spin-rotation constants and energy difference estimated using the second-order equation (4.5.2)

	$\varepsilon_{aa}/A \times 10^2$	$\varepsilon_{bb}/B \times 10^2$	$\varepsilon_{cc}/C \times 10^2$	A^a_{SO}/cm^{-1} [a]	$\Delta E_{calc}/cm^{-1}$ [e]	E_{obs}/cm^{-1}
HO_2	−8.1228	−1.2613	0.0272	158.5	7800	7029.48
HSO	−3.4613	−2.0808	0.0012	300 [b]	34700	14383.28
FO_2	−1.0473	-	-	158.5	60500	22500
FSO	−0.8774	0.3736	0.0248	300 [b]	136800	
NH_2	−1.3046	−0.3488	0.0049	80	24500	10249
PH_2	−3.0782	−1.0145	−0.0056	300	39000	18276.59
NF_2	−1.3501	−0.7822	0.0443	80	23700	
ClO_2	−2.6654	−2.1794	0.0552	587	88100	54689 ($\tilde{C}^2A_1$)
CH_2F	−0.4057	−0.6003	0.0051	65 [c]	64100	
CH_2Cl	−1.1478	−1.4900	0.0785	116 [d]	40400	
HCO	1.5938	0.0423	−0.4906	27.1	6800	9279
FCO	0.9576	-	-	27.1	11300	
NO_2	2.2536	0.0593	−0.7742	80	14200	
CH_3	-	−0.1219	-	27.1	88900	
CF_3	-	−0.3348	-	27.1	32400	

[a] Spin-orbit coupling constant of the most important atom
[b] $0.6\,A_{SO}(S) + 0.4\,A_{SO}(O)$
[c] $0.87\,A_{SO}(C) + 0.13\,A_{SO}(F)$
[d] $0.9\,A_{SO}(C) + 0.1\,A_{SO}(Cl)$
[e] ε_{aa} is used to estimate ΔE except for CH_3 and CF_3

As the second-order expression (4.5.2) indicates, the spin-rotation interaction constant is proportional to the rotational constant. Table 4.25 thus gives the diagonal components scaled by the appropriate rotational constants for ease of comparison between species. As *Curl* [4.328] has pointed out, these scaled second-order sums are related to the g-factor corrections due to mixing of excited electronic states through the electron orbital angular momentum. When we use the A_{SO} constants listed in Table 4.25 and assume $|\langle 0|\,L_a\,|n\rangle|^2 = 1$, we may estimate the excitation energy ΔE from ε_{aa}/A (ε_{bb}/B for CH_3 and CF_3), provided that only one excited state contributes to the second-order sum. For HO_2 the calculated ΔE agrees well with that for the first excited electronic state. For most other molecules, however, the calculated ΔE values are two to three times larger than the observed data, indicating that $|\langle 0|\,L_a\,|n\rangle|^2$ is less than unity. For HCO, the calculated value is smaller than the observed. This is probably explained by using a spin-orbit coupling constant which is a weighted average of those for the C and O atoms.

The vibrational change of the spin-rotation interaction constant has been discussed only in a few cases. Usually the bending vibration increases the ε_{aa} constant much more than the A rotational constant [4.336, 225]. This is presumably because the vibrational excitation decreases ΔE and increases $|\langle 0|\,L_a\,|n\rangle|^2$.

4.5.3 Hyperfine Interaction in Nonlinear Polyatomic Molecules

Both the magnetic interaction and nuclear electric quadrupole interaction tensors $\boldsymbol{T}$ and χ are normally determined in reference to the principal inertial axis system of the molecule. When one of the inertial axes coincides with the axis of the unpaired electron p orbital, the magnetic dipolar interaction tensor has diagonal components that approximately satisfy the ratios $4/5: -2/5: -2/5$, where the first number applies to the axial component and the second and third to the perpendicular components, as for a diatomic molecule. As shown in Table 4.26, for species with an unpaired electron in an out-of-plane p orbital, $T_{cc} = -2T_{aa} = -2T_{bb}$ is well satisfied; small discrepancies between T_{aa} and T_{bb} may be ascribed to coupling with neighboring atoms which modifies the unpaired electron distribution.

By assuming $T_{cc} = (4/5)\, g_S g_N \beta \beta_N \langle r^{-3} \rangle$, we may estimate the spin density on the atom involved in the hyperfine interaction, which is also included in Table 4.26, just as in Table 4.21 for diatomic free radicals. It is interesting to note that the unpaired electron of NO_2 occupies an in-plane A_1 symmetry orbital and thus has a positive T_{bb} constant roughly satisfying $T_{bb} = -2T_{aa} = -2T_{cc}$.

The hydrogen dipolar coupling constants are again exceptional, as they are in diatomic molecules, and must be discussed separately. The observed coupling constants are summarized in Table 4.27, where the Fermi coupling constants are also included. The coupling here is mainly ascribed to the interaction of the proton with the unpaired electron localized on the other atom X. Therefore, the tensor component along the $H-X$ bond is expected to have a large

Table 4.26. Magnetic hyperfine coupling constants of heavy nuclei (in MHz)

	a_F	T_{aa}	T_{bb}	T_{cc}	spin density [%]	Ref.
^{19}F						
FSO	67.23	−118.16	−117.06	235.22	6.7	[4.191]
NF_2	164.39	−241.75	−226.48	468.22	13.3	[4.332]
CH_2F	184.10	−255.21	−212.31	467.52	13.3	[4.286]
CF_3	408.5	$\lvert T_{aa} - T_{bb} \rvert = 40.1$		320.01	10.6	[4.275]
^{14}N						
NF_2	46.57	−47.72	−50.47	98.19	88.4	[4.332]
$NH_2\,(\tilde{X}^2 B_1)$	27.88	−43.38	−44.75	88.13	79.4	[4.331]
$NH_2\,(\tilde{A}^1 A_2)$	153.0	−39.5	76.9	−37.4	69.3	[4.337]
NO_2	147.26	−22.14	39.88	−17.74	35.9	[4.335]
^{31}P						
$PH_2\,(\tilde{X}^2 B_1)$	207.25	−300.24	−321.86	622.10	84.8	[4.233]
$PH_2\,(\tilde{A}^2 A_1)$	1747.2	−259.5	477.4	−217.9	65.1	[4.231]
^{35}Cl						
ClO_2	46.12	−77.74	−83.10	160.84	45.8	[4.333]
CH_2Cl	8.64	−32.29	−22.74	55.03	15.7	[4.288]

Table 4.27. Magnetic hyperfine coupling constants of proton (in MHz)

	a_F	T_{aa}	T_{bb}	T_{cc}	$\lvert T_{ab} \rvert$	Ref.
HO_2	−27.48	−8.34	19.68	−11.34	−18.3	[4.157]
HSO	−36.37	−11.96	10.44	1.52	−7.8	[4.185]
$NH_2\,(\tilde{X}^2\,B_1)$	−67.59	17.54	−12.75	−4.79		[4.331]
$PH_2\,(\tilde{X}^2\,B_1)$	−48.85	−1.00	−4.46	5.46		[4.233]
CH_2F	−60.73	−25.68	24.2	1.4		[4.286]
CH_2Cl	−61.40	−21.82	19.1	2.7		[4.288]
HCO	390.76	11.56	3.87	−15.43		[4.334]
$NH_2\,(\tilde{A}^2\,A_1)$	52.2	59.5	−9.5	−50.0		[4.337]
$PH_2\,(\tilde{A}^2\,A_1)$	190.06	12.3	−4.0	−8.3		[4.231]

Table 4.28. Principal values of the proton dipolar coupling tensors in HO_2 and HSO (in MHz)

	T_{XX} [a]	T_{YY}	$T_{ZZ} = T_{cc}$	θ [b] [°]	ϕ [c] [°]
HO_2	28.8	−17.4	−11.4	63.8	72.5
HSO	12.9	−14.4	1.52	72.5	70.5

[a] The X axis nearly coincides with the HX bond
[b] The angle between the a and X axes
[c] The angle between the a axis and HX bond

positive value and the component perpendicular to both the $H - X$ bond and the axis of the unpaired electron p orbital should take a negative value of similar magnitude. Unfortunately, such components are not obtainable directly from experiment. In two cases, HO_2 and HSO, however, the off-diagonal component T_{ab} has been determined, and the principal values of the $\boldsymbol{T}$ tensor have been evaluated, Table 4.28, where the angle of transformation from the inertial to the dipolar interaction principal axis systems is also given and is compared with the angle between the a inertial axis and the $H - X$ bond. (There are two possible ways of axis rotation, but it is not difficult to choose the correct one.)

The Fermi coupling constants listed in Tables 4.26, 27 are all much smaller than the corresponding atomic values [4.277], as expected since most of them are π radicals. The signs of the constant is positive for heavy nuclei, whereas it is negative for hydrogen, indicating the spin polarization effect to be important. The HCO and NO_2 radicals have the unpaired electrons in in-plane orbitals, and, in fact, the Fermi constant is large in these radicals. Large coupling constants have also been reported for NH_2 and PH_2 in the $\tilde{A}\,^2A_1$ excited states [4.337, 230, 231], where they are σ radicals with the unpaired electron orbital considerably $s - p$ hybridized. The CF_3 radical is similar; its Fermi coupling constant is much larger than those in other π radicals.

4.6 Molecules in Metastable Electronic States

Excited electronic states with spin multiplicity different from that of the ground state are usually difficult to observe by conventional spectroscopic methods mainly because the transition probabilities from the ground state are not large. These metastable states which thus often remain unknown are interesting and important from spectroscopic as well as chemical points of view. Variations in molecular structure and electronic properties of a molecule among different multiplet states may reflect characteristic interactions among electrons in the molecule. Spectroscopy of metastable states will also be indispensable in clarifying and understanding dynamical properties of a molecule such as relaxation and even chemical reactivities. A molecule may behave quite differently in different multiplet states, between which it is transferred through intersystem crossing. This section describes a laser-excited fluorescence (LEF) and microwave optical double resonance (MODR) study of H_2CS in the $\tilde{a}\,^3A_2$ state and also laser excitation spectroscopic studies of perturbations appearing in the singlet-singlet ($\tilde{A}\,^1A'' - \tilde{X}\,^1A'$) transition of a few simple carbenes, some of which have been ascribed to interactions with the low-lying triplet ($\tilde{a}\,^3A''$) state.

4.6.1 LEF and MODR Study of H_2CS in the $\tilde{a}\,^3A_2$ State

Thioformaldehyde has been subjected to a number of spectroscopic investigations since it was first unambiguously identified by a microwave study [4.338] in 1970, as recently reviewed by *Clouthier* and *Ramsay* [4.339]. *Fung* and *Ramsay* [4.340] have investigated the fine details of the $\tilde{A}\,^1A_2 - \tilde{X}\,^1A_1\; 4_0^1$ band using an intermodulated fluorescence technique and have confirmed that many small perturbations they observed can be ascribed to interactions with excited vibrational levels associated with the electronic ground state. *Suzuki* et al. [4.341] observed and analyzed the LEF and MODR spectrum of the $\tilde{a}\,^3A_2 - \tilde{X}\,^1A_1\; 3_0^1$ band of H_2CS. As described in detail below, this study furnishes the first gas phase MODR study of a small molecule in a triplet state, yielding the hyperfine coupling constants in the $\tilde{a}\,^3A_2$ state and also revealing the Coriolis interaction between the 3^1 and 6^1 levels. *Clouthier* et al. [4.342] recently analyzed perturbations observed in the $\tilde{A}\,^1A_1 - \tilde{X}\,^1A_1\; 0_0^0$ band of H_2CS in terms of the singlet-triplet interaction with the $4^3\,6^1$ level of the $\tilde{a}\,^3A_2$ state. The ν_6 frequency in the $\tilde{a}$ state reported by Suzuki et al. is consistent with this explanation.

Suzuki et al. [4.341] recorded the LEF $\tilde{a}\,^3A_2 - \tilde{X}\,^1A_1\; 3_0^1$ spectrum of H_2CS in the region 15 280–15 400 cm^{-1} at the Doppler-limited resolution and measured about 900 lines. The observed spectral features are consistent with *Hougen*'s theory of the singlet-triplet transition [4.343]. Figure 4.40 shows a part of the observed spectrum.

The observed spectrum was analyzed using the Hamiltonian

$$H = H_{\mathrm{ROT}} + H_{\mathrm{SR}} + H_{\mathrm{SS}}, \tag{4.6.1}$$

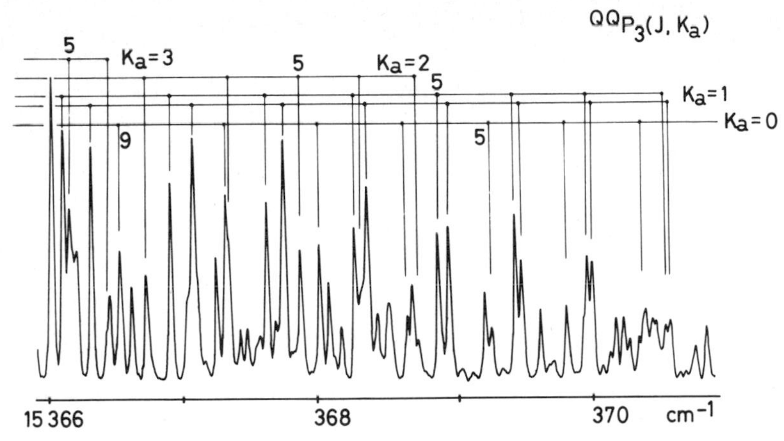

Fig. 4.40. A part of the excitation spectrum of the H_2CS $\tilde{a}^3A_2 - \tilde{X}^1A_1$ 3_0^1 band [4.341]

where $\boldsymbol{H}_{\mathrm{ROT}}$, $\boldsymbol{H}_{\mathrm{SR}}$, and $\boldsymbol{H}_{\mathrm{SS}}$ stand for the rotational, spin-rotation interaction, and spin-spin interaction Hamiltonians, which are all amply discussed in Chap. 2. For convenience, they are reproduced below:

$$\begin{aligned}\boldsymbol{H}_{\mathrm{ROT}} = {} & A N_a^2 + B N_b^2 + C N_c^2 - \Delta_N (N^2)^2 - \Delta_{NK} N^2 N_a^2 - \Delta_K N_a^4 \\ & - 2\delta_N N^2 (N_b^2 - N_c^2) - \delta_K [N_a^2 (N_b^2 - N_c^2) + (N_b^2 - N_c^2) N_a^2], \end{aligned} \quad (4.6.2)$$

$$\boldsymbol{H}_{\mathrm{SR}} = - a_0 \boldsymbol{N} \cdot \boldsymbol{S} - a(2 N_a S_a - N_b S_b - N_c S_c) - b(N_b S_b - N_c S_c), \quad (4.6.3)$$

$$\boldsymbol{H}_{\mathrm{SS}} = \alpha(3 S_a^2 - \boldsymbol{S}^2) + \beta(S_b^2 - S_c^2). \quad (4.6.4)$$

For the singlet ground state only (4.6.2) was retained.

The observed spectrum was well fitted for K_a' up to 3, but $K_a' = 4$ and 5 lines showed deviations because of the c-type Coriolis interaction between the 3^1 (a_1) and 6^1 (b_2) levels.

The Coriolis interaction was assumed to take the form

$$\boldsymbol{H}_{\mathrm{COR}} = i G_{36}^c N_c, \quad (4.6.5)$$

where the parameter G_{36}^c is related to the Coriolis coupling constant ζ_{36}^c as

$$G_{36}^c = C[(\omega_3/\omega_6)^{1/2} + (\omega_6/\omega_3)^{1/2}]\,\zeta_{36}^c. \quad (4.6.6)$$

All the 3^1 constants, G_{36}^c, the unperturbed energy difference between 3^1 and 6^1, and the $B + C$ constant of 6^1 were chosen as adjustable parameters, whereas other 6^1 constants were constrained to the $v = 0$ constants in the $\tilde{a}$ state [4.344] and the ground vibronic state constants to the microwave values [4.345]. Some of the principal constants for the 3^1 and 6^1 states are summarized in Table 4.29, where those of $v = 0$ are also included for comparison. When use

Table 4.29. Molecular constants of H_2CS in the $\tilde{a}^3 A_2$ state (in MHz)[a]

	$v_3 = 1$	$v = 0$[b]	$v_6 = 1$
A	280457.6 (87)	281289 (13)	
B	16401.74 (28)	16534.75 (90)	$B + C = 32189.5$ (24)
C	15488.75 (31)	15605.58 (90)	
a_0	4880.2 (45)	4495.7 (144)	
a	4685.9 (50)	4290.0 (152)	
b	−82.87	−86.6 (99)	
α	−43474.6 (61)	−40780 (170)	
β	2550.2 (61)	2827 (580)	
ν [cm^{-1}]	15369.0317 (37)	14507.3846 (45)	$\nu_3 - \nu_6 = 99.341$ (24)

[a] Values in parentheses denote three standard errrors and apply to the last digits of the constants. The Coriolis coupling constant between $v_3 = 1$ and $v_6 = 1$ was determined to be $G_{36}^c = 5307$ (41) MHz

[b] [4.344]

is made of the 3_0^0 and 3_0^1 band origins, the observed $\nu_3 - \nu_6$ value yields the ν_6 frequency in the $\tilde{a}\,^3A_2$ state as 762.3 cm^{-1}.

The molecular constants determined for an excited electronic state are often difficult to interpret because the excited electronic state is apt to be perturbed by other electronic states. This statement holds even when such constants reproduce the observed transition wave numbers well. It may thus be interesting to examine the physical significance of the molecular constants obtained from the present analysis. To do this, the centrifugal distortion constants and the Coriolis coupling constant were calculated from a harmonic force field and compared with the observed values.

The harmonic force field was calculated from the vibrational frequencies of H_2CS and D_2CS in $\tilde{a}\,^3A_2$, which are either the observed values or the ab initio calculated values, and the molecule was assumed to be planar, i.e., to be of C_{2v} symmetry. Although three F-matrix elements F_{12}, F_{13}, and F_{56} could not be determined, other elements readily led to the centrifugal distortion constants and the Coriolis coupling constant which agreed well with the observed values, except that the calculated ζ_{36}^c constant was slightly larger than the observed value. The discrepancy might be due to higher-order Coriolis terms which are neglected. The centrifugal distortion constants in the ground vibrational state, although less precise than the data on the ν_3 state, show excellent agreement with the calculated values. In conclusion, the molecular constants of H_2CS in $\tilde{a}\,^3A_2$ do not show any anomalous features except for the ν_3/ν_6 Coriolis interaction, suggesting that the $\tilde{a}\,^3A_2\,\nu_3$ state is almost completely free of perturbations by other vibronic states.

The MODR spectrum was observed for H_2CS in the $\tilde{a}\,^3A_2\; v_3 = 1$ state, in order to determine the proton hyperfine coupling constants. The spectroscopic arrangement used is described in Sect. 3.4. In the vibronic state considered,

$I_H = 1$ and 0 for rotational levels of K_a = odd and even, respectively. Therefore, two K-type transitions and two R-branch transitions, all $K_a = 1$, were selected for the observation; the F_3 components were chosen because they showed the largest splittings among the three spin components.

The hyperfine Hamiltonian used is discussed in detail in Sect. 2.3.7, and is reproduced here:

$$\boldsymbol{H}_{\mathrm{HFS}} = a_{\mathrm{F}}\boldsymbol{S} \cdot \boldsymbol{I} + T_{aa}S_a I_a + T_{bb}S_b I_b + T_{cc}S_c I_c. \tag{4.6.7}$$

In the least-squares analysis, only the hyperfine constants were chosen as parameters, whereas others were fixed to those determined by the LEF experiment. The results are included in Table 4.30.

Table 4.30. Hyperfine coupling constants of H_2CS in the $\tilde{a}^3 A_2\ v_3 = 1$ state (MHz)[a]

a_F	27.7 (25)
$T_{bb} + T_{cc}$	12.8 (44)
$T_{bb} - T_{cc}$	7.89 (19)

[a] Values in parentheses denote standard errors and apply to the last digits of the constants

Figure 4.41 illustrates the unpaired electron distribution in H_2CS in the $\tilde{a}\,^3A_2$ state; one unpaired electron remains in a nonbonding orbital which is predominantly the sulfur in-plane $3p_b$ orbital, and the second unpaired electron is transferred to an antibonding π^* orbital which approximates an "out-of-phase" linear combination of the sulfur $3p_c$ and carbon $2p_c$ orbitals. The former electron contributes positively to the spin density at the proton nuclei, whereas the latter electron contributes negatively through a spin polarization

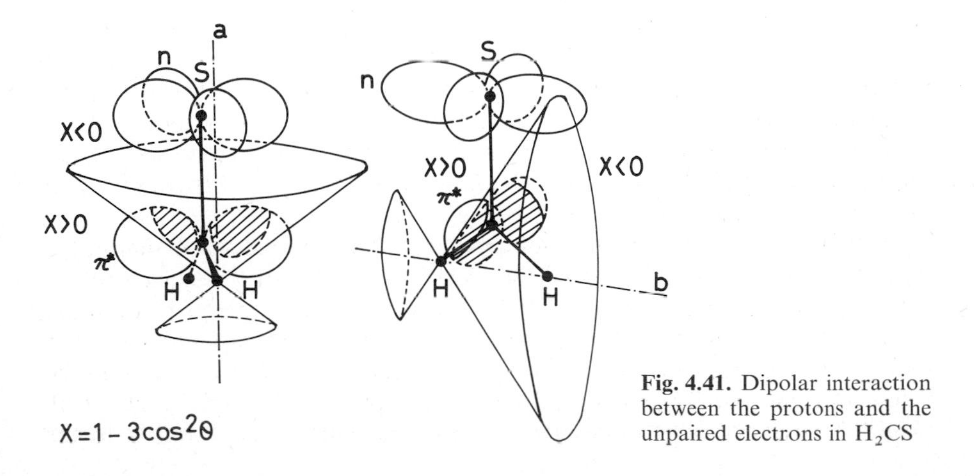

Fig. 4.41. Dipolar interaction between the protons and the unpaired electrons in H_2CS

effect on the sp^2 bounding orbital of the carbon atom. The positive sign of the observed Fermi term $a_F = +27.2$ MHz indicates that the former contribution predominates over the latter. On the other hand, the magnetic dipolar interaction mainly takes place between the second unpaired electron and the proton nuclear spins, because the dipolar interaction is proportional to the inverse cube of the distance between the two interacting dipoles. In fact, as shown in Fig. 4.41, the sign of $X = 1 - 3\cos^2\theta$ in the region where the π^* orbital density is large corresponds well to those of the $\boldsymbol{T}$ components, where θ denotes the angle between the radius vector pointing from the proton to the unpaired electron and one of the principal axes.

4.6.2 Perturbations in the Singlet-Singlet Transition of a Few Simple Carbenes

Carbenes play important roles in organic chemical reactions; a unique feature of these species is that they behave quite differently in the singlet and triplet states. It has been well established that the methylene radical has triplet ground state, whereas the ground state is singlet for CHF and CHCl. The singlet-triplet separation in a carbene is strongly dependent on the group substituted. For most simple carbenes, only the ground state, either singlet or triplet, has been identified and characterized by spectroscopic methods, while almost no information has been obtained on the metastable states which have multiplicity different from that of the ground state. It is of great significance to determine the singlet-triplet energy separation and to unravel the nature of the metastable state.

Methylene, the simplest carbene, is an exception to the above statement; *Herzberg* [4.200] and *Herzberg* and *Johns* [4.201] analyzed the triplet-triplet (Rydberg $n = 3 \sim 6 - \tilde{X}\,^3B_1$) and singlet-singlet ($\tilde{b}\,^1B_1 - \tilde{a}\,^1A_1$) transitions, respectively, and *McKellar* and co-workers [4.346] recently succeeded in determining the $\tilde{a}\,^1A_1 - \tilde{X}\,^3B_1$ separation to be $T_0 = 3165 \pm 20\ \mathrm{cm}^{-1}$ (the equilibrium separation $T_e = 2994 \pm 30\ \mathrm{cm}^{-1}$). *Ohashi* et al. [4.347] also attempted to refine Herzberg and Johns' data on the $\tilde{b}\,^1B_1 - \tilde{a}\,^1A_1$ transition, by applying Doppler-limited dye laser excitation spectroscopy to singlet methylene generated from ketene photolyzed with the 334 nm Ar^+ laser line. About 40 lines have thus been measured to a precision of $0.003\ \mathrm{cm}^{-1}$, which may be of some use in exploring the fine details of the level structure of the $\tilde{a}\,^1A_1$ state. (For other spectroscopic studies on methylene, see Sect. 4.3.2.)

Halogen-substituted methylenes are much less characterized than CH_2 itself. In 1966, *Merer* and *Travis* [4.348, 349] reported the first high-resolution spectroscopic investigations of HCF, HCCl, and DCCl, i.e., they observed the $\tilde{A}\,^1A'' - \tilde{X}\,^1A'$ transitions of these species in absorption. They pointed out that the upper electronic states are affected by perturbations and have discussed the origin of these perturbations. *Kakimoto* et al. [4.350, 351] and *Suzuki* et al. [4.352] have carried out more detailed measurements of a few bands in the

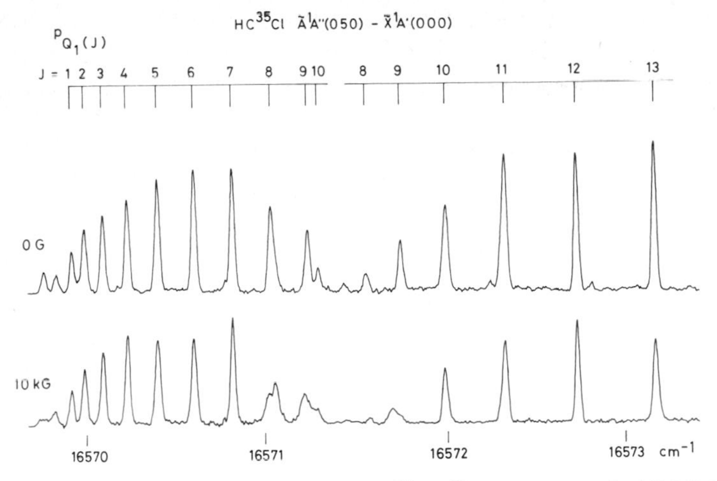

Fig. 4.42. ${}^{P}Q_1(J)$ transitions of the $HC^{35}Cl$ $\tilde{A}^1A''(050) - X\,{}^1A'(000)$ band [4.351]

$\tilde{A} - \tilde{X}$ system of the two species using Doppler-limited dye laser excitation spectroscopy.

Kakimoto et al. [4.351] focused on the perturbations observed for $J' = 8 - 10$, $K'_a = 0$ of the HCCl $\tilde{A}(0, 5, 0)$ state. Figure 4.42 reproduces the ${}^{P}Q_1(J)$ branch; it clearly shows that there are two lines observed for $J = 8 - 10$, one component being much stronger than the other for $J = 8$ and 10, while the two components are of almost equal intensity for $J = 9$. The lower trace of Fig. 4.42 shows that the perturbed lines bear some paramagnetic character. *Hirota* [4.353] has ascribed the perturbation to the interaction of the $\tilde{A}(0, 5, 0)$ state with a vibrational state associated with the low-lying $\tilde{a}\,{}^3A''$ state; the Zeeman effect observed for the $J' = 9$ levels was qualitatively explained in terms of the singlet-triplet mixing.

Suzuki et al. [4.354] investigated the $\tilde{A}\,{}^1A''(0, 1, 0) - \tilde{X}\,{}^1A'(0, 0, 0)$ vibronic band of HCF by paying special attention to the perturbations observed for $K'_a = 0$ and 1 levels in the upper state; they observed additional series of lines for the branches with $K'_a = 1$. In accordance with an argument by *Merer* and *Travis* [4.349], they interpreted the complications observed for $K'_a = 1$ levels in terms of electronic Coriolis interaction. Since both $\tilde{A}$ and $\tilde{X}$ states are singlet, the diagonal matrix element of the electronic orbital angular momentum $\boldsymbol{L}$ vanishes but, as suggested by the fact that two states may be considered to be derived from a ${}^1\Delta$ state in the limit of linearity, the L_a and L_b components have nonzero matrix elements between $\tilde{A}$ and $\tilde{X}$. When $\boldsymbol{R}$ is replaced by $\boldsymbol{J} - \boldsymbol{L}$, the rotational Hamiltonian $\boldsymbol{H}_{\mathrm{R}} = AR_a^2 + BR_b^2 + CR_c^2$ gives

$$\boldsymbol{H}_{\mathrm{COR}} = -2AJ_aL_a - 2BJ_bL_b, \tag{4.6.8}$$

which is referred to as the electronic Coriolis term. Using this model, *Suzuki* et al. [4.354] have satisfactorily explained the $K'_a = 1$ term values including the additional levels. The perturbing state is thus certainly one of the highly excited vibrational states associated with the $\tilde{X}\,^1A'$ state, but it is hard to make a vibrational assignment for it.

Suzuki et al. [4.354] have also found quite a large number of local perturbations and observed that some of these perturbed levels exhibit Zeeman effects as large as 4–14 GHz/T, i.e., 0.3–1.0 β, β denoting the Bohr magneton. Although the electronic Coriolis interaction predicts that the perturbed levels should bear some paramagnetic character, the Zeeman effect which is expected for this model is an order of magnitude smaller than the observed values. Therefore, these local perturbations can be explained only by taking into account singlet-triplet mixing, but a quantitative analysis has so far been unsuccessful, presumably because there are more than one perturbing level.

Butcher et al. [4.355] have carefully examined the HCF $\tilde{A}\,^1A''(0,0,0) - \tilde{X}\,^1A'(0,0,0)$ band, which has been known to be much less perturbed than the (0, 1, 0) – (0, 0, 0) band, using the technique of intermodulated fluorescence (Sect. 3.3), and have found that there are weak electronic Coriolis interactions in addition to some local singlet-triplet perturbations.

The analysis of local perturbations as described here does not provide any perspective for the perturbing levels including the low-lying triplet state which we are most interested in. However, it may be of some use in future in observing singlet-triplet transitions using some double-resonance techniques and in exploring the level structure of the triplet state.

5. Applications and Future Prospects

High-resolution spectroscopy is an ideal tool for identifying chemical species and monitoring their abundance not only in total but also in individual quantum states, and can thus provide very detailed information on various chemical systems. Three representative fields are considered in this chapter, and some speculative discussion on future prospects of high-resolution spectroscopic studies of unstable molecules follows.

5.1 Applications to Chemical Reactions

Even a qualitative observation of molecular spectra can often aid in elucidating the reaction mechanism. This particularly holds for high-resolution spectra of chemically active species. As mentioned in Sect. 3.5, a few phosphorus-containing molecules have recently been identified by high-resolution spectroscopy. The chemistry employed to generate such species, when combined with the previous reaction studies such as those described in [5.1], helps unravel the details of processes involved.

As in many other previous attempts, *Kawaguchi* et al. [5.2] failed to generate the PO radical when they simply passed oxygen microwave discharge products over red phosphorus. However, they noticed that when hydrogen atoms are present in the system, PO and also PO_2 are readily formed from P + O. This is similar to the situation *Uehara* [5.3] encountered in studying the AsO radical; he was able to generate this species when he passed microwave discharge products of a H_2/O_2 mixture over metallic arsenic.

These observations suggest that the initial step of the phosphorus oxidation reaction is either (5.1.1 or 2):

$$H + P\ (\text{solid}) \rightarrow PH_n\ (n = 1 - 3), \tag{5.1.1}$$

$$OH + P\ (\text{solid}) \rightarrow PO + H. \tag{5.1.2}$$

In (5.1.2) the OH radical was assumed to be made by the discharge in a H_2/O_2 mixture. However, when OH prepared by the reaction of NO_2 with a small amount H atoms was passed over red phosphorus, no PO spectrum was observed. Therefore, the reaction must primarily proceed as (5.1.1) at its initial stage. In fact, the PH and PH_2 radicals were confirmed to exist in the system by FIR LMR and PH_3 by diode laser spectroscopy. This initial step will then probably be followed by these following reactions:

$$PH + H = PH_2, \tag{5.1.3}$$

$$PH + O = PO + H + 52\ \text{kcal/mol}, \tag{5.1.4}$$

$$= P\ (\text{gas}) + OH + 30\ \text{kcal/mol}, \tag{5.1.5}$$

$$PH + OH = PO + H_2, \tag{5.1.6}$$

$$P\ (\text{gas}) + O_2 = PO + O + 28\ \text{kcal/mol}, \tag{5.1.7}$$

$$PH_2 + O = PH + OH + 21\ \text{kcal/mol}, \tag{5.1.8}$$

$$= HPO + H + 29\ \text{kcal/mol}, \tag{5.1.9}$$

$$O + HPO = OH + PO + 44\ \text{kcal/mol}, \tag{5.1.10}$$

$$PO + O = PO_2, \tag{5.1.11}$$

$$PO + OH = PO_2 + H, \tag{5.1.12}$$

$$PO + O_2 = PO_2 + O + 9\ \text{kcal/mol}. \tag{5.1.13}$$

The spectrum of PH_2 disappeared when oxygen was added, indicating that reaction (5.1.3) is slower than (5.1.4–6) or PH_2 is rapidly removed by (5.1.8 or 9). When a fraction of microwave-discharged hydrogen (without any oxygen) was converted to OH by the reaction with NO_2 and then was passed over red phosphorus, the spectrum of PO was about one third as intense as that observed for an optimum mixture: O_2 of 30 mTorr and H_2 of 350 mTorr. These observations show reaction (5.1.6) to be less important than (5.1.4, 7). When oxygen was added in excess, the PO spectrum was finally quenched whereas the PO_2 spectrum increased in intensity; these observations are accounted for by reactions (5.1.11–13). Bright green chemiluminescence emitted from the reaction system was found to behave parallel with PO and PO_2 IR and/or microwave spectra and may thus be employed to monitor the abundance of the two species.

Both PO and PO_2 radicals have also been detected in the reaction system of PH_3 and microwave discharge products of a H_2 and O_2 mixture. Here again hydrogen was indispensable. Because the $O + PH_3$ reaction is known to be slow [5.1], the initial step of this reaction is probably

$$H + PH_3 = PH_2 + H_2 + 23\ \text{kcal/mol} \quad \text{or} \tag{5.1.14}$$

$$OH + PH_3 = PH_2 + H_2O + 38\ \text{kcal/mol}, \tag{5.1.15}$$

and then reactions (5.1.8, 9 and 6, 10) follow to generate PO, which is further converted to PO_2 by reactions (5.1.11–13), as mentioned above. As inferred from reaction (5.1.9), the microwave spectrum of HPO has also been observed in this reaction system [5.4].

In this connection it is worth mentioning that *Clyne* and *Heaven* [5.5] observed the dye laser excitation spectrum of PO by generating the radical using the PH_3 + (O and N) reaction. They inferred that the initial step of the reaction is $PH_3 + N \rightarrow PH_2 + NH$, because $PH_3 + O$ is slow.

In any quantitative applications of spectroscopic data to chemical analysis, the observed spectral intensity must be converted to the abundance of the molecular species responsible for the spectrum. This is by no means easy, especially for unstable molecules.

We shall restrict ourselves here to linear spectroscopy only, because nonlinear effects are dependent on too many parameters, making calibration even harder. Let the intensity of an absorption line be expressed by the absorption coefficient $\alpha(\nu)$ as a function of the frequency ν of the incident radiation as follows:

$$I(\nu) = I_0(\nu)\exp[-\alpha(\nu)\,l], \tag{5.1.16}$$

where $I_0(\nu)$ and $I(\nu)$ denote the intensities of the radiation before and after passing through an absoption cell l long, respectively. According to the time-dependent perturbation theory $\alpha(\nu)$ is expressed by

$$\alpha(\nu) = Sf(\nu - \nu_0), \quad \text{where} \tag{5.1.17}$$

$$S = [8\pi^3 N f_i/3hc]\,|\langle i|\,\mu\,|f\rangle|^2\,\nu[1 - \exp(-h\nu_0/kT)]. \tag{5.1.18}$$

In (5.1.18) N means the number of molecules in the unit volume, f_i the fraction of molecules present in the lower state of the transition, $\langle i|\,\mu\,|f\rangle$ the dipole moment of the i to f transition, and ν_0 the molecular transition frequency. The last factor in (5.1.18) represents the difference in population between the lower (i) and the upper (f) states. Equation (5.1.18) applies to the absorption spectrum, whereas the intensity of an emission line is given by a formula similar to (5.1.18) except for the last factor and inclusion of the quantum yield, i.e., corrections for nonradiative and other radiative processes.

The factor $f(\nu - \nu_0)$ is called the line-shape function. In the microwave region it takes the form [5.6]

$$f(\nu - \nu_0) = (\Delta\nu_C/\pi)\,\{[(\nu - \nu_0)^2 + (\Delta\nu_C)^2]^{-1} + [(\nu + \nu_0)^2 + (\Delta\nu_C)^2]^{-1}\}, \tag{5.1.19}$$

when the sample pressure is not too low, in other words, the pressure broadening dominates over the Doppler broadening. The linewidth parameter $\Delta\nu_C$ (half-width at half-maximum, HWHM) is related to the average time interval τ between two successive collisions by $\Delta\nu_C = 1/(2\pi\tau)$. Because $\Delta\nu_C$ is much smaller than ν_0 in most cases, the second term of (5.1.19) may be neglected, leaving

$$f(\nu - \nu_0) = (\Delta\nu_C/\pi)/[(\nu - \nu_0)^2 + (\Delta\nu_C)^2]. \tag{5.1.20}$$

In the wavelength regions shorter than IR, the line shape is governed by the Doppler effect, and the following expression holds for $f(\nu - \nu_0)$:

$$f(\nu - \nu_0) = (\ln 2/\pi\,\Delta\nu_D^2)^{1/2} \exp\{-(\ln 2)\,[(\nu - \nu_0)/\Delta\nu_D]^2\}, \tag{5.1.21}$$

where $\Delta\nu_D$ is given by

$$\Delta\nu_D = (\nu_0/c)\,(2\,N_0\,k\,T\,\ln 2/M)^{1/2}, \tag{5.1.22}$$

where N_0 denotes Avogadro's number and M the molecular weight. In the millimeter to IR region the line shape will be of the form which convolutes (5.1.20, 21) and is referred to as the Voigt function.

In the microwave region it is customary to measure the peak intensity, while in the IR and optical regions an integrated intensity is more common. In any case, to derive Nf_i (the number of molecules giving the absorption spectrum) from the observed intensity, we need to know the transition moment and the linewidth parameter $\Delta\nu_C$, if the collision broadening is more dominant. For transient molecules, it is fairly difficult to measure the dipole moment, because the surface of the Stark electrodes, which are normally made of metal plates or metal-coated glass plates, easily deteriorates most of the free radicals and unstable molecules. Although the linewidth parameter may be read directly from the observed spectrum, the sample pressure or the partial pressure of a transient molecule (which needs to be known to determine the transition moment experimentally) is extremely difficult to measure in a discharge plasma. Therefore we often have to be content with some estimated value. Furthermore, the distribution of transient molecules in the absorption cell is not necessarily homogeneous, rather the concentration gradient can be appreciable. Therefore, the observed intensity corresponds to no more than a distribution averaged over the cell. In view of these difficulties inherent in free radical studies, it might be sensible to calculate the dipole moment including the vibrational and electronic transition moments by ab initio MO methods, provided that they give results on related molecules which agree well with experimental data.

Some chemical methods may be employed to estimate the concentrations of transient species and to calibrate the observed spectral intensities, e.g., the chemical titration technique. *Thrush* and *Tyndall* [5.7, 8] showed that the concentration of the HO_2 radical may be determined by examining the change in that of formaldehyde. They employed two reactions: (i) $Cl_2 + CH_3OH + O_2$ and (ii) $Cl_2 + H_2CO + O_2$, and in both cases they initiated the reactions by flash photolyzing the Cl_2 molecule as follows:

$$Cl_2 + h\nu \rightarrow Cl + Cl.$$

Then in case (i)

$$Cl + CH_3OH \rightarrow HCl + CH_2OH$$

$$CH_2OH + O_2 \rightarrow H_2CO + HO_2$$

and in case (ii)

$$Cl + H_2CO \rightarrow HCl + HCO$$

$$HCO + O_2 \rightarrow CO + HO_2$$

reactions follow, and in each case an increase (i) or a decrease (ii) in the H_2CO amount is exactly converted to increase in HO_2 abundance. There are perhaps a number of similar techniques. It is interesting to note that if we wish to determine the first-order or pseudo-first-order reaction rate, we need to measure only the relative intensity.

An interesting application was made using the diode laser spectra of the CH_3 radical to measure the recombination rate constant of this molecule:

$$CH_3 + CH_3 \xrightarrow{k_R} C_2H_6 .$$

Yamada and *Hirota* [5.9] employed the discharge current modulation method (Sect. 3.2.1) and confirmed that the $v_2 = 1 \leftarrow 0$ $Q(4, 4)$ line intensity falls off according to

$$[CH_3] = [CH_3]_0/\{1 + 2[CH_3]_0 k_R t\}, \tag{5.1.23}$$

where $[CH_3]_0$ and $[CH_3]$ denote, respectively, the initial concentration of the CH_3 radical and its concentration at t after the start of the reaction. In this experiment di-*tert*-butylperoxide was used as a starting material. On the other hand, *Laguna* and *Baughcum* [5.10] generated the CH_3 radical by excimer laser (KrF, 248 nm) photolysis of CH_3I. To avoid excitation due to excess energy, they diluted CH_3I to 0.5 and 1% using Ar and He, respectively, making the total pressure 2–20 Torr. From a plot of $[CH_3]_0/[CH_3]$ versus t, we may determine $2\,[CH_3]_0 k_R$ or $|\langle i|\,\mu\,|f\rangle|^2 k_R$. Both sets of data were consistent with $k_R = 1.55 \times 10^{-11}$ cm^{3}molecule^{-1}s^{-1} and $|\langle i|\,\mu\,|f\rangle| = 0.280$ D.

Laguna and Baughcum's experiment may indicate that high-resolution spectra of free radicals such as the diode laser spectrum of CH_3 provide nascent distribution of the products at the initial stage of photolysis. According to the results of *Leone* and *Hermann* [5.11, 12], the CH_3I photolysis system contains a large amount of excess energy and, in fact, methyl radicals thus generated are distributed over vibrational states of up to $v_2 = 10$ with the maximum at $v_2 = 2$. A similar study [5.13] was made of $(CH_3)_2Hg$ photolyzed by the 248 and 193 nm laser lines and both the ν_2 and ν_3 bands of CH_3 were monitored. It would be interesting to diagnose these systems right after the photolysis with diode laser spectroscopy.

Kanamori et al. [5.14] recently investigated SO_2 photolysis with the 193 nm ArF laser line by monitoring the vibration-rotation spectra of SO generated. The SO_2 molecule is excited to high vibrational levels associated with the $\tilde{C}\,{}^1B_2$ state, which does not correlate with the "ground-state" channel, $SO(X\,{}^3\Sigma^-) + O({}^3P)$, located at about 6560 cm^{-1} lower than the excimer energy, but with an excited-state channel, $SO(a\,{}^1\Delta) + O({}^1D)$. So the potential curves must cross. They have found that 70–80% of $SO(X\,{}^3\Sigma^-)$ are prepared in the $v = 2$ state and some in the $v = 1$ and 5 states, but the $v = 0$, 3, and 4 states are essentially vacant. The initial population over the rotational levels is quite different for the three vibrational states, $N = 32$ to 42, 17 to 43, and 2 to 15 for $v = 1$, 2, and 5, which may be compared with $N_{max} = 12$ for the thermal distribution at $T = 300$ K. A more interesting observation is that the initial distribution among the three spin states deviates much from the thermal distribution, although the deviation disappears after a few collisions; the F_1 $(J = N + 1)$ and F_3 $(J = N - 1)$ components dominate over the F_2 $(J = N)$ component. In classical terminology, this means that the spin is aligned either parallel or antiparallel to the rotational angular momentum which is always directed perpendicular to the molecular axis.

The SO_2 molecule excited to the $\tilde{C}$ state may bear some triplet nature by spin-orbit interaction. A consideration of the correlation diagram suggests that there is a repulsive triplet state crossing with the $\tilde{C}$ state and converging to the "ground-state" dissociation channel mentioned above. An argument based upon coupling of the rotational and spin angular momenta suggests that two of the three spin components F_1 and F_3 of the triplet repulsive state can be mixed more with the $\tilde{C}$ state than F_2, and the F_1 and F_3 spin populations thus generated in the repulsive state are transferred to SO without being affected by any serious perturbations.

5.2 Applications to Atmospheric Chemistry

There seem to be two types of approaches for investigating chemistry in the Earth's atmosphere. In the first type of approach chemical species constituting the atmosphere are directly observed and monitored using equipment specially designed for the field work. The second type of research is primarily based upon simulation, supported by a large amount of data taken from atmospheric molecules in laboratories [5.15]. High-resolution spectroscopy contributes to both types of approaches.

Howard and co-workers [5.16–19] have carried out systematic applications of FIR LMR spectroscopy to determine the rate constants of the free radical reactions which are of some interest for atmospheric chemistry. One of the largest impacts arising from these studies is a high value [$(8.1 \pm 1.5) \times 10^{-12}$ $cm^3 molecule^{-1} s^{-1}$ at 296 K] for the $HO_2 + NO \rightarrow OH + NO_2$ reaction [5.17]. Additional measurements [5.18, 20, 21] have supported that this value is correct within the uncertainty stated. If this is really so high, addition of nitrogen oxides to the stratosphere might even increase the amount of ozone [5.22], although NO_x has been thought to catalyze ozone decomposition by the following cycle

$$O + NO_2 \rightarrow NO + O_2$$

$$NO + O_3 \rightarrow NO_2 + O_2,$$

resulting in $O + O_3 \rightarrow 2O_2$. The rapid consumption of HO_2 by the above reaction $HO_2 + NO \rightarrow OH + NO_2$ may replace the two famous ozone consumption routes:

$$\text{(i)} \quad \begin{aligned} HO_2 + O &\rightarrow HO + O_2 \\ HO + O_3 &\rightarrow HO_2 + O_2 \end{aligned}$$

$$\text{(ii)} \quad \begin{aligned} HO_2 + O_3 &\rightarrow HO + 2O_2 \\ HO + O_3 &\rightarrow HO_2 + O_2 \end{aligned}$$

by

$$HO_2 + NO \rightarrow HO + NO_2$$

$$NO_2 + h\nu \rightarrow NO + O$$

$$O + O_2 + M \rightarrow O_3 + M$$

$$HO + O_3 \rightarrow HO_2 + O_2$$

which leaves the O_3 concentration practically unchanged.

The field measurements have already been carried out for a number of major constituents of the atmosphere. In future, such observations will be extended to minor components some of which must be free radicals. In laboratories such species have been detected by absorption spectroscopy to an abundance as small as 10^8 molecules cm^{-3} when the path length was 10 m. Therefore, in principle, a field observation using a path length of 1 km will detect free radicals of concentration as small as 10^6 molecules cm^{-3}. As already mentioned in Chap. 1, OH, HO_2, ClO, SH, and NO_3 are candidates to be detected among atmospheric free radicals that are of special importance in the chemistry of the troposphere and stratosphere [5.23].

5.3 Applications to Astronomy

Table 5.1 summarizes interstellar molecules, 55 in total, so far detected by radiotelescopic observations. It is remarkable that nearly 40% of them are transient molecules. This fact is not so surprising if one considers the low

Table 5.1. Interstellar molecules observed by radio telescopes

Ion	HCO^+, $[HOC^+]^a$, HCS^+, HN_2^+, $[CO^+]^a$, HCO_2^+ (or HOCN)
Free radical	CH, OH, CN, NO, NS, SO, HCO, C_2H, $[C_3H]^a$, C_4H, C_3N
Unstable molecule	CS, SiO, SiS, $[HNO]^a$, H_2CS
Inorganic compound	CO, H_2O, H_2S, SO_2, OCS, NH_3, HNCO, HNCS
Organic compound	
alcohol, thiol	CH_3OH, C_2H_5OH, CH_3SH
ether	$(CH_3)_2O$
aldehyde, ketone	H_2CO, H_2CCO, CH_3CHO
acid, ester	HCOOH, CH_3OCHO
amide, imide	NH_2CN, CH_2NH, CH_3NH_2, NH_2CHO
cyanide, isocyanide	HCN, HNC, CH_3CN, C_2H_5CN, CH_2CHCN
polyyne	CH_3CCH, HC_3N, HC_5N, HC_7N, HC_9N, $HC_{11}N$

[a] Detection to be confirmed

temperature and the low density of interstellar space; transient molecules are even more abundant (10^{-7}–10^{-8} relative to H_2) in dense molecular clouds than medium-size organic molecules which are only 10^{-9}–10^{-10} abundant [5.24]. It has been well established that the ion-molecule reaction is the main route of forming interstellar molecules [5.25], and, as Table 5.1 shows, 3 to 6 molecular ions have, in fact, been identified in space, of which HCO^+ and HN_2^+ play important roles in interstellar chemistry. Free radicals such as OH, CH, and CN are likely to result from the principal molecular ions in space such as H_2O^+, H_3O^+, CH_2^+, CH_3^+, and H_2CN^+ through dissociative ion-electron recombination reactions [5.26], and their existence is more support for the ion-molecule reaction theory.

The characteristic conditions of interstellar space have enabled certain molecular species to exist as stable molecules, whereas they are very difficult to generate in terrestrial laboratories because of their high chemical activities. As a result, some transient molecules have been identified in space prior to laboratory observations, e.g., HCO^+, HN_2^+, HCS^+, CH, CN, C_2H, C_4H, and C_3N. The assignments werc based on chemical reasoning, ab initio calculations, and fine and hyperfine structures when observed. It is quite difficult to observe the microwave spectra of these species in laboratories; for example, it took *Gottlieb* et al. [5.27] 60 min to record microwave lines of C_4H and C_3N using their highly sensitive spectrometer.

The recent development of the radio telescope has dramatically increased the sensitivity, especially in high-frequency regions, allowing more lines to be detected, since the molecular absorption increases in proportion to the square or cube of the transition frequency. As a result, many more lines remain to be assigned in future. Even now in 1984, there are more than one hundred unidentified lines (referred to as *U* lines) [5.28]. Probably many *U* lines are ascribed to transient molecules that have not been detected by laboratory microwave spectroscopy. In view of the importance of transient species in astronomy, in particular when choosing correct models for interstellar chemistry, further laboratory studies then should be forwarded using high-resolution and high-sensitivity spectroscopic tools.

Of the transient species mentioned in Chap. 4, the molecules SiN, CCN, CH_3, CH_3O, FeO, PO, and HPO may exist in interstellar space [5.29]. Here the HCCN radical and a few phosphorus-containing molecules are discussed in some detail.

5.3.1 HCCN

The HCCN radical is of some atronomical importance, because, as Table 5.1 shows, several polyyne cyanides and related molecules have been found in interstellar space. The chemistry of polyynes in space has been discussed by several authors, based on ion-molecule reaction models. *Schiff* and *Bohme* [5.30] and *Mitchell* et al. [5.31] have proposed mechanisms for the formation of polyynes in dense molecular clouds such that the carbon chain grows by ion-molecule reactions involving $C_2H_2^+$, $C_2H_3^+$, or C_2H_2, thus predicting that only $HC_{2m}CN$ (m denoting an integer) type molecules are produced. On the other hand, *Suzuki* [5.32, 33], focusing on ion-molecule reactions in regions where carbon atoms are partially ionized, proposed that the carbon chain can be lengthened by the reaction with C^+, predicting that $HC_{2m+1}CN$ molecules are produced in as nearly equal amount as $HC_{2m}CN$ species. In the laboratory, the $HC_{2m+1}CN$ molecules will be chemically active, and, in fact, only HCCN has been identified in the gas phase [5.34]. The lack of experimental data might have hampered the identification of the $HC_{2m+1}CN$ family in space, but the HCCN spectrum will provide critical information on polyyne formation in interstellar space.

5.3.2 Phosphorus-Containing Compounds

The cosmic abundance is smaller and perhaps the depletion ratio is greater for phosphorus than for the atoms contained in the interstellar molecules listed in Table 5.1 [5.35], but, because it is as important for life as Na, Mg, S, Cl, K, and Ca, it is worth trying to detect interstellar molecules containing phosphorus. In the past, PN [5.36], PH_3 [5.36], HCP [5.37], and NCCP [5.38] have been searched for in interstellar space without success. These molecules are not necessarily easy to detect astronomically; PH_3 does not give any strong transitions in the millimeter wave region, the dipole moments of PH_3 and HCP are 0.57397 D [5.39] and 0.390 D [5.40], respectively, and the chemical processes in interstellar space do not seem to generate these species efficiently.

Recently *Huntress* and his co-workers [5.41, 42] examined the reactions of the PH_n^+ ions with a number of neutral molecules using ion cyclotron resonance and concluded that neither P^+ nor PH^+ reacts with H_2, N_2, and CO, but does with O_2 and other simple molecules to form PO^+, HPO^+, PNH_2^+, and others. These results indicate that it is unlikely that phosphorus exists in the form of PH_2 or PH_3, and, since the main repositories for C and N in dense clouds are CO and N_2, respectively, this leaves molecules containing a P – O bond as the most likely candidates for phosphorus-containing interstellar molecules. Among the molecules discussed in Chap. 4, it is promising that PO [5.2], HPO [5.4], and PO_2 [5.43] could be observed in interstellar space. Although not experimentally measured, the dipole moments of these molecules seem to be quite large, as judged from the intensities of their rotational spectra.

5.4 Future Developments

5.4.1 Possible Improvements of Spectroscopic Techniques

There are two wavelength regions where the development of light sources has been delayed, in comparison with other regions, namely the far and near IR regions.

Klystrons are commercially available up to about 200 GHz, but beyond this frequency only a few backward wave oscillators (BWO) may be purchased in some specified frequency regions; one has to have recourse to a frequency multiplication technique. In view of the recent progress in semiconductor technology, we may expect to get nonlinear elements of efficiency much higher than hitherto available.

Evenson and co-workers [5.44] have succeeded in generating difference frequency radiation in FIR by mixing the output of two CO_2 lasers in a nonlinear element which consists of three layers, metal, insulator, and metal, and is thus called MIM. The output power they obtained is a fraction of μW, but it may be increased by a factor of ten or so in the near future by improving the quality of MIM. By replacing a CO_2 laser by a waveguide laser and by selecting appropriate combinations of two lines, nearly the entire FIR region 0.5–3.5 THz may be scanned continuously. Spectroscopy in the FIR region allows us to detect molecules of very low concentration through observing the rotational spectrum, as demonstrated by FIR LMR.

On the other hand, near IR spectroscopy can provide us with information which is quite different from that obtainable from FIR spectroscopy. A few simple molecules like HCF and HNO may show weak absorption spectra which are ascribed to transitions from the ground state to excited states of different multiplicity (i.e., metastable states). High sensitivity is necessary to observe such spectra, and a wide wavelength region must be scannned because of large uncertainties in their transition frequencies.

The color-center laser is a promising source; *Tittel* and co-workers [5.45] have demonstrated that Tl-doped crystals pumped by a Nd : YAG laser oscillate even at room temperature. The wavelength region 0.9–3.3 μm can be covered with spectral purity better than 3 MHz and with a single frequency output of 1–500 mW. The diode laser is also undergoing rapid development in the near IR region, mainly in an attempt to cope with communication needs. The InP/InGaAsP diode is representative for the 1.3–1.5 μm region. Diodes may require some modification when applied to spectroscopy, however.

As already discussed in Sect. 3.2.1, IR diodes have been employed in observing vibration-rotation spectra of transient molecules. They are still subjected to various shortcomings: low power, multimode oscillation, narrow mode, short lifetime, and so on. However, they are being improved, and in the near future will make it possible to detect free radicals which are more reactive and thus shorter lived than those hitherto known.

5.4.2 Future Trends in High-Resolution Spectroscopic Studies of Transient Molecules

A number of free radicals and short-lived molecules still remain to be investigated by high-resolution spectroscopy. For example, the discharge plasma in silane, which generates amorphous silicon, contains SiH_3, SiH_2, SiH, Si, and species involving two or more Si atoms. So far no high-resolution spectroscopic observations has been reported for SiH_3, only the IR spectra in an Ar matrix being published [5.46].

High-resolution spectroscopy has been mainly directed to transient molecules consisting of two to five atoms. It is certainly difficult to investigate more complicated molecules, but some of them may have characteristic structures of special importance for the study of molecular structure. Two examples can be mentioned here, the ethyl radical and the allyl radical. *Pacansky* and *Schrader* [5.47] have reported that the CH_2 out-of-plane mode of the ethyl radical is the strongest band in an Ar matrix. The molecule may show internal rotation effects, and, when the CCH_2 group is planar, the V_3 term in the internal rotation potential function vanishes, leaving the V_6 term as the most dominant, but it is usually very small (a few cal/mol). Almost no spectroscopic information has been obtained on the allyl radical.

Special attention has been paid to charged molecular species, i.e., molecular ions, presumably because of their importance in many fields including astronomy (Sect. 5.3) [5.48]. *Woods* and his students [5.48–50] have observed the microwave spectra of several molecular ions such as CO^+, HCO^+, HN_2^+, HCS^+, and HOC^+. Oka and co-workers have applied either difference frequency laser spectroscopy or color-center laser spectroscopy (also diode laser spectroscopy in a few cases) to simple protonated ions MH^+, e.g., H_3^+ ν_2 [5.51], HeH^+ [5.52], NeH^+ [5.53], ArH^+ [5.54], HCO^+ ν_1 [5.55, 56], HN_2^+ ν_1 [5.57], NH_4^+ ν_3 [5.58, 59], H_3O^+ $\nu_2(1^- - 0^+)$ [5.60] and ν_3 [5.61], $HCNH^+$ [5.62], $DCNH^+$ [5.63], H_2D^+ ν_1 [5.64] and ν_2/ν_3 [5.65], HeH^+ [5.66], and H_3O^+ $\nu_2(1^+ - 0^-, 1^- - 1^+)$ [5.67]. *Evenson* and co-workers employed FIR LMR to observe the pure rotational spectra of three diatomic ions HCl^+ [5.68], HBr^+ [5.69] and HF^+ [5.70], which are radical ions and may thus be more reactive than the MH^+ type ions mentioned above.

The ion-beam technique is very attractive for the spectroscopic study of molecular ions. *Wing* et al. [5.71] employed ion beams accelerated for Doppler tuning and crossed with a CO laser beam. *Carrington* and co-workers [5.48, 72] have also exploited the ion-beam technique and obtained information on high vibrational states of ions; a beautiful example is the case of HD^+ [5.72]. It is certain that many more results will appear in the near future on spectroscopy of molecular ions, not only positive but also negative in charge.

High-resolution spectroscopy will find another important direction to follow, namely chemical reactions, in particular photochemical reactions. As briefly discussed in Sect. 5.1, reaction products seem to have nascent distributions over quantum states that are considerably different from the Boltzmann distribution. For a low-pressure gas sample (a few mTorr), the intermolecular collision will not affect the distribution very much for a few μs after the reaction has started. So there are good chances to examine molecular distributions over fine structure and hyperfine structure levels for which almost on detailed data have been reported.

References

Chapter 1

1.1 G. Herzberg: *The Spectra and Structures of Simple Free Radicals, An Introduction to Molecular Spectroscopy* (Cornell University Press, New York 1971)
1.2 G. Herzberg: *Molecular Spectra and Molecular Structure I. Spectra of Diatomic Molecules* (Van Nostrand, New York 1950)
1.3 G. Herzberg: *Molecular Spectra and Molecular Structure III. Electronic Spectra and Electronic Structure of Polyatomic Molecules* (Van Nostrand, Princeton 1966)
1.4 A. Carrington: *Microwave Spectroscopy of Free Radicals* (Academic, London 1974)
1.5 C. H. Townes, A. L. Schawlow: *Microwave Spectroscopy* (McGraw-Hill, New York 1955)
1.6 W. Gordy, R. L. Cook: *Microwave Molecular Spectra* (Interscience, New York 1970)
1.7 G. C. Dousmanis, T. M. Sanders, C. H. Townes: Phys. Rev. **100,** 1735 (1955)
1.8 F. X. Powell, D. R. Lide: J. Chem. Phys. **41,** 1413 (1964)
1.9 M. Winnewisser, K. V. L. N. Sastry, R. L. Cook, W. Gordy: J. Chem. Phys. **41,** 1687 (1964)
1.10 K. C. Herr, G. C. Pimentel: Appl. Opt. **4,** 25 (1965)
1.11 D. E. Milligan, M. E. Jacox: J. Chem. Phys. **47,** 5157 (1967); M. E. Jacox: J. Phys. Chem. **87,** 3126 (1983)
1.12 J. I. Raymond, L. Andrews: J. Phys. Chem. **75,** 3235 (1971)
1.13 M. E. Jacox: J. Phys. Chem. Ref. Data (to be published)
1.14 K. M. Evenson, H. P. Broida, J. S. Wells, R. J. Mahler, M. Mizushima: Phys. Rev. Lett. **21,** 1038 (1968)
1.15 K. M. Evenson, R. J. Saykally, D. A. Jennings, R. F. Curl, Jr., J. M. Brown: "Far Infrared Laser Magnetic Resonance", in *Chemical and Biochemical Applications of Lasers,* Vol. V, ed. by C. B. Moore (Academic, New York 1980)
1.16 K. Sakurai, S. E. Johnson, H. P. Broida: J. Chem. Phys. **52,** 1625 (1970)
1.17 M. M. Hessel: Phys. Rev. Lett. **26,** 215 (1971)
1.18 G. W. Hills, C. R. Brazier, J. M. Brown, J. M. Cook, R. F. Curl, Jr.: J. Chem. Phys. **76,** 240 (1982) and references cited therein
1.19 R. W. Field, A. D. English, T. Tanaka, D. O. Harris, D. A. Jennings: J. Chem. Phys. **59,** 2191 (1973) and references cited therein
1.20 M. S. Sorem, A. L. Schawlow: Opt. Commun. **5,** 148 (1972)
1.21 R. S. Low, H. Gerhardt, W. Dillenschneider, R. F. Curl, Jr., F. K. Tittel: J. Chem. Phys. **70,** 42 (1979)
1.22 G. W. Hills, D. L. Philen, R. F. Curl, Jr., F. K. Tittel: Chem. Phys. **12,** 107 (1976)

Chapter 2

2.1 C. H. Townes, A. L. Schawlow: *Microwave Spectroscopy* (McGraw-Hill, New York 1955)
2.2 W. Gordy, R. L. Cook: *Microwave Molecular Spectra* (Interscience, New York 1970)

2.3 H. W. Kroto: *Molecular Rotation Spectra* (Wiley, New York 1975)
2.4 J. H. Van Vleck: Rev. Mod. Phys. **23,** 213 (1951)
2.5 E. U. Condon, G. H. Shortley: *The Theory of Atomic Spectra* (Cambridge Univ. Press, Cambridge 1935)
2.6 A. R. Edmonds: *Angular Momentum in Quantum Mechanics* (Princeton Univ. Press, Princeton 1957)
2.7 M. E. Rose: *Elementary Theory of Angular Momentum* (Wiley, New York 1957)
2.8 D. M. Brink, G. R. Satchler: *Angular Momentum* (Oxford Univ. Press, Oxford 1962)
2.9 T. Oka, Y. Morino: J. Mol. Spectrosc. **6,** 472 (1961)
2.10 E. B. Wilson, Jr., J. C. Decius, P. C. Cross: *Molecular Vibrations*, Chap. 11, (McGraw-Hill, New York 1955) pp. 273–284
2.11 H. H. Nielsen: Rev. Mod. Phys. **23,** 90 (1951)
2.12 H. H. Nielsen: "The Vibration-Rotation Energies of Molecules and Their Spectra in the Infrared", in *Handbuch der Physik*, Vol. 37/1, ed. by S. Flügge (Springer, Berlin, Heidelberg, New York 1959) pp. 173–313
2.13 J. K. G. Watson: "Aspects of Quartic and Sextic Centrifugal Effects on Rotational Energy Levels", in *Vibrational Spectra and Structure,* Vol. 6, Chap. 1, ed. by J. R. Durig (Elsevier, Amsterdam 1977) pp. 1–89
2.14 T. Oka: J. Chem. Phys. **47,** 5410 (1967)
2.15 D. K. Russell, H. E. Radford: J. Chem. Phys. **72,** 2750 (1980)
2.16 C. C. Lin: Phys. Rev. **116,** 903 (1959)
2.17 K. Kayama, J. C. Baird: J. Chem. Phys. **46,** 2604 (1967)
2.18 W. T. Raynes: J. Chem. Phys. **41,** 3020 (1964)
2.19 M. H. Hebbs: Phys. Rev. **49,** 610 (1936)
2.20 E. Hill, J. H. Van Vleck: Phys. Rev. **32,** 250 (1928)
2.21 R. S. Mulliken, A. Christy: Phys. Rev. **38,** 87 (1931)
2.22 M. Tinkham, M. W. P. Strandberg: Phys. Rev. **97,** 937 (1955)
2.23 J. M. Brown, D. J. Milton, J. K. G. Watson, R. N. Zare, D. L. Albritton, M. Horani, J. Rostas: J. Mol. Spectrosc. **90,** 139 (1981)
2.24 D. W. Lepard: Can. J. Phys. **48,** 1664 (1970)
2.25 J. H. Van Vleck: Phys. Rev. **33,** 467 (1929)
2.26 I. C. Bowater, J. M. Brown, A. Carrington: Proc. R. Soc. London Ser. **A 333,** 265 (1973)
2.27 J. M. Brown, T. J. Sears: J. Mol. Spectrosc. **75,** 111 (1979)
2.28 K.-E. J. Hallin, Y. Hamada, A. J. Merer: Can. J. Phys. **54,** 2118 (1976)
2.29 H. C. Longuet-Higgins, U. Öpik, M. H. L. Pryce, R. A. Sack: Proc. R. Soc. London Ser. **A 244,** 1 (1958)
2.30 J. M. Brown: Mol. Phys. **20,** 817 (1971)
2.31 J. T. Hougen: J. Mol. Spectrosc. **81,** 73 (1980)
2.32 M. S. Child: Mol. Phys. **5,** 391 (1962)
2.33 M. S. Child, H. C. Longuet-Higgins: Philos. Trans. R. Soc. London Sect. **A 254,** 259 (1961)
2.34 Y. Endo, S. Saito, E. Hirota: J. Chem. Phys. **81,** 122 (1984)
2.35 J. K. G. Watson: J. Mol. Spectrosc. **103,** 125 (1984)
2.36 A. Carrington, D. H. Levy, T. A. Miller: Adv. Chem. Phys. **18,** 149 (1970)
2.37 R. A. Frosch, H. M. Foley: Phys. Rev. **88,** 1337 (1952)
2.38 J. M. Brown, M. Kaise, C. M. L. Kerr, D. J. Milton: Mol. Phys. **36,** 553 (1978)
2.39 J. M. Cook, G. W. Hills, R. F. Curl, Jr.: J. Chem. Phys. **67,** 1450 (1977)
2.40 C. Yamada, Y. Endo, E. Hirota: J. Chem. Phys. **79,** 4159 (1983)
2.41 Y. Endo, S. Saito, E. Hirota: J. Mol. Spectrosc. **97,** 204 (1983)
2.42 Y. Endo, C. Yamada, S. Saito, E. Hirota: J. Chem. Phys. **77,** 3376 (1982)
2.43 M. Born, J. R. Oppenheimer: Ann. Phys. **84,** 457 (1927)
2.44 A. Messiah: *Quantum Mechanics,* translated by G. M. Temmer, Chap. 18 (North-Holland, Amsterdam 1961) pp. 781–783
2.45 R. Renner: Z. Phys. **92,** 172 (1934)

2.46 H. A. Jahn, E. Teller: Proc. R. Soc. London Ser. **A 161,** 220 (1937)
2.47 H. C. Longuet-Higgins: Adv. Spectrosc. **2,** 429 (1961)
2.48 G. Herzberg, E. Teller: Z. Physik. Chem. **B 21,** 410 (1933)
2.49 C. Jungen, A. J. Merer: "The Renner-Teller Effect", in *Molecular Spectroscopy: Modern Research,* Vol. II, Chap. 3, ed. by K. Narahari Rao (Academic, New York 1976) pp. 127–164
2.50 C. Di Lauro, I. M. Mills: J. Mol. Spectrosc. **21,** 386 (1966)
2.51 L. D. Landau, E. M. Lifshitz: *Quantum Mechanics,* translated by J. B. Sykes, J. S. Bell, Chap. 4, Sect. 28 (Pergamon, London 1958)
2.52 J. T. Hougen: J. Chem. Phys. **36,** 519 (1962)
2.53 J. M. Brown: J. Mol. Spectrosc. **68,** 412 (1977)
2.54 A. D. Walsh: J. Chem. Soc. 2266 (1953)
2.55 J. W. C. Johns: Can. J. Phys. **42,** 1004 (1964)
2.56 D. Gauyacq, C. Larcher, J. Rostas: Can. J. Phys. **57,** 1634 (1979)
2.57 P. S. H. Bolman, J. M. Brown, A. Carrington, I. Kopp, D. A. Ramsay: Proc. R. Soc. London Ser. **A 343,** 17 (1975)
2.58 F. Dorman, C. C. Lin: J. Mol. Spectrosc. **12,** 119 (1964)
2.59 K. Kawaguchi, S. Saito, E. Hirota: Mol. Phys. **49,** 663 (1983)
2.60 J. T. Hougen: J. Chem. Phys. **37,** 403 (1962)
2.61 R. F. Curl, Jr.: Mol. Phys. **9,** 585 (1965)
2.62 C. E. Barnes, J. M. Brown, A. Carrington, J. Pinkstone, T. J. Sears, P. J. Thistlethwaite: J. Mol. Spectrosc. **72,** 86 (1978)
2.63 N. Ohashi, K. Kawaguchi, E. Hirota: J. Mol. Spectrosc. **103,** 337 (1984)

Chapter 3

3.1 C. H. Townes, A. L. Schawlow: *Microwave Spectroscopy* (McGraw-Hill, New York 1955)
3.2 S. Saito: Pure Appl. Chem. **50,** 1239 (1978)
3.3 G. C. Dousmanis, T. M. Sanders, Jr., C. H. Townes: Phys. Rev. **100,** 1735 (1955)
3.4 F. X. Powell, D. R. Lide, Jr.: J. Chem. Phys. **41,** 1413 (1964)
3.5 M. Winnewisser, K. V. L. N. Sastry, R. L. Cook, W. Gordy: J. Chem. Phys. **41,** 1687 (1964)
3.6 R. Kewley, K. V. L. N. Sastry, M. Winnewisser, W. Gordy: J. Chem. Phys. **39,** 2856 (1963)
3.7 A. Carrington: *Microwave Spectroscopy of Free Radicals* (Academic, London 1974)
3.8 T. Amano, E. Hirota, Y. Morino: J. Mol. Spectrosc. **27,** 257 (1968)
3.9 F. X. Powell, D. R. Johnson: J. Chem. Phys. **50,** 4596 (1969)
3.10 T. Amano, S. Saito, E. Hirota, Y. Morino: J. Mol. Spectrosc. **32,** 97 (1969)
3.11 S. Saito: J. Chem. Phys. **53,** 2544 (1970)
3.12 S. Saito, T. Amano: J. Mol. Spectrosc. **34,** 383 (1970)
3.13 K. Takagi, S. Saito: J. Mol. Spectrosc. **44,** 81 (1972)
3.14 S. Saito: Astrophys. J. **178,** L 95 (1972)
3.15 S. Saito: J. Mol. Spectrosc. **48,** 530 (1973)
3.16 T. Amano, E. Hirota: J. Mol. Spectrosc. **45,** 417 (1973)
3.17 R. D. Brown, F. R. Burden, P. D. Godfrey, I. R. Gillard: J. Mol. Spectrosc. **52,** 301 (1974)
3.18 Y. Beers, C. J. Howard: J. Chem. Phys. **63,** 4212 (1975)
3.19 R. C. Woods: Rev. Sci. Instrum. **44,** 282 (1973)
3.20 T. A. Dixon, R. C. Woods: Phys. Rev. Lett. **34,** 61 (1975)
3.21 R. C. Woods, T. A. Dixon, R. J. Saykally, P. G. Szanto: Phys. Rev. Lett. **35,** 1269 (1975)
3.22 R. J. Saykally, P. G. Szanto, T. G. Anderson, R. C. Woods: Astrophys. J. **204,** L 143 (1976)

3.23 R. J. Saykally, T. A. Dixon, T. G. Anderson, P. G. Szanto, R. C. Woods: Astrophys. J. **205,** L 101 (1976)
3.24 T. A. Dixon, R. C. Woods: J. Chem. Phys. **67,** 3956 (1977)
3.25 C. S. Gudeman, N. N. Haese, N. D. Piltch, R. C. Woods: Astrophys. J. **246,** L 47 (1981)
3.26 C. S. Gudeman, R. C. Woods: Phys. Rev. Lett. **48,** 1344 (1982)
3.27 Y. Endo, S. Saito, E. Hirota: J. Chem. Phys. **75,** 4379 (1981)
3.28 Y. Endo, S. Saito, E. Hirota: Astrophys. J. **278,** L 131 (1984)
3.29 S. Saito, Y. Endo, M. Takami, E. Hirota: J. Chem. Phys. **78,** 116 (1983)
3.30 E. Hirota, M. Imachi: Can. J. Phys. **53,** 2023 (1975)
3.31 Z. Kisiel, D. J. Millen: J. Phys. Chem. Ref. Data **11,** 101 (1982)
3.32 C. Yamada, K. Nagai, E. Hirota: J. Mol. Spectrosc. **85,** 416 (1981)
3.33 K. Nagai, K. Kawaguchi, C. Yamada, K. Hayakawa, Y. Takagi, E. Hirota: J. Mol. Spectrosc. **84,** 197 (1980)
3.34 J. L. Hall, S. A. Lee: Appl. Phys. Lett. **29,** 367 (1976)
3.35 C. Yamada, E. Hirota: J. Chem. Phys. **78,** 669 (1983)
3.36 K. Kawaguchi, C. Yamada, Y. Hamada, E. Hirota: J. Mol. Spectrosc. **86,** 136 (1981)
3.37 C. Yamada, E. Hirota, K. Kawaguchi: J. Chem. Phys. **75,** 5256 (1981)
3.38 Y. Endo, K. Nagai, C. Yamada, E. Hirota: J. Mol. Spectrosc. **97,** 213 (1983)
3.39 C. Yamada, E. Hirota: Unpublished
3.40 C. Yamada, Y. Endo, E. Hirota: J. Chem. Phys. **78,** 4379 (1983)
3.41 C. S. Gudeman, M. H. Begemann, J. Pfaff, R. J. Saykally: Phys. Rev. Lett. **50,** 727 (1983)
3.42 N. N. Haese, F.-S. Pan, T. Oka: Phys. Rev. Lett. **50,** 1575 (1983)
3.43 N. N. Haese, T. Oka: J. Chem. Phys. **80,** 572 (1984)
3.44 A. R. W. McKellar, C. Yamada, E. Hirota: J. Chem. Phys. **79,** 1220 (1983)
3.45 H. Kanamori, C. Yamada, J. E. Butler, K. Kawaguchi, E. Hirota: To be published
3.46 H. Kanamori, J. E. Butler, K. Kawaguchi, C. Yamada, E. Hirota: J. Chem. Phys. submitted
3.47 K. M. Evenson, R. J. Saykally, D. A. Jennings, R. F. Curl, J. M. Brown: "Far Infrared Laser Magnetic Resonance", in *Chem. Biochem. Appl. Lasers,* Vol. 5, ed. by C. B. Moore (Academic, New York 1980) pp. 95–138
3.48 K. Kawaguchi, C. Yamada, E. Hirota, J. M. Brown, J. Buttenshaw, C. R. Parent, T. J. Sears: J. Mol. Spectrosc. **81,** 60 (1980)
3.49 C. Freed, L. C. Bradley, R. G. O'Donnell: IEEE J. Quantum Electron. QE-**16,** 1195 (1980)
3.50 F. R. Petersen, J. S. Wells, A. G. Maki, K. J. Siemsen: Appl. Opt. **20,** 3635 (1981)
3.51 D. M. Dale, M. Herman, J. W. C. Johns, A. R. W. McKellar, S. Nagler, I. K. M. Strathy: Can. J. Phys. **57,** 677 (1979)
3.52 C. R. Pollock, F. R. Petersen, D. A. Jennings, J. S. Wells, A. G. Maki: J. Mol. Spectrosc. **99,** 357 (1983)
3.53 K. M. Evenson, H. P. Broida, J. S. Wells, R. J. Mahler, M. Mizushima: Phys. Rev. Lett. **21,** 1038 (1968)
3.54 T. Y. Chang, T. J. Bridges: Opt. Commun. **1,** 423 (1970)
3.55 D. J. E. Knight: *NPL Report QU-45* (1st Revision) (1981)
3.56 F. R. Petersen, K. M. Evenson, D. A. Jennings, J. S. Wells, K. Goto, J. J. Jiménez: IEEE J. Quantum Electron. QE-**11,** 838 (1975)
3.57 H. E. Radford, M. M. Litvak: Chem. Phys. Lett. **34,** 561 (1975)
3.58 K. M. Evenson, D. A. Jennings, F. R. Petersen, J. A. Mucha, J. J. Jiménez, R. M. Charlton, C. J. Howard: IEEE J. Quantum Electron. QE-**13,** 442 (1977)
3.59 K. M. Evenson: Faraday Discuss. Chem. Soc. **71,** 7 (1981)
3.60 K. Kawaguchi, S. Saito, E. Hirota: J. Chem. Phys. **79,** 629 (1983)
3.61 N. Ohashi, K. Kawaguchi, E. Hirota. J. Mol. Spectrosc. **103,** 337 (1984)
3.62 A. S. Pine: J. Opt. Soc. Am. **64,** 1683 (1974); **66,** 97 (1976)

3.63 T. Amano, P. F. Bernath, C. Yamada, Y. Endo, E. Hirota: J. Chem. Phys. **77,** 5284 (1982)
3.64 C. R. Vidal: Appl. Opt. **19,** 3897 (1980);
E. E. Marinero, C. T. Rettner, R. N. Zare, A. H. Kung: Chem. Phys. Lett. **95,** 486 (1983)
3.65 S. Blit, E. G. Weaver, T. A. Rabson, F. K. Tittel: Appl. Opt. **17,** 721 (1978)
3.66 T. F. Johnston, Jr., R. H. Brady, W. Proffitt: Appl. Opt. **21,** 2307 (1982)
3.67 L. A. Bloomfield, H. Gerhardt, T. W. Hänsch, S. C. Rand: Opt. Commun. **42,** 247 (1982)
3.68 S. Gerstenkorn, P. Luc: *Atlas du Spectre d'Absorption de la Molécule d'Iode,* 14000–15600 cm^{-1}; 14800–20000 cm^{-1} (ed. du CNRS, Orsay 1978); Rev. Phys. Appl. **14,** 791 (1979)
3.69 J. Cariou, P. Luc: *Atlas du Spectre d'Absorption de la Molécule de Tellure,* Partie 2: 18500–21200 cm^{-1}; Partie 5: 21100–23800 cm^{-1} (ed. du CNRS, Orsay 1980)
3.70 M. Kakimoto, S. Saito, E. Hirota: J. Mol. Spectrosc. **80,** 334 (1980)
3.71 M. S. Sorem, A. L. Schawlow: Opt. Commun. **5,** 148 (1972)
3.72 A. Muirhead, K. V. L. N. Sastry, R. F. Curl, Jr., J. Cook, F. K. Tittel: Chem. Phys. Lett. **24,** 208–211 (1974);
R. S. Lowe, H. Gerhardt, W. Dillenschneider, R. F. Curl, Jr., F. K. Tittel: J. Chem. Phys. **70,** 42 (1979)
3.73 G. W. Hills, D. L. Philen, R. F. Curl, Jr., F. K. Tittel: Chem. Phys. **12,** 107 (1976)
3.74 M. Kakimoto, E. Hirota: J. Mol. Spectrosc. **94,** 173 (1982)
3.75 P. F. Bernath, P. G. Cummins, R. W. Field: Chem. Phys. Lett. **70,** 618 (1980)
3.76 T. Tanaka, D. O. Harris: J. Mol. Spectrosc. **59,** 413 (1976)
3.77 G. R. Bird, J. C. Baird, A. W. Jache, J. A. Hodgeson, R. F. Curl, Jr., A. C. Kunkle, J. W. Bransford, J. Rastrup-Andersen, J. Rosenthal: J. Chem. Phys. **40,** 3378 (1964)
3.78 R. W. Field, R. S. Bradford, D. O. Harris, H. P. Broida: J. Chem. Phys. **56,** 4712 (1972)
3.79 R. W. Field, R. S. Bradford, H. P. Broida, D. O. Harris: J. Chem. Phys. **57,** 2209 (1972)
3.80 R. W. Field, A. D. English, T. Tanaka, D. O. Harris, D. A. Jennings: J. Chem. Phys. **59,** 2191 (1973)
3.81 P. J. Domaille, T. C. Steimle, D. O. Harris: J. Mol. Spectrosc. **66,** 503 (1977)
3.82 P. J. Domaille, T. C. Steimle, D. O. Harris: J. Mol. Spectrosc. **68,** 146 (1977)
3.83 J. Nakagawa, P. T. Domaille, T. C. Steimle, D. O. Harris: J. Mol. Spectrosc. **70,** 374 (1978)
3.84 J. M. Cook, G. W. Hills, R. F. Curl, Jr.: Astrophys. J. **207,** L 139 (1976)
3.85 G. W. Hills, J. M. Cook, R. F. Curl, Jr., F. K. Tittel: J. Chem. Phys. **65,** 823 (1976)
3.86 G. W. Hills, J. M. Cook: Astrophys. J. **209,** L 157 (1976)
3.87 G. W. Hills, R. F. Curl, Jr.: J. Chem. Phys. **66,** 1507 (1977)
3.88 J. M. Cook, G. W. Hills, R. F. Curl, Jr.: J. Chem. Phys. **67,** 1450 (1977)
3.89 G. W. Hills, R. S. Lowe, J. M. Cook, R. F. Curl, Jr.: J. Chem. Phys. **68,** 4073 (1978)
3.90 J. V. V. Kasper, R. S. Lowe, R. F Curl, Jr.: J. Chem. Phys. **70,** 3350 (1979)
3.91 R. S. Lowe, J. V. V. Kasper, G. W. Hills, W. Dillenschneider, R. F. Curl, Jr.: J. Chem. Phys. **70,** 3356 (1979)
3.92 J. M. Brown, T. C. Steimle: Astrophys. J. **236,** L 101 (1980)
3.93 T. C. Steimle, J. M. Brown, R. F. Curl, Jr.: J. Chem. Phys. **73,** 2552 (1980)
3.94 G. W. Hills, C. R. Brazier, J. M. Brown, J. M. Cook, R. F. Curl, Jr.: J. Chem. Phys. **76,** 240 (1982)
3.95 J. M. Cook, G. W. Hills: J. Chem. Phys. **78,** 2144 (1983)
3.96 C. R. Brazier, J. M. Brown: J. Chem. Phys. **78,** 1608 (1983)
3.97 K. Takagi, S. Saito, M. Kakimoto, E. Hirota: J. Chem. Phys. **73,** 2570 (1980)
3.98 T. Suzuki, S. Saito, E. Hirota: J. Chem. Phys. **79,** 1641 (1983)
3.99 R. H. Judge, G. W. King: Can. J. Phys. **53,** 1927 (1975)
3.100 T. Suzuki, S. Saito, E. Hirota: J. Mol. Spectrosc. **110,** (1985)

3.101 W. Kaiser, J. P. Maier, A. Seilmeier: J. Mol. Struct. **59,** 249 (1980)
3.102 A. S. Sudbo, M. M. T. Loy: Chem. Phys. Lett. **82,** 135 (1981)
3.103 M. L. Lesiecki, G. R. Smith, J. A. Stewart, W. A. Guillory: J. Mol. Struct. **59,** 237 (1980)
3.104 B. J. Orr, G. F. Nutt: J. Mol. Spectrosc. **84,** 272 (1980)
3.105 B. J. Orr, J. G. Haub: J. Mol. Spectrosc. **103,** 1 (1984)
3.106 T. Amano, K. Kawaguchi, M. Kakimoto, S. Saito, E. Hirota: J. Chem. Phys. **77,** 159 (1982)
3.107 R. E. Muenchausen, G. W. Hills: Chem. Phys. Lett. **99,** 335 (1983)
3.108 K. Kawaguchi: Thesis, Kyushu University (1981)
3.109 C. Yamada, E. Hirota: J. Mol. Spectrosc. **74,** 203 (1979)
3.110 K. Matsumura, K. Kawaguchi, K. Nagai, C. Yamada, E. Hirota: J. Mol. Spectrosc. **84,** 68 (1980)
3.111 C. Yamada, K. Kawaguchi, E. Hirota: J. Chem. Phys. **69,** 1942 (1978)
3.112 K. Kawaguchi, C. Yamada, E. Hirota: J. Chem. Phys. **71,** 3338 (1979)
3.113 Y. Endo, S. Saito, E. Hirota: J. Mol. Spectrosc. **92,** 443 (1982)
3.114 K. Kawaguchi, S. Saito, E. Hirota: J. Chem. Phys. **79,** 629 (1983)
3.115 K. Kawaguchi, E. Hirota: J. Mol. Spectrosc. **106,** 423 (1984)
3.116 S. Saito: J. Mol. Spectrosc. **65,** 229 (1977)
3.117 Y. Endo, S. Saito, E. Hirota, T. Chikaraishi: J. Mol. Spectrosc. **77,** 222 (1979)
3.118 Y. Endo, S. Saito, E. Hirota: J. Chem. Phys. **74,** 1568 (1981)
3.119 K. Kawaguchi, S. Saito, E. Hirota: Mol. Phys. **49,** 663 (1983)
3.120 K. Kawaguchi, S. Saito, E. Hirota, N. Ohashi: J. Chem. Phys. (in press)
3.121 S. Saito, Y. Endo, E. Hirota: To be published
3.122 R. F. Curl, Y. Endo, M. Kakimoto, S. Saito, E. Hirota: Chem. Phys. Lett. **53,** 536 (1978)
3.123 M. Kakimoto, S. Saito, E. Hirota: J. Mol. Spectrosc. **88,** 300 (1981)
3.124 M. Kakimoto, S. Saito, E. Hirota: J. Mol. Spectrosc. **97,** 194 (1983)
3.125 T. Suzuki, S. Saito, E. Hirota: To be published
3.126 K. Kawaguchi, T. Suzuki, S. Saito, E. Hirota, T. Kasuya: J. Mol. Spectrosc. **106,** 320 (1984)
3.127 S. Saito, Y. Endo, E. Hirota: J. Chem. Phys. **80,** 1427 (1984)
3.128 Y. Endo, C. Yamada, S. Saito, E. Hirota: J. Chem. Phys. **79,** 1605 (1983)
3.129 Y. Endo, S. Saito, E. Hirota: Can. J. Phys. **62,** 1347 (1984)
3.130 Y. Endo, S. Saito, E. Hirota: J. Chem. Phys. **81,** 122 (1984)
3.131 Y. Endo, S. Saito, E. Hirota: To be published
3.132 Y. Endo, S. Saito, E. Hirota: To be published
3.133 S. Saito, Y. Endo, E. Hirota: J. Chem. Phys. **78,** 6447 (1983)
3.134 M. Tanimoto, S. Saito, Y. Endo, E. Hirota: J. Mol. Spectrosc. **100,** 205 (1983)
3.135 M. Tanimoto, S. Saito, Y. Endo, E. Hirota: J. Mol. Spectrosc. **103,** 330 (1984)
3.136 M. Tanimoto, S. Saito, E. Hirota: To be published
3.137 Y. Endo, S. Saito, E. Hirota: Bull. Chem. Soc. Jpn. **56,** 3410 (1983)
3.138 C. Yamada, Y. Endo, E. Hirota: J. Chem. Phys. **79,** 4159 (1983)
3.139 S. Saito, Y. Endo, E. Hirota: J. Chem. Phys. **82,** No. 7 (1985)
3.140 T. Minowa, S. Saito, E. Hirota: To be published
3.141 A. R. W. McKellar, C. Yamada, E. Hirota: J. Mol. Spectrosc. **97,** 425 (1983)
3.142 J. E. Butler, K. Kawaguchi, E. Hirota: J. Mol. Spectrosc. **104,** 372 (1984)
3.143 S. Saito, Y. Endo, E. Hirota: J. Mol. Spectrosc. **98,** 138 (1983)
3.144 S. Saito, Y. Endo, E. Hirota: To be published
3.145 Y. Endo, S. Saito, E. Hirota: J. Mol. Spectrosc. **97,** 204 (1983)
3.146 S. Saito, Y. Endo, E. Hirota: To be published
3.147 K. Nagai, C. Yamada, Y. Endo, E. Hirota: J. Mol. Spectrosc. **90,** 249 (1981)
3.148 K. Kawaguchi, E. Hirota, C. Yamada: Mol. Phys. **44,** 509 (1981)
3.149 K. Kawaguchi, Y. Endo, E. Hirota: J. Mol. Spectrosc. **93,** 381 (1982)
3.150 C. Yamada, E. Hirota: J. Chem. Phys. **80,** 4694 (1984)

3.151 C. Yamada, E. Hirota: J. Chem. Phys. **78,** 1703 (1983)
3.152 Y. Endo, C. Yamada, S. Saito, E. Hirota: J. Chem. Phys. **77,** 3376 (1982)
3.153 N. Ohashi, S. Saito, T. Suzuki, E. Hirota: To be published
3.154 T. Ishiwata, T. Tanaka, K. Kawaguchi, E. Hirota: J. Chem. Phys. **82,** No. 5 (1985)
3.155 H. E. Radford, K. M. Evenson, C. J. Howard: J. Chem. Phys. **60,** 3178 (1974)
3.156 U. Schurath, M. Weber, K. H. Becker: J. Chem. Phys. **67,** 110 (1977)
3.157 A. Carrington, G. N. Currie, T. A. Miller, D. H. Levy: J. Chem. Phys. **50,** 2726 (1969)
3.158 D. R. Johnson, F. X. Powell: Science **164,** 950 (1969)
3.159 H. E. Radford, F. D. Wayne, J. M. Brown: J. Mol. Spectrosc. **99,** 209 (1983)

Chapter 4

4.1 C. Yamada, E. Hirota: J. Mol. Spectrosc. **74,** 203 (1979)
4.2 T. R. Todd, W. B. Olson: J. Mol. Spectrosc. **74,** 190 (1979)
4.3 J. F. Dolan: Astrophys. J. **142,** 1621 (1965)
4.4 W. Jevons: Proc. R. Soc. London Ser. **A 89,** 187 (1913)
4.5 R. S. Mulliken: Phys. Rev. **26,** 319 (1925)
4.6 H. Bredohl, I. Dubois, Y. Houbrechts, M. Singh: Can. J. Phys. **54,** 680 (1976)
4.7 S. Saito, Y. Endo, E. Hirota: J. Chem. Phys. **78,** 6447 (1983)
4.8 C. Yamada, E. Hirota: J. Chem. Phys. **82,** No. 6 (1985)
4.9 J. C. Knights, J. P. M. Schmitt, J. Perrin, G. Guelachvili: J. Chem. Phys. **76,** 3414 (1982)
4.10 C. Linton: J. Mol. Spectrosc. **55,** 108 (1975)
4.11 S. C. Foster: J. Mol. Spectrosc. **106,** 369 (1984)
4.12 A. B. Meinel: Astrophys. J. **112,** 562 (1950)
4.13 A. E. Douglas: Astrophys. J. **117,** 380 (1953)
4.14 W. Benesch, R. Rivers, J. Moore: J. Opt. Soc. Am. **70,** 792 (1980)
4.15 J. M. Cook, T. A. Miller, V. E. Bondybey: J. Chem. Phys. **69,** 2562 (1978)
4.16 F. J. Grieman, J. C. Hansen, J. T. Moseley: Chem. Phys. Lett. **85,** 53 (1982)
4.17 J. C. Hansen, C. H. Kuo, F. J. Grieman, J. T. Moseley: J. Chem. Phys. **79,** 1111 (1983)
4.18 T. A. Miller, T. Suzuki, E. Hirota: J. Chem. Phys. **80,** 4671 (1984)
4.19 F. X. Powell, D. R. Lide, Jr.: J. Chem. Phys. **41,** 1413 (1964)
4.20 M. Bogey, C. Demuynck, J. L. Destombes: Chem. Phys. **66,** 99 (1982)
4.21 C. A. Gottlieb, J. A. Ball: Astrophys. J. **184,** L 59 (1973)
4.22 K. Kawaguchi, C. Yamada, E. Hirota: J. Chem. Phys. **71,** 3338 (1979)
4.23 M. Wong, T. Amano, P. Bernath: J. Chem. Phys. **77,** 2211 (1982)
4.24 H. Kanamori, J. E. Butler, K. Kawaguchi, C. Yamada, E. Hirota: J. Chem. Phys. submitted
4.25 D. E. Milligan: J. Chem. Phys. **35,** 372 (1961)
4.26 D. E. Milligan, M. E. Jacox: J. Chem. Phys. **40,** 2461 (1964)
4.27 A. G. Briggs, R. G. W. Norrish: Proc. R. Soc. London Ser. **A 278,** 27 (1964)
4.28 R. Colin, W. E. Jones: Can. J. Phys. **45,** 301 (1967)
4.29 C. Yamada, Y. Endo, E. Hirota: To be published
4.30 C. Yamada, Y. Endo, E. Hirota: J. Chem. Phys. **79,** 4159 (1983)
4.31 J. K. G. Watson: J. Mol. Spectrosc. **45,** 99 (1973)
4.32 J. K. G. Watson: J. Mol. Spectrosc. **80,** 411 (1980)
4.33 A. E. Douglas, M. Frackowiak: Can. J. Phys. **40,** 832 (1962)
4.34 R. Colin, J. Devillers, F. Prevot: J. Mol. Spectrosc. **44,** 230 (1972)
4.35 S. Saito, Y. Endo, E. Hirota: J. Chem. Phys. **82,** No. 7 (1985)
4.36 C. Ryzlewicz, H.-U. Schütze-Pahlmann, J. Hoeft, T. Törring: Chem. Phys. **71,** 389 (1982)
4.37 K. P. Huber, G. Herzberg: *Molecular Spectra and Molecular Structure IV. Constants of Diatomic Molecules* (Van Nostrand-Reinhold, New York 1979)

4.38 T. Minowa, S. Saito, E. Hirota: To be published
4.39 H. Kanamori, J. E. Butler, K. Kawaguchi, C. Yamada, E. Hirota: To be published
4.40 M. Ishaque, R. W. B. Pearse: Proc. R. Soc. London Ser. **A 173,** 265 (1939)
4.41 F. Legay: Can. J. Phys. **38,** 797 (1960)
4.42 J. Rostas, D. Cossart, J. R. Bastien: Can. J. Phys. **52,** 1274 (1974)
4.43 P. B. Davies, D. K. Russell, B. A. Thrush: Chem. Phys. Lett. **36,** 280 (1975)
4.44 P. B. Davies, D. K. Russell, D. R. Smith, B. A. Thrush: Can. J. Phys. **57,** 522 (1979)
4.45 H. Uehara, K. Hakuta: J. Chem. Phys. **74,** 4326 (1981)
4.46 N. Ohashi, K. Kawaguchi, E. Hirota: J. Mol. Spectrosc. **103,** 337 (1984)
4.47 J. R. Anacona, P. B. Davies, P. A. Hamilton: Chem. Phys. Lett. **104,** 269 (1984)
4.48 R. N. Dixon, H. M. Lamberton: J. Mol. Spectrosc. **25,** 12 (1968)
4.49 K. Kawaguchi, E. Hirota: J. Mol. Spectrosc. **106,** 423 (1984)
4.50 R. N. Dixon, H. W. Kroto: Trans. Faraday Soc. **59,** 1484 (1963)
4.51 K. M. Evenson, H. E. Radford, M. M. Moran, Jr.: Appl. Phys. Lett. **18,** 426 (1971)
4.52 J. T. Hougen, J. A. Mucha, D. A. Jennings, K. M. Evenson: J. Mol. Spectrosc. **72,** 463 (1978)
4.53 J. M. Brown, K. M. Evenson: J. Mol. Spectrosc. **98,** 392 (1983)
4.54 O. E. H. Rydbeck, J. Elldér, W. M. Irvine, A. Sume, Å. Hjalmarson: Astron. Astrophys. **34,** 479 (1974)
4.55 C. R. Brazier, J. M. Brown: J. Chem. Phys. **78,** 1608 (1983)
4.56 M. Bogey, C. Demuynck, J. L. Destombes: Chem. Phys. Lett. **100,** 105 (1983)
4.57 K. Kawaguchi, C. Yamada, Y. Hamada, E. Hirota: J. Mol. Spectrosc. **86,** 136 (1981)
4.58 R. J. Saykally, K. G. Lubic, A. Scalabrin, K. M. Evenson: J. Chem. Phys. **77,** 58 (1982)
4.59 F. C. Van den Heuvel, W. L. Meerts, A. Dymanus: Chem. Phys. Lett. **88,** 59 (1982)
4.60 S. Saito, Y. Endo, M. Takami, E. Hirota: J. Chem. Phys. **78,** 116 (1983)
4.61 C. Yamada, K. Nagai, E. Hirota: J. Mol. Spectrosc. **85,** 416 (1981)
4.62 Y. Endo, S. Saito, E. Hirota: J. Mol. Spectrosc. **92,** 443 (1982)
4.63 Y. Endo, S. Saito, E. Hirota: J. Mol. Spectrosc. **94,** 199 (1982)
4.64 M. Tanimoto, S. Saito, Y. Endo, E. Hirota: J. Mol. Spectrosc. **100,** 205 (1983)
4.65 M. Tanimoto, S. Saito, Y. Endo, E. Hirota: J. Mol. Spectrosc. **103,** 330 (1984)
4.66 M. Tanimoto, S. Saito, Y. Endo, E. Hirota: To be published
4.67 K. Kawaguchi, S. Saito, E. Hirota: J. Chem. Phys. **79,** 629 (1983)
4.68 J. E. Butler, K. Kawaguchi, E. Hirota: J. Mol. Spectrosc. **101,** 161 (1983)
4.69 H. Uehara: Chem. Phys. Lett. **84,** 539 (1981)
4.70 A. R. W. McKellar: Can. J. Phys. **57,** 2106 (1979)
4.71 A. R. W. McKellar, C. Yamada, E. Hirota: J. Mol. Spectrosc. **97,** 425 (1983)
4.72 A. R. W. McKellar: J. Mol. Spectrosc. **101,** 186 (1983)
4.73 P. B. Davies, F. Temps, H. G. Wagner, D. P. Stern: Max-Planck-Institute Report 19, December 1982
4.74 Y. Endo: Unpublished
4.75 A. R. W. McKellar: J. Mol. Spectrosc. **86,** 43 (1981)
4.76 E. A. Cohen, H. M. Pickett, M. Geller: J. Mol. Spectrosc. **87,** 459 (1981)
4.77 M. Barnett, E. A. Cohen, D. A. Ramsay: Can. J. Phys. **59,** 1908 (1981)
4.78 J. E. Butler, K. Kawaguchi, E. Hirota: J. Mol. Spectrosc. **104,** 372 (1984)
4.79 E. A. Cohen, H. M. Pickett, M. Geller: J. Mol. Spectrosc. **106,** 430 (1984)
4.80 Y. Endo, K. Nagai, C. Yamada, E. Hirota: J. Mol. Spectrosc. **97,** 213 (1983)
4.81 C. Yamada, J. E. Butler, K. Kawaguchi, H. Kanamori, E. Hirota: To be published
4.82 L. Herzberg, G. Herzberg: Astrophys. J. **105,** 353 (1947)
4.83 A. M. Falick, B. H. Mahan, R. J. Myers: J. Chem. Phys. **42,** 1837 (1965)
4.84 A. Scalabrin, R. J. Saykally, K. M. Evenson, H. E. Radford, M. Mizushima: J. Mol. Spectrosc. **89,** 344 (1981)
4.85 G. Cazzoli, C. D. Esposti, P. G. Favero: Chem. Phys. Lett. **100,** 99 (1983)
4.86 A. Carrington, D. H. Levy, T. A. Miller: Proc. R. Soc. London Ser. **A 293,** 108 (1966)
4.87 S. Saito: J. Chem. Phys. **53,** 2544 (1970)

4.88 C. Yamada, K. Kawaguchi, E. Hirota: J. Chem. Phys. **69,** 1942 (1978)
4.89 A. H. Curran, R. G. MacDonald, A. J. Stone, B. A. Thrush: Chem. Phys. Lett. **8,** 451 (1971)
4.90 P. B. Davies, P. A. Hamilton, M. Okumura: J. Chem. Phys. **75,** 4294 (1981)
4.91 P. B. Davies, W. J. Rothwell: Proc. R. Soc. London Ser. **A 389,** 205 (1983)
4.92 H. Schall, C. Linton, R. W. Field: J. Mol. Spectrosc. **100,** 437 (1983)
4.93 C. Linton, S. McDonald, S. Rice, M. Dulick, Y. C. Liu, R. W. Field: J. Mol. Spectrosc. **101,** 332 (1983)
4.94 A. S.-C. Cheung, R. M. Gordon, A. J. Merer: J. Mol. Spectrosc. **87,** 289 (1981)
4.95 A. S.-C. Cheung, N. Lee, A. M. Lyyra, A. J. Merer, A. W. Taylor: J. Mol. Spectrosc. **95,** 213 (1982)
4.96 Y. Endo, S. Saito, E. Hirota: Astrophys. J. **278,** L 131 (1984)
4.97 D. Buhl, L. E. Snyder: Nature **228,** 267 (1970)
4.98 W. Klemperer: Nature **227,** 1230 (1970)
4.99 R. C. Woods, T. A. Dixon, R. J. Saykally, P. G. Szanto: Phys. Rev. Lett. **35,** 1269 (1975)
4.100 M. Guélin, P. Thaddeus: Astrophys. J. **212,** L 81 (1977)
4.101 M. Guélin, S. Green, P. Thaddeus: Astrophys. J. **224,** L 27 (1978)
4.102 P. Friberg, Å. Hjalmarson, W. M. Irvine, M. Guélin: Astrophys. J. **241,** L 99 (1980)
4.103 M. B. Bell, T. J. Sears, H. E. Matthews: Astrophys. J. **255,** L 75 (1982)
4.104 C. A. Gottlieb, E. W. Gottlieb, P. Thaddeus, H. Kawamura: Astrophys. J. **275,** 916 (1983)
4.105 P. G. Carrick, A. J. Merer, R. F. Curl, Jr.: J. Chem. Phys. **78,** 3652 (1983)
4.106 A. J. Merer, D. N. Travis: Can. J. Phys. **43,** 1795 (1965)
4.107 A. J. Merer, D. N. Travis: Can. J. Phys. **44,** 353 (1966)
4.108 S. Saito, Y. Endo, E. Hirota: J. Chem. Phys. **80,** 1427 (1984)
4.109 R. A. Bernheim, R. J. Kempf, P. W. Humer, P. S. Skell: J. Chem. Phys. **41,** 1156 (1964)
4.110 R. A. Bernheim, R. J. Kempf, J. V. Gramas, P. S. Skell: J. Chem. Phys. **43,** 196 (1965)
4.111 R. A. Bernheim, R. J. Kempf, E. F. Reichenbecher: J. Magn. Reson. **3,** 5 (1970)
4.112 D. E. Milligan, M. E. Jacox: J. Chem. Phys. **47,** 5157 (1967)
4.113 K. Kawaguchi, E. Hirota: To be published
4.114 J. W. C. Johns: Can. J. Phys. **39,** 1738 (1961); ibid. **42,** 1004 (1964)
4.115 D. K. Russell, M. Kroll, R. A. Beaudet: J. Chem. Phys. **66,** 1999 (1977)
4.116 R. N. Dixon, D. Field, M. Noble: Chem. Phys. Lett. **50,** 1 (1977)
4.117 I. Suzuki: J. Mol. Spectrosc. **25,** 479 (1968)
4.118 I. Suzuki: Bull. Chem. Soc. Jpn. **48,** 1685 (1975)
4.119 K. G. Weyer, R. A. Beaudet, R. Straubinger, H. Walther: Chem. Phys. **47,** 171 (1980)
4.120 M. S. Kim, R. E. Smalley, D. H. Levy: J. Mol. Spectrosc. **71,** 458 (1978)
4.121 A. Muirhead, K. V. L. N. Sastry, R. F. Curl, J. Cook, F. K. Tittel: Chem. Phys. Lett. **24,** 208 (1974)
4.122 R. S. Lowe, H. Gerhardt, W. Dillenschneider, R. F. Curl, Jr., F. K. Tittel: J. Chem. Phys. **70,** 42 (1979)
4.123 W. Schulz, K. G. Weyer, R. A. Beaudet, H. Walther: J. Chem. Phys. **72,** 589 (1980)
4.124 K. Kawaguchi, E. Hirota, C. Yamada: Mol. Phys. **44,** 509 (1981)
4.125 U. Öpik, M. H. L. Pryce: Proc. R. Soc. London **A 238,** 425 (1957)
4.126 L. Gausset, G. Herzberg, A. Lagerqvist, B. Rosen: Astrophys. J. **142,** 45 (1965)
4.127 D. Gauyacq, C. Larcher, J. Rostas: Can. J. Phys. **57,** 1634 (1979)
4.128 K. Kawaguchi, S. Saito, E. Hirota: Mol. Phys. **49,** 663 (1983)
4.129 K. Kawaguchi, T. Suzuki, S. Saito, E. Hirota, T. Kasuya: J. Mol. Spectrosc. **106,** 320 (1984)
4.130 K. Kawaguchi, Y. Endo, E. Hirota: J. Mol. Spectrosc. **93,** 381 (1982)
4.131 R. N. Dixon: Philos. Trans. R. Soc. London Ser. **A 252,** 165 (1960)
4.132 R. N. Dixon: Can. J. Phys. **38,** 10 (1960)
4.133 A. Carrington, A. R. Fabris, B. J. Howard, N. J. D. Lucas: Mol. Phys. **20,** 961 (1971)

4.134 P. S. H. Bolman, J. M. Brown: Chem. Phys. Lett. **21**, 213 (1973)
4.135 P. S. H. Bolman, J. M. Brown, A. Carrington, I. Kopp, D. A. Ramsay: Proc. R. Soc. London **A 343**, 17 (1975)
4.136 C. E. Barnes, J. M. Brown, A. D. Fackerell, T. J. Sears: J. Mol. Spectrosc. **92**, 485 (1982)
4.137 S. Saito, T. Amano: J. Mol. Spectrosc. **34**, 383 (1970)
4.138 T. Amano, E. Hirota: J. Chem. Phys. **57**, 5608 (1972)
4.139 K. Kawaguchi, S. Saito, E. Hirota: Mol. Phys. (in press)
4.140 L. Veseth: J. Mol. Spectrosc. **38**, 228 (1971)
4.141 J. M. Brown, J. K. G. Watson: J. Mol. Spectrosc. **65**, 65 (1977)
4.142 M. Kakimoto, T. Kasuya: J. Mol. Spectrosc. **94**, 380 (1982)
4.143 K. Hakuta, H. Uehara: J. Chem. Phys. **78**, 6484 (1983)
4.144 K. Hakuta, H. Uehara, K. Kawaguchi, T. Suzuki, T. Kasuya: J. Chem. Phys. **79**, 1094 (1983)
4.145 A. C. Lloyd: Int. J. Chem. Kinet. **6**, 169 (1974)
4.146 N. N. Semenov: "Modern Concepts of the Mechanism of Hydrocarbon Oxidation in Gas Phase", in *Photochemistry and Reaction Kinetics,* ed. by P. G. Ashmore, F. S. Dainton, T. M. Sugden (Cambridge Univ. Press, Cambridge 1967) pp. 229–249
4.147 D. E. Milligan, M. E. Jacox: J. Chem. Phys. **38**, 2627 (1963)
4.148 M. E. Jacox, D. E. Milligan: J. Mol. Spectrosc. **42**, 495 (1972)
4.149 T. T. Paukert, H. S. Johnston: J. Chem. Phys. **56**, 2824 (1972)
4.150 H. E. Radford, K. M. Evenson, C. J. Howard: J. Chem. Phys. **60**, 3178 (1974)
4.151 J. T. Hougen: J. Mol. Spectrosc. **54**, 447 (1975)
4.152 J. T. Hougen, H. E. Radford, K. M. Evenson, C. J. Howard: J. Mol. Spectrosc. **56**, 210 (1975)
4.153 Y. Beers, C. J. Howard: J. Chem. Phys. **63**, 4212 (1975)
4.154 Y. Beers, C. J. Howard: J. Chem. Phys. **64**, 1541 (1976)
4.155 S. Saito: J. Mol. Spectrosc. **65**, 229 (1977)
4.156 F. J. Adrian, E. L. Cochran, V. A. Bowers: J. Chem. Phys. **47**, 5441 (1967)
4.157 C. E. Barnes, J. M. Brown, A. Carrington, J. Pinkstone, T. J. Sears, P. J. Thislethwaite: J. Mol. Spectrosc. **72**, 86 (1978)
4.158 J. M. Brown, T. J. Sears: J. Mol. Spectrosc. **75**, 111 (1979)
4.159 A. Charo, F. C. De Lucia: J. Mol. Spectrosc. **94**, 426 (1982)
4.160 S. Saito, C. Matsumura: J. Mol. Spectrosc. **80**, 34 (1980)
4.161 J. W. C. Johns, A. R. W. McKellar, M. Riggin: J. Chem. Phys. **68**, 3957 (1978)
4.162 A. R. W. McKellar: Faraday Discuss. **71**, 63 (1981)
4.163 K. Nagai, Y. Endo, E. Hirota: J. Mol. Spectrosc. **89**, 520 (1981)
4.164 C. Yamada, Y. Endo, E. Hirota: J. Chem. Phys. **78**, 4379 (1983)
4.165 P. A. Preedman, W. J. Jones: J. Chem. Soc. Faraday Trans. II **72**, 207 (1976)
4.166 R. P. Tuckett, P. A. Freedman, W. J. Jones: Mol. Phys. **37**, 379 (1979)
4.167 C. E. Barnes, J. M. Brown, H. E. Radford: J. Mol. Spectrosc. **84**, 179 (1980)
4.168 S. Saito, Y. Endo, E. Hirota: J. Mol. Spectrosc. **98**, 138 (1983)
4.169 A. R. W. McKellar: J. Chem. Phys. **71**, 81 (1979)
4.170 H. Uehara, K. Kawaguchi, E. Hirota: To be published
4.171 K. G. Lubic, T. Amano, H. Uehara, K. Kawaguchi, E. Hirota: J. Chem. Phys. **81**, 4826 (1984)
4.172 A. Arkell: J. Am. Chem. Soc. **87**, 4057 (1965)
4.173 R. D. Spratley, J. J. Turner, G. C. Pimentel: J. Chem. Phys. **44**, 2063 (1966)
4.174 P. N. Noble, G. C. Pimentel: J. Chem. Phys. **44**, 3641 (1966)
4.175 M. E. Jacox: J. Mol. Spectrosc. **84**, 74 (1980)
4.176 P. H. Kasai, A. D. Kirshenbaum: J. Am. Chem. Soc. **87**, 3069 (1965)
4.177 R. W. Fessenden, R. H. Schuler: J. Chem. Phys. **44**, 434 (1966)
4.178 F. J. Adrian: J. Chem. Phys. **46**, 1543 (1967)
4.179 C. Yamada, E. Hirota: J. Chem. Phys. **80**, 4694 (1984)
4.180 U. Schurath, M. Weber, K. H. Becker: J. Chem. Phys. **67**, 110 (1977)

4.181 M. Kakimoto, S. Saito, E. Hirota: J. Mol. Spectrosc. **80,** 334 (1980)
4.182 N. Ohashi, M. Kakimoto, S. Saito, E. Hirota: J. Mol. Spectrosc. **84,** 204 (1980)
4.183 M. Satoh, N. Ohashi, S. Matsuoka: Bull. Chem. Soc. Jpn. **56,** 2545 (1983)
4.184 K. Tsukiyama, I. Tanaka, S. Saito, T. Suzuki, E. Hirota: To be published
4.185 Y. Endo, S. Saito, E. Hirota: J. Chem. Phys. **75,** 4379 (1981)
4.186 C. W. Webster, P. J. Brucat, R. N. Zare: J. Mol. Spectrosc. **92,** 184 (1982)
4.187 T. J. Sears, A. R. W. McKellar: Mol. Phys. **49,** 25 (1983)
4.188 A. B. Sannigrahi, K. H. Thunemann, S. D. Peyerimhoff, R. J. Buenker: Chem. Phys. **20,** 25 (1977)
4.189 A. B. Sannigrahi, S. D. Peyerimhoff, R. J. Buenker: Chem. Phys. **20,** 381 (1977)
4.190 H. E. Radford, F. D. Wayne, J. M. Brown: J. Mol. Spectrosc. **99,** 209 (1983)
4.191 Y. Endo, S. Saito, E. Hirota: J. Chem. Phys. **74,** 1568 (1981)
4.192 S. Saito, Y. Endo, E. Hirota: To be published
4.193 D. E. Milligan, M. E. Jacox, A. M. Bass, J. J. Comeford, D. E. Mann: J. Chem. Phys. **42,** 3187 (1965)
4.194 M. E. Jacox: J. Mol. Spectrosc. **80,** 257 (1980)
4.195 D. K. W. Wang, W. E. Jones: J. Photochem. **1,** 147 (1972/1973)
4.196 F. J. Adrian, E. L. Cochran, V. A. Bowers: J. Chem. Phys. **43,** 462 (1965)
4.197 E. L. Cochran, F. J. Adrian, V. A. Bowers: J. Chem. Phys. **44,** 4626 (1966)
4.198 K. Nagai, C. Yamada, Y. Endo, E. Hirota: J. Mol. Spectrosc. **90,** 249 (1981)
4.199 G. Herzberg, J. Shoosmith: Nature **183,** 1801 (1959)
4.200 G. Herzberg: Proc. R. Soc. London Ser. **A 262,** 291 (1961)
4.201 G. Herzberg, J. W. C. Johns: Proc. R. Soc. London Ser. **A 295,** 107 (1966)
4.202 R. A. Bernheim, H. W. Bernard, P. S. Wang, L. S. Wood, P. S. Skell: J. Chem. Phys. **53,** 1280 (1970)
4.203 R. A. Bernheim, H. W. Bernard, P. S. Wang, L. S. Wood, P. S. Skell: J. Chem. Phys. **54,** 3223 (1971)
4.204 E. Wasserman, W. A. Yager, V. J. Kuck: Chem. Phys. Lett. **7,** 409 (1970)
4.205 E. Wasserman, V. J. Kuck, R. S. Hutton, W. A. Yager: J. Am. Chem. Soc. **92,** 7491 (1970)
4.206 J. A. Mucha, K. M. Evenson, D. A. Jennings, G. B. Ellison, C. J. Howard: Chem. Phys. Lett. **66,** 244 (1979)
4.207 T. J. Sears, P. R. Bunker, A. R. W. McKellar: J. Chem. Phys. **75,** 4731 (1981)
4.208 T. J. Sears, P. R. Bunker, A. R. W. McKellar: J. Chem. Phys. **77,** 5363 (1982)
4.209 T. J. Sears, P. R. Bunker, A. R. W. McKellar, K. M. Evenson, D. A. Jennings, J. M. Brown: J. Chem. Phys. **77,** 5348 (1982)
4.210 A. R. W. McKellar, T. J. Sears: Can. J. Phys. **61,** 480 (1983)
4.211 P. R. Bunker, T. J. Sears, A. R. W. McKellar, K. M. Evenson, F. J. Lovas: J. Chem. Phys. **79,** 1211 (1983)
4.212 F. J. Lovas, R. D. Suenram, K. M. Evenson: Astrophys. J. **267,** L 131 (1983)
4.213 A. R. W. McKellar, C. Yamada, E. Hirota: J. Chem. Phys. **79,** 1220 (1983)
4.214 P. Jensen, P. R. Bunker, A. R. Hoy: J. Chem. Phys. **77,** 5370 (1982)
4.215 K. M. Evenson, T. J. Sears, A. R. W. McKellar: J. Opt. Soc. Am. **B 1,** 15 (1984)
4.216 P. R. Bunker, P. Jensen: J. Chem. Phys. **79,** 1224 (1983)
4.217 C. Jungen, K.-E. J. Hallin, A. J. Merer: Mol. Phys. **40,** 25 (1980)
4.218 D. A. Ramsay: J. Chem. Phys. **25,** 188 (1956)
4.219 K. Dressler, D. A. Ramsay: Philos. Trans. R. Soc. London Ser. **A 251,** 553 (1959)
4.220 J. W. C. Johns, D. A. Ramsay, S. C. Ross: Can. J. Phys. **54,** 1804 (1976)
4.221 F. W. Birss, M.-F. Merienne-Lafore, D. A. Ramsay, M. Vervloet: J. Mol. Spectrosc. **85,** 493 (1981)
4.222 P. B. Davies, D. K. Russell, B. A. Thrush, H. E. Radford: Proc. R. Soc. London Ser. **A 353,** 299 (1977)
4.223 J. M. Cook, G. W. Hills, R. F. Curl, Jr.: J. Chem. Phys. **67,** 1450 (1977)
4.224 G. W. Hills, A. R. W. McKellar: J. Mol. Spectrosc. **74,** 224 (1979)

4.225 K. Kawaguchi, C. Yamada, E. Hirota, J. M. Brown, J. Buttenshaw, C. R. Parent, T. J. Sears: J. Mol. Spectrosc. **81**, 60 (1980)
4.226 T. Amano, P. F. Bernath, A. R. W. McKellar: J. Mol. Spectrosc. **94**, 100 (1982)
4.227 T. Amano, K. Kawaguchi, M. Kakimoto, S. Saito, E. Hirota: J. Chem. Phys. **77**, 159 (1982)
4.228 P. B. Davies, D. K. Russell, B. A. Thrush, H. E. Radford: Chem. Phys. **44**, 421 (1979)
4.229 G. W. Hills, A. R. W. McKellar: J. Chem. Phys. **71**, 1141 (1979)
4.230 R. F. Curl, Jr., Y. Endo, M. Kakimoto, S. Saito, E. Hirota: Chem. Phys. Lett. **53**, 536 (1978)
4.231 M. Kakimoto, E. Hirota: J. Mol. Spectrosc. **94**, 173 (1982)
4.232 M. Kakimoto, E. Hirota: To be published
4.233 Y. Endo, S. Saito, E. Hirota: J. Mol. Spectrosc. **97**, 204 (1983)
4.234 R. D. Verma, C. F. McCarthy: Can. J. Phys. **61**, 1149 (1983)
4.235 K. Kawaguchi, S. Saito, E. Hirota, N. Ohashi: J. Chem. Phys. (in press)
4.236 P. F. Bernath, T. Amano: J. Mol. Spectrosc. **95**, 359 (1982)
4.237 J. M. Berthou, B. Pascat, H. Guenebaut, D. A. Ramsay: Can. J. Phys. **50**, 2265 (1972)
4.238 C. Camy-Peyret, J.-M. Flaud, A. Perrin, K. N. Rao: J. Mol. Spectrosc. **95**, 72 (1982)
4.239 Y. Morino, M. Tanimoto, S. Saito, E. Hirota, R. Awata, T. Tanaka: J. Mol. Spectrosc. **98**, 331 (1983)
4.240 J. W. C. Johns, A. R. W. McKellar, E. Weinberger: Can. J. Phys. **61**, 1106 (1983)
4.241 E. Hirota: To be published
4.242 M. Larzillière, M. E. Jacox: J. Mol. Spectrosc. **79**, 132 (1980)
4.243 M. Larzillière, N. Damany, Lam Thanh My: Chem. Phys. **46**, 401 (1980)
4.244 E. Hirota: Unpublished
4.245 J. F. Ogilvie: J. Mol. Structure **31**, 407 (1976)
4.246 A. R. Hoy, I. M. Mills, G. Strey: Mol. Phys. **24**, 1265 (1972)
4.247 P. Botschwina: Chem. Phys. **40**, 33 (1979)
4.248 F. W. Dalby: Can. J. Phys. **36**, 1336 (1958)
4.249 J. M. Brown, D. A. Ramsay: Can. J. Phys. **53**, 2232 (1975)
4.250 G. Herzberg: *Molecular Spectra and Molecular Structure III. Electronic Spectra and Electronic Structure of Polyatomic Molecules* (Van Nostrand, New York 1966)
4.251 G. Herzberg, J. Shoosmith: Can. J. Phys. **34**, 523 (1956)
4.252 A. D. Walsh: J. Chem. Soc. 2296 (1953)
4.253 L. Y. Tan, A. M. Winer, G. C. Pimentel: J. Chem. Phys. **57**, 4028 (1972)
4.254 M. Karplus: J. Chem. Phys. **30**, 15 (1959)
4.255 R. W. Fessenden: J. Phys. Chem. **71**, 74 (1967)
4.256 J. M. Riveros: J. Chem. Phys. **51**, 1269 (1969)
4.257 A. Snelson: J. Phys. Chem. **74**, 537 (1970)
4.258 J. Pacansky, J. Bargon: J. Am. Chem. Soc. **97**, 6896 (1975)
4.259 M. E. Jacox: J. Mol. Spectrosc. **66**, 272 (1977)
4.260 Y. Ishikawa, R. C. Binning, Jr.: Chem. Phys. Lett. **40**, 342 (1976)
4.261 C. Yamada, E. Hirota, K. Kawaguchi: J. Chem. Phys. **75**, 5256 (1981)
4.262 T. Amano, P. F. Bernath, C. Yamada, Y. Endo, E. Hirota: J. Chem. Phys. **77**, 5284 (1982)
4.263 C. Yamada, E. Hirota: J. Chem. Phys. **78**, 669 (1983)
4.264 G. A. Laguna, S. L. Baughcum: Chem. Phys. Lett. **88**, 568 (1982)
4.265 R. E. Florin, D. W. Brown, L. A. Wall: J. Phys. Chem. **66**, 2672 (1962)
4.266 R. W. Fessenden, R. H. Schuler: J. Chem. Phys. **43**, 2704 (1965)
4.267 M. T. Rogers, L. D. Kispert: J. Chem. Phys. **46**, 3193 (1967)
4.268 J. Maruani, C. A. McDowell, H. Nakajima, P. Raghunathan: Mol. Phys. **14**, 349 (1968)
4.269 J. Maruani, J. A. R. Coope, C. A. McDowell: Mol. Phys. **18**, 165 (1970)
4.270 M. Karplus, G. K. Fraenkel: J. Chem. Phys. **35**, 1312 (1961)
4.271 D. M. Shrader, M. Karplus: J. Chem. Phys. **40**, 1953 (1964)
4.272 G. A. Carlson, G. C. Pimentel: J. Chem. Phys. **44**, 4053 (1966)

4.273 D. E. Milligan, M. E. Jacox, J. J. Comeford: J. Chem. Phys. **44,** 4058 (1966)
4.274 D. E. Milligan, M. E. Jacox; J. Chem. Phys. **48,** 2265 (1968)
4.275 Y. Endo, C. Yamada, S. Saito, E. Hirota: J. Chem. Phys. **77,** 3376 (1982)
4.276 C. Yamada, E. Hirota: J. Chem. Phys. **78,** 1703 (1983)
4.277 J. R. Morton, K. F. Preston: J. Magn. Res. **30,** 577 (1978)
4.278 J. Dyke, N. Jonathan, E. Lee, A. Morris: J. Chem. Soc. Faraday Trans. II **72,** 1385 (1976)
4.279 H. W. Hermann, S. R. Leone: J. Chem. Phys. **76,** 4759 (1982)
4.280 E. Hirota, C. Yamada: J. Mol. Spectrosc. **96,** 175 (1982)
4.281 A. B. Callear, M. P. Metcalfe: Chem. Phys. **14,** 275 (1976)
4.282 M. E. Jacox, D. E. Milligan: J. Chem. Phys. **50,** 3252 (1969)
4.283 M. E. Jacox: Chem. Phys. **59,** 199 (1981)
4.284 J. I. Raymond, L. Andrews: J. Phys. Chem. **75,** 3235 (1971)
4.285 J. A. Mucha, D. A. Jennings, K. M. Evenson, J. T. Hougen: J. Mol. Spectrosc. **68,** 122 (1977)
4.286 Y. Endo, C. Yamada, S. Saito, E. Hirota: J. Chem. Phys. **79,** 1605 (1983)
4.287 C. Yamada, E. Hirota: To be published
4.288 Y. Endo, S. Saito, E. Hirota: Can. J. Phys. **62,** 1347 (1984)
4.289 M. E. Jacox, D. E. Milligan: J. Chem. Phys. **53,** 2688 (1970)
4.290 L. Andrews, D. W. Smith: J. Chem. Phys. **53,** 2956 (1970)
4.291 K. Morokuma, L. Pedersen, M. Karplus: J. Chem. Phys. **48,** 4801 (1968)
4.292 P. Botchwina, J. Flesch, W. Meyer: Chem. Phys. **74** 321 (1983)
4.293 W. J. Marinelli, D. M. Swanson, H. S. Johnston: J. Chem. Phys. **76,** 2864 (1982)
4.294 D. A. Ramsay: Proc. Colloq. Spectrosc. Int. 10th, 583 (1962)
4.295 H. H. Nelson, L. Pasternack, J. R. McDonald: J. Phys. Chem. **87,** 1286 (1983)
4.296 T. Ishiwata, I. Fujiwara, Y. Naruge, K. Obi, I. Tanaka: J. Phys. Chem. **87,** 1349 (1983)
4.297 H. H. Nelson, L. Pasternack, J. R. McDonald: J. Chem. Phys. **79,** 4279 (1983)
4.298 A. D. Walsh: J. Chem. Soc. 2301 (1953)
4.299 T. E. H. Walker, J. A. Horseley: Mol. Phys. **21,** 939 (1971)
4.300 T. Ishiwata, I. Tanaka, K. Kawaguchi, E. Hirota: J. Chem. Phys. **82,** No. 5 (1985)
4.301 Y. Kawashima, A. P. Cox: J. Mol. Spectrosc. **65,** 319 (1977)
4.302 Y. Kawashima, A. P. Cox: J. Mol. Spectrosc. **72,** 423 (1978)
4.303 H. A. Jahn, E. Teller: Proc. R. Soc. London Ser. **A 161,** 220 (1937)
4.304 H. E. Radford: Chem. Phys. Lett. **71,** 195 (1980)
4.305 D. W. G. Style, J. C. Ward: Trans. Faraday Soc. **49,** 999 (1953)
4.306 K. Ohbayashi, H. Akimoto, I. Tanaka: J. Phys. Chem. **81,** 798 (1977)
4.307 H. E. Radford, D. K. Russell: J. Chem. Phys. **66,** 2222 (1977)
4.308 D. K. Russell, H. E. Radford: J. Chem. Phys. **72,** 2750 (1980)
4.309 Y. Endo, S. Saito, E. Hirota: J. Chem. Phys. **81,** 122 (1984)
4.310 J. T. Hougen: J. Mol. Spectrosc. **81,** 73 (1980)
4.311 J. K. G. Watson: J. Mol. Spectrosc. **103,** 125 (1984)
4.312 J. H. Callomon, E. Hirota, K. Kuchitsu, W. J. Lafferty, A. G. Maki, C. S. Pote (eds.): *Landolt-Börnstein Tables,* New Series, Group II, Vol. 7 (Springer, Berlin, Heidelberg, New York 1976)
4.313 D. R. Yarkony, H. F. Schaefer III, S. Rothenberg: J. Am. Chem. Soc. **96,** 656 (1974)
4.314 G. F. Adams, G. D. Bent, G. D. Purvis, R. J. Bartlett: Chem. Phys. Lett. **81,** 461 (1981)
4.315 C. F. Jackels: J. Chem. Phys. **76,** 505 (1982)
4.316 S. Saebø, L. Radom, H. F. Schaefer III: J. Chem. Phys. **78,** 845 (1983)
4.317 C. H. Towns, A. L. Schawlow: *Microwave Spectroscopy* (McGraw-Hill, New York 1955)
4.318 W. Gordy, R. L. Cook: *Microwave Molecular Spectra* (Interscience, New York 1970)
4.319 F. C. Van den Heuvel, W. L. Meerts, A. Dymanus: Chem. Phys. Lett. **92,** 215 (1982)
4.320 J. M. Brown, M. Kaise, C. M. L. Kerr, D. J. Milton: Mol. Phys. **36,** 553 (1978)

4.321 W. L. Meerts, A. Dymanus: Can. J. Phys. **53,** 2123 (1975)
4.322 D. D. Skatrud, F. C. De Lucia, G. A. Blake, K. V. L. N. Sastry: J. Mol. Spectrosc. **99,** 35 (1983)
4.323 P. Kristiansen: J. Mol. Spectrosc. **66,** 177 (1977)
4.324 F. J. Lovas, R. D. Suenram: J. Mol. Spectrosc. **93,** 416 (1982)
4.325 J. R. Morton: Chem. Rev. **64,** 453 (1964)
4.326 Y. Endo, S. Saito, E. Hirota: Bull. Chem. Soc. Jpn. **56,** 3410 (1983)
4.327 B. Fabricant, J. S. Muenter: J. Chem. Phys. **66,** 5274 (1977)
4.328 R. F. Curl, Jr.: Mol. Phys. **9,** 585 (1965)
4.329 J. H. Van Vleck: Rev. Mod. Phys. **23,** 213 (1951)
4.330 J. M. Brown, T. J. Sears, J. K. G. Watson: Mol. Phys. **41,** 173 (1980)
4.331 G. W. Hills, J. M. Cook: J. Mol. Spectrosc. **94,** 456 (1982)
4.332 R. D. Brown, F. R. Burden, P. D. Godfrey, I. R. Gillard: J. Mol. Spectrosc. **52,** 301 (1974)
4.333 M. Tanoura, K. Chiba, K. Tanaka, T. Tanaka: J. Mol. Spectrosc. **95,** 157 (1982)
4.334 J. M. Brown, K. Dumper, R. S. Lowe: J. Mol. Spectrosc. **97,** 441 (1983)
4.335 W. C. Bowman, F. C. De Lucia: J. Chem. Phys. **77,** 92 (1982)
4.336 B. M Landsberg, A. J. Merer, T. Oka: J. Mol. Spectrosc. **67,** 459 (1977)
4.337 G. W. Hills, C. R. Brazier, J. M. Brown, J. M. Cook, R. F. Curl, Jr.: J. Chem. Phys. **76,** 240 (1982)
4.338 D. R. Johnson, F. X. Powell: Science **169,** 679 (1970)
4.339 D. J. Clouthier, D. A. Ramsay: Ann. Rev. Phys. Chem. **34,** 31 (1983)
4.340 K. H. Fung, D. A. Ramsay: J. Phys. Chem. **88,** 395 (1984)
4.341 T. Suzuki, S. Saito, E. Hirota: J. Chem. Phys. **79,** 1641 (1983)
4.342 D. J. Clouthier, D. A. Ramsay, F. W. Birss: J. Chem. Phys. **79,** 5851 (1983)
4.343 J. T. Hougen: Can. J. Phys. **42,** 433 (1964)
4.344 R. H. Judge, D. C. Moule, G. W. King: J. Mol. Spectrosc. **81,** 37 (1980)
4.345 Y. Beers, G. P. Klein, W. H. Kirchhoff, D. R. Johnson: J. Mol. Spectrosc. **44,** 553 (1972)
4.346 A. R. W. McKellar, P. R. Bunker, T. J. Sears, K. M. Evenson, R. J. Saykally, S. R. Langhoff: J. Chem. Phys. **79,** 5251 (1983)
4.347 N. Ohashi, S. Saito, T. Suzuki, E. Hirota: To be published
4.348 A. J. Merer, D. N. Travis: Can. J. Phys. **44,** 525 (1966)
4.349 A. J. Merer, D. N. Travis: Can. J. Phys. **44,** 1541 (1966)
4.350 M. Kakimoto, S. Saito, E. Hirota: J. Mol. Spectrosc. **88,** 300 (1981)
4.351 M. Kakimoto, S. Saito, E. Hirota: J. Mol. Spectrosc. **97,** 194 (1983)
4.352 T. Suzuki, S. Saito, E. Hirota: J. Mol. Spectrosc. **90,** 447 (1981)
4.353 E. Hirota: Discuss. Faraday Soc. **71,** 87 (1981)
4.354 T. Suzuki, S. Saito, E. Hirota: Can. J. Phys. **62,** 1328 (1984)
4.355 R. J. Butcher, S. Saito, E. Hirota: J. Chem. Phys. **80,** 4000 (1984)

Chapter 5

5.1 P. B. Davies, B. A. Thrush: Proc. R. Soc. London Ser. **A 302,** 243 (1968)
5.2 K. Kawaguchi, S. Saito, E. Hirota: J. Chem. Phys. **79,** 629 (1983)
5.3 H. Uehara: Chem. Phys. Lett. **84,** 539 (1981)
5.4 S. Saito, Y. Endo, E. Hirota: To be published
5.5 M. A. A. Clyne, M. C. Heaven: Chem. Phys. **58,** 145 (1981)
5.6 J. H. Van Vleck, V. F. Weisskopf: Rev. Mod. Phys. **17,** 227 (1945)
5.7 B. A. Thrush, G. S. Tyndall: J. Chem. Soc. Faraday Trans. 2 **78,** 1469 (1982)
5.8 B. A. Thrush, G. S. Tyndall: Chem. Phys. Lett. **92,** 232 (1982)
5.9 C. Yamada, E. Hirota: J. Chem. Phys. **78,** 669 (1983)
5.10 G. A. Laguna, S. L. Baughcum: Chem. Phys. Lett. **88,** 568 (1982)
5.11 H. W. Hermann, S. R. Leone: J. Chem. Phys. **76,** 4759 (1982)

5.12 H. W. Hermann, S. R. Leone: J. Chem. Phys. **76**, 4766 (1982)
5.13 S. L. Baughcum, S. R. Leone: Chem. Phys. Lett. **89**, 183 (1982)
5.14 H. Kanamori, J. E. Butler, K. Kawaguchi, C. Yamada, E. Hirota: J. Chem Phys. submitted
5.15 See, for example, a compilation by NASA and the Jet Propulsion Laboratory: Chemical Kinetics and Photochemical Data for Use in Stratospheric Modeling, 1982
5.16 C. J. Howard, K. M. Evenson: J. Chem. Phys. **61**, 1943 (1974)
5.17 C. J. Howard, K. M. Evenson: Geophys. Res. Lett. **4**, 437 (1977)
5.18 C. J. Howard: J. Chem. Phys. **71**, 2352 (1979)
5.19 Y.-P. Lee, C. J. Howard: J. Chem. Phys. **77**, 756 (1982) and references cited therein
5.20 J. P. Burrows, D. I. Cliff, G. W. Harris, B. A. Thrush, J. P. T. Wilkinson: Proc. R. Soc. London Ser. **A 368**, 463 (1979)
5.21 B. A. Thrush, J. P. T. Wilkinson: Chem. Phys. Lett. **81**, 1 (1981)
5.22 B. A. Thrush: Acc. Chem. Res. **14**, 116 (1981)
5.23 See, for example, two recent review articles: W. L. Chameides, D. D. Davis: Chem. Eng. News 39 (1982);
R. M. Baum: Chem. Eng. News 21 (1982)
5.24 W. T. Huntress: Chem. Soc. Rev. 295 (1977)
5.25 E. Herbst, W. Klemperer: Astrophys. J. **185**, 505 (1973)
5.26 D. Smith, N. G. Adams: Int. Rev. Phys. Chem. **1**, 271 (1981)
5.27 C. A. Gottlieb, E. W. Gottlieb, P. Thaddeus, H. Kawamura: Astrophys. J. **275**, 916 (1983)
5.28 B. E. Turner: Bull. Am. Astron. Soc. **10**, 627 (1979)
5.29 H. Suzuki: Prog. Theor. Phys. **62**, 936 (1979)
5.30 H. I. Schiff, D. K. Bohme: Astrophys. J. **232**, 740 (1979)
5.31 G. F. Mitchell, W. T. Huntress, S. S. Prasad: Astrophys. J. **233**, 102 (1979)
5.32 H. Suzuki: Astrophys. J. **272**, 579 (1983)
5.33 K. F. Freed, T. Oka, H. Suzuki: Astrophys. J. **263**, 718 (1982)
5.34 S. Saito, Y. Endo, E. Hirota: J. Chem. Phys. **80**, 1427 (1984)
5.35 C. W. Allen: *Astrophysical Quantities* 3rd ed. (Athlone, London 1973)
5.36 J. M. Hollis, L. E. Snyder, F. J. Lovas, B. L. Ulich: Astrophys. J. **241**, 158 (1980)
5.37 J. M. Hollis, L. E. Snyder, D. H. Blake, F. J. Lovas, R. D. Suenram, B. L. Ulich: Astrophys. J. **251**, 541 (1982)
5.38 H. W. Kroto: Private communication
5.39 P. B. Davies, R. M. Neumann, S. C. Wofsy, W. Klemperer: J. Chem. Phys. **55**, 3564 (1971)
5.40 J. K. Tyler: J. Chem. Phys. **40**, 1170 (1964)
5.41 L. R. Thorne, V. G. Anicich, W. T. Huntress: Chem. Phys. Lett. **98**, 162 (1983)
5.42 L. R. Thorne, V. G. Anicich, S. S. Prasad, W. T. Huntress, Jr.: Astrophys. J. **280**, 139 (1984)
5.43 K. Kawaguchi, S. Saito, E. Hirota: J. Chem. Phys. (in press)
5.44 K. M. Evenson: 16th Int. Symp. on Free Radicals, Brussels, September, 1983
5.45 F. K. Tittel: Private communication
5.46 D. E. Milligan, M. E. Jacox: J. Chem. Phys. **52**, 2594 (1970)
5.47 J. Pacansky, B. Schrader: J. Chem. Phys. **78**, 1033 (1983)
5.48 T. A. Miller, V. E. Bondybey (eds.): *Molecular Ions: Spectroscopy, Structure and Chemistry* (North-Holland, Amsterdam, 1983)
5.49 R. J. Saykally, R. C. Woods: Ann. Rev. Phys. Chem. **32**, 403 (1981)
5.50 R. C. Woods: J. Mol. Struct. **97**, 195 (1983)
5.51 T. Oka: Phys. Rev. Lett. **45**, 531 (1980)
5.52 P. Bernath, T. Amano: Phys. Rev. Lett. **48**, 20 (1982)
5.53 M. Wong, P. Bernath, T. Amano: J. Chem. Phys. **77**, 693 (1982)
5.54 N. N. Haese, F.-S. Pan, T. Oka: Phys. Rev. Lett. **50**, 1575 (1983)
5.55 C. S. Gudeman, M. H. Begemann, J. Pfaff, R. J. Saykally: Phys. Rev. Lett. **50**, 727 (1983)

5.56 T. Amano: J. Chem. Phys. **79,** 3595 (1983)
5.57 C. S. Gudeman, M. H. Begemann, J. Pfaff, R. J. Saykally: J. Chem. Phys. **78,** 5837 (1983)
5.58 M. W. Crofton, T. Oka: J. Chem. Phys. **79,** 3157 (1983)
5.59 E. Schafer, M. H. Begemann, C. S. Gudeman, R. J. Saykally: J. Chem. Phys. **79,** 3159 (1983)
5.60 N. N. Haese, T. Oka: J. Chem. Phys. **80,** 572 (1984)
5.61 M. H. Begemann, C. S. Gudeman, J. Pfaff, R. J. Saykally: Phys. Rev. Lett. **51,** 554 (1983)
5.62 R. S. Altman, M. W. Crofton, T. Oka: J. Chem. Phys. **80,** 3911 (1984)
5.63 T. Amano: J. Chem. Phys. **81,** 3350 (1984)
5.64 T. Amano, J. K. G. Watson: J. Chem. Phys. **81,** 2869 (1984)
5.65 S. C. Foster, A. R. W. McKellar: J. Chem. Phys. (in press)
5.66 M. W. Crofton, R. S. Altman, N. N. Haese, T. Oka: 39th Symp. on Molecular Spectroscopy, Columbus, June 1984
5.67 P. B. Davies, P. A. Hamilton: Private communication
5.68 D. Ray, K. G. Lubic, R. J. Saykally: Mol. Phys. **46,** 217 (1982)
5.69 R. J. Saykally, K. M. Evenson: Phys. Rev. Lett. **43,** 515 (1979)
5.70 D. C. Hovde, E. Schäfer, S. E. Strahan, C. A. Ferrari, D. Ray, K. G. Lubic, R. J. Saykally: Mol. Phys. **52,** 245 (1984)
5.71 J.-T. Shy, J. W. Farley, W. H. Wing: Phys. Rev. **A 24,** 1146 (1981) and references cited therein
5.72 A. Carrington, J. Buttenshaw, R. A. Kennedy: Mol. Phys. **48,** 775 (1983) and references cited therein and also in [5.48]

Subject Index

A (asymmetric) reduction 28
Absorption coefficient 93, 203
Angular momentum 5–8, 30, 35, 57, 63
– –, electron orbital 4, 17, 18, 19, 30, 34, 47, 48, 50, 62, 70
– –, electron spin 4, 17, 30, 62
– –, electronic 47
– –, nuclear spin 17, 33, 34, 63
– –, rotational 18, 24, 30. 41, 63
– –, total 6, 18, 34, 41
– –, vibrational 14, 18, 30, 47, 51
Anharmonic potential 59
– – constant 120, 164–169
Anomalous sign 7
ArH^+ 211
AsH 114, 128–129
AsO 130, 134
Asymmetric rotor, *see* Asymmetric top
Asymmetric top 10, 14, 18, 24

Backward wave oscillator (BWO) 210
BaO 2, 105
BCl 115, 188, 189
BClO (ClBO) 115, 147
BH_2 90
BH_3 90
BO_2 53, 56, 102, 115, 141–147, 148
Bohr magneton 4, 63, 91
Born-Oppenheimer approximation 47–48
BrD^+ (DBr^+), *see also* BrH^+ 131
BrH^+ (HBr^+) 131, 212
BrO 77, 115, 130, 136

C_2 1
C_3 146, 148
CaF 102
Carbene 194, 198–200
CCl 8, 113, 114, 133, 185, 186, 188, 189
CClD (CDCl), *see also* CHCl 198–200
CD_2, *see also* CH_2 88, 114, 156–158
$CDHN^+$ ($DCNH^+$) 211
Centrifugal distortion
– – constant 15
– – effect 15
– – term 28, 57
CF 77, 80, 81, 85, 114, 117, 118, 131, 132, 133, 185, 186
CF_3 42, 45, 81, 115, 116, 172–175, 190–193
CFO (FCO) 115, 155, 190–191
CH 1, 93, 105, 131, 185, 187, 208
CH_2 88, 113, 115, 156–158, 198
CH_3 85, 86, 87, 90, 98, 113, 115, 116, 169–171, 190–191, 205
C_2H 140, 208
C_2H_5 (ethyl) 211
C_3H 208
C_3H_5 (allyl) 211
C_4H 140, 208
CHCl 114, 198–200
CH_2Cl 81, 114, 178, 190–193
Chemical titration 204
CHF (HCF) 101, 114, 118, 198–200
CH_2F 81, 114, 118, 178, 190–193
CHF_2 178
CHN (HNC) 77
CH_2N^+ ($HCNH^+$) 211
C_2HN (HCCN) 81, 114, 118, 140, 209
C_2H_2N (CH_2CN) 118
CHO (HCO) 77, 90, 93, 155, 166–168, 190–193, 208
CHO^+ (HCO^+) 77, 139, 208, 211
CHO^+ (HOC^+) 77, 208
CHO_2^+ (HCO_2^+) 208
CH_3O 28, 42, 46, 47, 81, 113, 114, 180–183
C_2H_3O (CH_2CHO) 81, 114
CHS^+ (HCS^+) 77, 139, 208
CH_2S (H_2CS) 104, 105, 107, 108, 113, 114, 194–198, 208
CH_3S 81, 114
Cl 90
ClD^+ (DCl^+), *see also* ClH^+ 131
ClF 188, 189
ClH^+ (HCl^+) 131, 212

ClN (NCl) 81, 115, 125–127, 185, 186, 187, 188, 189
ClO 4, 77, 136, 185, 186, 207
ClO_2 190–192
ClOS (ClSO) 81, 115, 154–155
ClP (PCl) 81, 89, 90, 115, 127
ClS (SCl) 115, 137
ClSi (SiCl) 81, 114, 117, 133, 185, 186
CN 1, 77, 120, 185, 186, 188, 208
C_2N (CCN) 56, 114, 118, 140, 146, 148, 149–150
C_2N (CNC) 56, 140, 146, 148
C_3N 140, 208
CNO (NCO) 56, 77, 81, 113, 114, 146, 147–149
CNS (NCS) 77
CO laser 82, 90, 92, 93
CO^+ 77, 208
CO_2^+ 56, 144, 146, 148
CO_2 laser 82, 90, 92, 93, 109
– – pumped (far infrared) laser, *see* Far infrared laser
C_2O (CCO) 81, 90
Color center laser 2, 98
Commutation relation 6, 7
Coriolis coupling constant 13, 32, 58
– interaction 13, 17, 195–196
– –, electronic 20, 189, 199
– –, first order 14
– –, vibronic 30
– term 13
– –, electronic 66
CS 90, 113, 114, 119–120, 208
Cubic anharmonic potential 58, 59
– – – constant 15
Cylindrical resonant cavity 105

Decoupled representation 18
Density
–, orbital 131
–, spin 131, 185–187
Detector 79, 83
Diatomic molecule 18, 19, 21, 24, 40, 57, 64, 119–139
– –, Δ 137–139
– –, Π 129–137
– –, Σ 119–129
Difference frequency laser 2, 96, 98
– – – spectroscopic system 96–98
– – – spectroscopy 88
Diode laser 82–90, 109, 131
– – spectrometer 82–90
– – spectroscopic system 82–90
– – spectroscopy 2, 85, 88, 89, 116, 117, 135

Dipolar interaction, *see* Dipole-dipole interaction
Dipole-dipole interaction 20, 34, 36, 39, 42, 43, 44, 46, 185–187, 192–193
– – tensor, off-diagonal component of 152
Dipole moment 71, 72, 107, 135
– –, electric 62
– –, magnetic 62, 70
Direction cosine 5
Discharge 113
–, AC 113, 117
–, DC 113, 117
D_2N (ND_2), *see also* H_2N 109
DO (OD), *see also* HO 131
DO_2, *see also* HO_2 115, 116
Doppler effect 91
– broadening 203
Doppler-limited excitation spectroscopy 99–102
– – spectrum 99, 101
– resolution 99, 102, 104
– width 97, 100, 104, 113
DOS (DSO), *see also* HOS 153–154
double resonance (DR) 98, 103–112
– –, infrared optical (IODR) 108–112, 159–161
– –, microwave optical (MODR) 2, 104–105, 194
– –, rf optical 104–105
– – spectrometer, microwave optical 105–108
– – spectrum, microwave optical 107
Doublet Π state 21, 22–24, 48–57, 61–62
–, inverted 129
–, regular 129
DP (PD), *see also* HP 114, 127–128, 134
DS (SD), *see also* HS 131
DSe (SeD), *see also* HSe 130, 131
Dye laser 2, 98, 99, 102, 109
– – excitation spectrometer 98–103
– – spectroscopic system 98–103

Eckart condition 12, 13
Electron paramagnetic resonance (EPR) 1, 2, 67, 77, 93
– – – spectrometer 77
Electron spin resonance (ESR) 1, 93, 113
Electron spin-rotation interaction, *see* Spin-rotation interaction
Equilibrium structure 164–169
Etalon 83
Euler angle 51

Excimer laser photolysis 89, 206
Excitation spectrum, *see* Laser excitation spectrum

Far infrared (FIR) laser 82
– – –, optically pumped 94
FeO 81, 113, 114, 139
Fermi contact term 42, 44
– coupling constant 46, 187–188, 192–193
– interaction 17, 34, 36, 41
– term, *see* Fermi contact term
FGe (GeF) 81, 114, 133
FH^+ (HF^+) 131, 212
Field gradient tensor 38
Fine structure 5, 17–47, 100
– – interaction 19, 184–193
– – Hamiltonian 19–20
– – transition 93
FIR LMR, *see* Laser magnetic resonance
Flash photolysis 1, 113
Fluorescence cell 99
FN (NF) 77, 138, 185, 186, 187
F_2N (NF_2) 77, 190–192
FO 115, 135, 185
FO_2 89, 115, 153, 169, 190–191
FOS (FSO) 81, 113, 114, 117, 154–155, 190–192
Four group 10, 11
FP (PF) 81, 115, 118, 127, 185
F_2P (PF_2) 81, 115, 118, 127
Free-space absorption cell 76, 77, 78
Free spectral range 83
Frosch and Foley hyperfine constant 38, 40
FS (SF) 77, 81, 87, 114, 117, 133, 137, 185, 186
F_2S (SF_2) 81, 114, 117
FSe (SeF) 77
FSi (SiF) 81, 114, 117, 133, 185, 186
F_2Si (SiF_2) 117

g factor 1, 62, 66, 91, 132
– –, nuclear rotational 63
– –, rotational 63, 66
g tensor
– –, anisotropic 66
GaAs Schottky diode 79

Harmonic generator 78, 80, 99
H_3^+ 87, 211
HD^+ 211
H_2D^+ 81, 211
Herzberg-Teller effect 48, 70
– interaction 54, 61, 147

HHe^+ (HeH^+) 211
HN (NH) 163, 185, 186, 187
HN_2^+ (N_2H^+) 77, 139, 208, 211
H_2N (NH_2) 42, 56, 102, 105, 110, 111, 114, 158–161, 163, 190–193
H_4N^+ (NH_4^+) 211
HNe^+ (NeH^+) 211
HNO 77, 105, 106, 113, 114, 163, 166–168, 208
HO (OH) 1, 4, 76, 77, 93, 131, 185, 187, 207, 208
HO_2 4, 77, 81, 88, 93, 97, 114, 115, 116, 150–153, 165–167, 169, 190–193, 204–205, 206–207
H_3O^+ 88, 211
H_2O laser 82, 93
Hollow cathode 78
HOP (HPO) 81, 114, 118, 163, 202–203, 210
HOS (HSO) 81, 101, 113, 114, 115, 116, 153–154, 190–193
HP (PH) 127–128, 163, 185, 187, 201–202
H_2P 42, 56, 81, 102, 103, 114, 115, 161–162, 163, 190–193, 201–202
HS (SH) 4, 131, 185, 187, 207
HSe (SeH) 130, 131
HSi (SiH) 121, 131
H_2Si (SiH_2) 114
Hund's case (a) 18, 19, 30, 37, 51
– – $(a)_\beta$ 34–38, 46
– – (b) 18, 19, 20, 24, 27, 60, 64, 69
– – $(b)_{\beta J}$ 39, 43, 46
– – (c) 64
Hyperfine coupling constant 1, 38, 40, 61, 130, 175, 184–189, 197
– interaction 33–34, 41–47, 184–193
– splitting 111
– structure 4, 5, 17–47, 61–62, 102

I_2 100
Infrared (IR) detector 92
– diode laser spectroscopic system 82–90
– laser spectrometer 81–98
InSb photoconductive detector 79
Intensity 56, 72
Intermediate case 131
Intermodulated fluorescence (IMF) 2, 102
– – spectroscopy 98, 102–103, 104
Intersteller space 77
Inversion operator 51
IO 77
IODR, *see* Double resonance spectroscopy
IR LMR, *see* Laser magnetic resonance

Jahn-Teller coupling 32
– effect 30, 46, 47, 180
– interaction 30

K degeneracy 9
Klystron 210
K-type doublet 72

L-uncoupling 63
– term 66
Ladder operator 7, 31
Lamb dip 91
Λ doublet 1, 22, 37, 70, 72
Λ doubling 19, 21, 22–24, 31, 58, 64, 76, 130, 131, 138
– – constant 133, 135
Laser excitation spectrometer 99
– – spectrum 116
Laser magnetic resonance (LMR) 2, 62, 67, 68, 82, 90, 91, 109
– – –, far infrared (FIR) 2, 82, 92, 115, 131
– – –, infrared (IR) 2, 93, 116
– – – spectrometer 90–95
– – – spectrometer, far infrared 93–95
– – – spectrometer, infrared 90–93
Laser Stark spectroscopy 62, 72, 82
LiNa 2
$LiNbO_3$ (lithium niobate) 96, 109
Line shape function 203
Linear molecule 9, 18, 19, 20, 24, 40, 47, 48, 50, 57–61, 66, 139–150
– – in Σ state 139–140
– – in Π state 141–150
LMR, *see* Laser magnetic resonance
l-type constant 16
– doubling 16
l-type doublet 16, 72

Magnetic dipole transition 69, 105
Magnetic moment, *see* Dipole moment
Metastable electronic state 194–200
Microwave spectrometer 75–81
Microwave spectroscopy 1, 77, 116, 117
Millimeter-wave spectroscopy, *see* Microwave spectroscopy
Minimum detectable absorption coefficient 76, 78, 80
Minimum detectable number of molecules 85, 92
MODR, *see* Double resonance spectroscopy
Modulation
–, discharge current 87, 137
–, source frequency 76, 77, 85, 86
–, Stark 62, 76
–, velocity 88
–, Zeeman 62, 86, 97, 133
Molecular rotation 5–11, 17, 19
Molecule-fixed coordinate 8, 10, 27, 35, 51
– – system 5, 8, 11, 13
Moment of inertia 8, 9, 13, 59
Multiple-reflection cell 88, 89

N_2^+ 122–123
NO 130, 133, 135, 163, 185, 186, 208
N_2O laser 82, 90, 92, 93, 109
NO_2 56, 163, 190–193
NO_3 4, 113, 115, 179–180, 207
Nonlinear spectroscopy 98, 102
Nonlinear triatomic molecule 150–169
– –, XY_2 type 156–164
– –, XYZ type 150–155, 164–169
Normal coordinate 12
– –, dimensionless 14
NS 77, 113, 114, 120–122, 135, 185, 186, 208
NSi (SiN) 81, 113, 114, 185, 186, 188
Nuclear electric quadrupole interaction 17, 33, 34, 37, 39, 188–189
– – – coupling tensor 39, 40
– – – coupling tensor, non-axially symmetric component of 188
– – – moment 37
Nuclear magneton 63
Nuclear spin rotation interaction 17, 34, 40
Nuclear spin-spin interaction 34

O_2 137, 138
Oblate top 9, 11
OP (PO) 81, 114, 118, 133, 134, 135, 163, 201–212
O_2P (PO_2) 81, 114, 118, 134, 162–164, 201–202, 210
Optical isolator 102
Orbital density, *see* Density
OS (SO) 1, 76, 77, 90, 114, 117, 123–124, 138, 206, 208
OSi (SiO) 208
Out-of-plane bending vibration 175–178

Parallel-plate cell 75, 76, 77, 78
Parity 11, 22, 28, 31, 58, 61, 130
Paschen-Back effect 68, 72
Passive enhancement cavity 99
Planarity 176
Polarization spectroscopy 62, 99
Pressure broadening 113, 203

Pressure width 113
Principal axis of inertia 8
Principal moment of inertia, *see* Moment of inertia
Prolate top 9, 10
Protonated ion 211
Pulsed dye laser 99
Pure precession 130
Pyrolysis 13

Radioastronomy 4
Ray's parameter 11
Renner effect 18, 47–62, 141, 146
– interaction 60, 147
– parameter 54, 55
Renner-Teller effect, *see* Renner effect
Resolution 98
Rigid body 8, 11
Ring laser 99
Rotation, *see* Molecular rotation
Rotation matrix 27
Rotational constant 9, 17
Rotational energy 8–11
Rotational Hamiltonian 9, 10, 13, 20, 30, 50, 59
Rotational transition 93
Rotational wavefunction 19

S (symmetric) reduction 28
s character 131, 187
Sensitivity 75, 76, 80, 85, 92, 94
Signal averaging 79, 89
Space-fixed coordinate 8, 35
– – system 5, 8
Spectrum analyzer 100
Spherical harmonics 9, 28, 35, 49, 50
Spherical tensor 24, 26, 34, 43, 45
Spherical top 14
Spin density, *see* Density
Spin polarization 188
Spin uncoupling 130
Spin-orbit interaction 19, 20, 31, 50, 52, 53, 55, 57, 63, 66, 190
– coupling constant 50, 55, 67, 130, 131
Spin-rotation coupling constant 26, 66, 67, 68, 72
– – –, off-diagonal component of 151
– interaction 19, 21, 24, 27, 28, 31, 40, 66, 189–192
Spin-spin interaction 19, 20, 28
SSi (SiS) 208
Stark cell 76, 107
– effect 62–73, 107
– –, first order 72
– modulation, *see* Modulation
– – spectrometer 76
sub-Doppler resolution 91, 98, 104, 109
Symmetric top 8, 9, 14, 16, 18, 19, 24, 28, 30, 42, 64, 71, 72, 169–183
– –, oblate 9
– –, prolate 9
– – wavefunction 9, 10, 27, 65
symmetry axis 8, 9, 15, 30

$^{130}Te_2$ 100
Teflon lens 78, 79
Transition moment 85
(2, 2) interaction 16, 31
(2, −1) interaction 17, 31
2*f* detection 85

Unpaired electron 4, 17, 19, 20, 23, 28

van Vleck transformation 23
Vibrational potential function 13, 17, 48
Vibration rotation constant 14–15, 57, 58, 165
– interaction 11–17, 47
Vibronic interaction 47–48, 57, 144–145, 176–177

Walsh diagram 55
Wang wavefunction, *see* Symmetric top wavefunction
Wavelength meter 84
Wavenumber marker 100
Wavenumber standard 100
White-type multiple-reflection cell, *see* Multiple-reflection cell
Wigner-Eckart theorem 37, 44, 65

Zeeman component 90
– effect 62–73, 91, 97
– Hamiltonian 63, 64
– modulation, *see* Modulation

K. Shimoda

Introduction to Laser Physics

1984. 87 figures. IX, 211 pages
(Springer Series in Optical Sciences,
Volume 44). ISBN 3-540-13430-1

Contents: The Laser - An Unprecedented Light Source. - The Coherence of Light. - Electromagnetic Theory of Light. - Emission and Absorption of Light. - Principle of the Laser. - Output Characteristics of the Laser. - Coherent Interaction. - Nonlinear Coherent Effects. - Theory of Laser Oscillation. - References. - Subject Index.

K. Uehara, H. Sasada

High Resolution Spectral Atlas of Nitrogen Dioxide 559–597 nm

1985. 7 figures and 179 charts. VII, 226 pages
(Springer Series in Chemical Physics,
Volume 41). ISBN 3-540-15027-7

Contents: Introduction. - High Resolution Spectral Atlas of No_2 Between 559 and 597 nm. - Progress in Spectroscopic Studies of NO_2 Using Lasers. - Selected Bibliography 1978–1983. - Atlas of Absorption and Stark Modulation Spectra. - References. - Subject Index.

Springer-Verlag
Berlin
Heidelberg
New York
Tokyo